This book provides a broad grounding for the design of instruments for use in scientific and other spacecraft.

The first chapter introduces the basic principles of designing for the space environment. Following chapters discuss mechanical, structural, thermal and electronic design including the problems that are frequently encountered in the testing and verification of spacecraft subsystems. Important topics are described, including stress analysis, multilayer insulation, two dimensional sensor systems, mechanisms, the structure of space optics, and project management and control. A final chapter looks towards future developments of space instrument design and addresses issues arising from financial constraints. The book contains lists of symbols, acronyms and units and a comprehensive reference list and bibliography. Worked examples are found throughout the text.

Principles of Space Instrument Design is suitable for researchers and engineers interested in spacecraft and space instrument design. It will also be valuable to graduate students of physics, space science, spacecraft engineering and astronautics.

T0211267

Principles of Space Instrument Design

CAMBRIDGE AEROSPACE SERIES

Principles of Space Instrument Design

A. M. Cruise
School of Physics and Space Research,
University of Birmingham

J. A. Bowles
Mullard Space Science Laboratory,
University College London

T. J. Patrick
Mullard Space Science Laboratory,
University College London

C. V. Goodall
School of Physics and Space Research,
University of Birmingham

CAMBRIDGE
UNIVERSITY PRESS

CAMBRIDGE UNIVERSITY PRESS
Cambridge, New York, Melbourne, Madrid, Cape Town, Singapore, São Paulo

Cambridge University Press
The Edinburgh Building, Cambridge CB2 2RU, UK

Published in the United States of America by Cambridge University Press, New York

www.cambridge.org
Information on this title: www.cambridge.org/9780521451642

First published 1998
This digitally printed first paperback version 2006

A catalogue record for this publication is available from the British Library

Library of Congress Cataloguing in Publication data

Principles of space instrument design / A. M. Cruise . . . [et al.].
p.   cm. – (Cambridge aerospace series ; 9)
Includes bibliographical references and index.
ISBN 0-521-45164-7 (hc)
1. Astronautical instruments–Design and construction.
I. Cruise, A. M. (Adrian Michael), 1947–   . II. Series.
TL1082.P75   1998
629.47′4–dc21   97-16356 CIP

ISBN-13 978-0-521-45164-2 hardback
ISBN-10 0-521-45164-7 hardback

ISBN-13 978-0-521-02594-2 paperback
ISBN-10 0-521-02594-X paperback

Contents

Preface

Scientific observations from space require instruments which can operate in the orbital environment. The skills needed to design such special instruments span many disciplines. This book aims to bring together the elements of the design process. It is, first, a manual for the newly graduated engineer or physicist involved with the design of instruments for a space project. Secondly the book is a text to support the increasing number of undergraduate and MSc courses which offer, as part of a degree in space science and technology, lecture courses in space engineering and management. To these ends, the book demands no more than the usual educational background required for such students.

Following their diverse experience, the authors outline a wide range of topics from space environment physics and system design, to mechanisms, some space optics, project management and finally small science spacecraft. Problems frequently met in design and verification are addressed. The treatment of electronics and mechanical design is based on taught courses wide enough for students with a minimum background in these subjects, but in a book of this length and cost, we have been unable to cover all aspects of spacecraft design. Hence topics such as the study of attitude control and spacecraft propulsion for inflight manœuvres, with which most instrument designers would not be directly involved, must be found elsewhere.

The authors are all associated with University groups having a long tradition of space hardware construction, and between them, they possess over a century of personal experience in this relatively young discipline. One of us started his career in the aerospace industry, but we all learned new and evolving skills from research group leaders and colleagues who gave us our chances to develop and practise space techniques on real projects; there was little formal training in the early days .

Lecture courses, both undergraduate and postgraduate in their various fields of expertise, have been given by the authors who have found it very difficult to recommend a text book to cover the required topics with sufficient detail to enable the reader to feel he had sufficient knowledge and confidence to start work on a space project. Hence we felt, collectively, that it would be useful to put down in a single volume the principles that underlie the design and preparation of space instruments, so that others might benefit from using this as a starting point for a challenging and sometimes difficult endeavour.

We gratefully acknowledge the enormous debt we owe to those pioneers who guided us early in our careers, and who had the vision to promote and develope space as a scientific tool. We recall especially Professor Sir Harrie Massey, Professor Sir Robert Boyd, Professor J L Culhane, Mr Peter Barker, Mr Peter Sheather, Dr Eric Dorling and many colleagues in our home institutes, in NASA and ESA.

J.A. Bowles
A.M. Cruise
C.V. Goodall
T.J. Patrick

October 1997

1

Designing for space

1.1   The challenge of space

The dawn of the space age in 1957 was as historic in world terms as the discovery of the Americas, the voyage of the Beagle or the first flight by the Wright brothers. Indeed, the space age contains elements of each of these events. It is an age of exploration of new places, an opportunity to acquire new knowledge and ideas and the start of a technological revolution whose future benefits can only be guessed at. For these reasons, and others, space travel has captured the public's imagination.

Unfortunately, access to the space environment is not cheap either in terms of money or in fractions of the working life of an engineer or scientist. The design of space instruments should therefore only be undertaken if the scientific or engineering need cannot be met by other means, or, as sometimes happens, if instruments in space are actually the cheapest way to proceed, despite their cost. Designing instruments, or spacecraft for that matter, to work in the space environment places exacting requirements on those involved. Three issues arise in this kind of activity which add to the difficulty, challenge and excitement of carrying out science and engineering in space. First, it is by no means straightforward to design highly sophisticated instruments to work in the very hostile physical environment experienced in orbit. Secondly, since the instruments will work remotely from the design team, the processes of design, build, test, calibrate, launch and operate, must have an extremely high probability of producing the performance required for the mission to be regarded as a success. Thirdly, the design and manufacture must achieve the performance requirements with access to a very strictly limited range of resources such as mass, power and size.

It should also be stated, in case prospective space scientists and engineers are discouraged by the enormity of the technical challenge, that access to space has already proved immensely valuable in improving our understanding of the Universe, in crucial studies of the Earth's atmosphere and in monitoring the health of our planet. In addition to the direct benefits from space research, the engineering discipline and technological stimulus which the requirements of space programmes have generated, make space projects a most effective attractor of young and talented people into the physical sciences and an excellent training ground for scientists and engineers who later seek careers in other branches of science. More benefits can be expected in the future when space technology could become critical to Mankind's survival.

This first chapter addresses two of the issues already mentioned as major determinants of the special nature of designing for space, namely the physical environment experienced by space instruments and the philosophy or methodology of creating a complex and sophisticated instrument which meets its performance goals reliably while remote from direct human intervention.

1.2   The physical environment in space

The environment in which space instruments are operated is hostile, remote and limited in resources. Many of the resources on which we depend in our everyday lives are absent in Earth orbit and therefore have to be provided by design if they are believed necessary or, in the contrary situation, the space instrument has to be designed to work in their absence. Compared to the situation on Earth, though, the environment is fairly predictable. Indeed, most space instrument designers, feel convinced that their brain–child is far safer in orbit than on the ground where fingers can poke and accidents can occur.

The first environment experienced by the instrument is that of the launch phase itself but since this is such a transient and specialised situation it will be discussed in Chapter 2 under the topic of Mechanical Design where one of the main objectives is to survive the launch environment, intact and functioning.

The orbital space environment comprises the full set of physical conditions which a space instrument will be exposed to and for which the instrument designer must cater.

1.2.1   Pressure

The most obvious difference between conditions in space and those on Earth is that the ambient pressure is very low. Depending on the altitude and the phase of the 11 year solar cycle the pressure experienced by space hardware can approach that of perfect vacuum.

Fig. 1.1 shows the variation of pressure with height above the surface of the Earth. The pressure is given in pascals, 10^5 pascals being roughly equivalent to normal atmospheric pressure, and 133 pascals being equal to the frequently used cgs unit of pressure, the torr. There are three principal implications for the design engineer of the very low gas pressure. The low ambient pressure allows volatile materials to 'outgas' from spacecraft components. The condensible fraction of such material may be deposited where it is not wanted (on optics or other cold surfaces) with considerable damage to the instrument performance. The loss of water vapour or waxes from composite materials such as carbon fibre reinforced plastic (CFRP) may alter their structural properties and size. It is customary to avoid the use in space of materials which outgas to any great extent, and tables of fractional mass–loss under suitable vacuum conditions are available from most space agencies to assist in material selection.

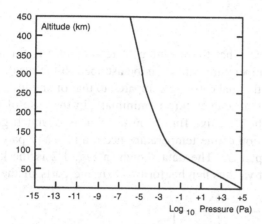

Fig. 1.1.   Variation of pressure with height.

The second effect of low pressure is that the mechanism for heat transfer by convection, which is so efficient at equalising temperatures between various components under normal atmospheric pressure, ceases to be effective if the gas pressure falls below 10^{-6} pascals. This issue will be dealt with in Chapter 3 on thermal design.

The third effect of the low pressure is that there are regimes of pressure under which electrical high voltage breakdown may occur. A particularly serious problem arises in the design of equipment which includes detectors requiring high voltages for their operation. Ways of delivering high voltages to separate subsystems without the danger of electrical breakdown are discussed in Chapter 4, section 4.6.6. The pressure data presented show the, so –called, 'Standard Atmosphere', relatively undisturbed by violent conditions on the Sun.

In addition to the overall number density or pressure as a function of height it is useful to notice that the composition of the atmosphere changes as one goes higher and this aspect is especially sensitive to solar conditions since it is the interaction of both solar radiation and the solar wind with the outer layers of our atmosphere which dominates the processes of dissociation, controlling the composition. Of particular interest to space instrument designers is the fact that the concentration of atomic oxygen (which is extremely reactive with hydrocarbons) increases at low Earth orbit altitudes over the more benign molecular component. Atomic oxygen can be extremely corrosive of thermal surfaces and their selection must bear this requirement in mind.

The COSPAR International Reference Atmosphere provides models for these aspects of atmospheric behaviour (Kallmann – Bijl, [1961]).

1.2.2   Temperature

It is not immediately clear what the meaning of temperature is for a near vacuum. There is a gas kinetic temperature which can be ascribed to the residual gas atoms but these have such a small thermal capacity compared to that of any instrument component that the relevance of that temperature is minimal. In the case of space instruments in the solar system the radiative flux from the Sun is of much greater importance. The detailed calculation of the temperature taken up by any part of an instrument is the subject of Chapter 3. The data shown in Fig. 1.2 is the kinetic temperature for the residual gas at various heights for two extreme parts of the solar cycle.

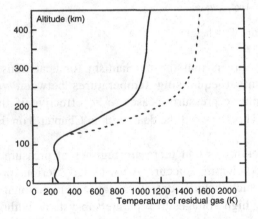

Fig. 1.2. Variation of temperature with height.

1.2.3   Radiation

The Sun dominates the radiation environment of the solar system at most energies even though it is a rather ordinary star with a surface temperature of only about 6000 K. Fig. 1.3 gives an overall impression of the distribution with wavelength of the radiation from the Sun. There are regions of the spectrum which do not approximate well to a blackbody and these show great variability, again often linked to both the solar cycle and periods of intense solar activity.

Space instrument designers have to take account of the radiation environment in several ways. Radiation outside the wavelengths being studied can penetrate sensitive

detectors either by means of imperfect absorption of high energy photons by screening
material, or by scattering off parts of the instrument or spacecraft which are within the
instrument field of view. Detectors which are sensitive in the desired wavelength
range can also exhibit low, but compromising, sensitivity at wavelengths where solar
radiation is intense. The full spectrum of the Sun in any likely flaring state has to be
considered as a background to be suppressed in any instrument design.

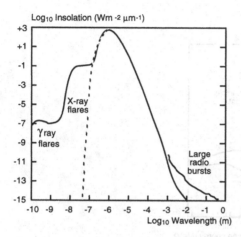

Fig. 1.3.   The solar spectrum.

Apart from the needs of thermal analysis and the effects of light on detectors and
surfaces the other important kind of solar radiation is the outflow of energetic particles
in the form of electrons (which can generally be shielded against quite easily) and
protons which can cause considerable damage. Fig. 1.4 shows the instantaneous fluxes

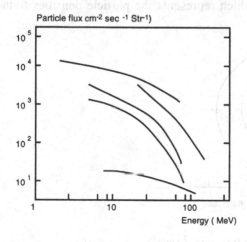

Fig. 1.4.   Instantaneous solar proton fluxes.

of energetic protons from separate solar particle events as a function of energy for discrete solar outbursts. Section 4.7 of this book provides guidance on the design of electronic systems and the choice of components to withstand the effects of radiation damage.

It is immediately clear that these fluxes are very high and greatly exceed the levels of particle radiation from outside the solar system which are shown in Fig. 1.5 as time integrated fluxes.

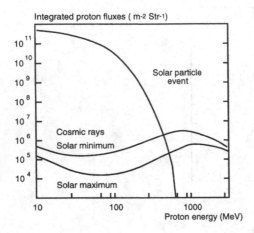

Fig. 1.5.   Time integrated solar proton fluxes.

For spacecraft in near Earth orbit the effect of particle contamination can be serious and this is more pronounced when the instrument is in an orbit which takes it through the belts of trapped particles confined by the Earth's magnetic field. The structure of these belts is indicated in Fig. 1.6 which represents the particle densities found in a

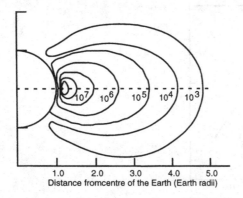

Fig. 1.6.   Contours of electron flux (cm^{-2} sec^{-1}).

plane at right angles to an equatorial orbit. Detailed three dimensional, energy dependent models of these particle distributions are now available in computer form so that the instrument designer can calculate total doses as well as instantaneous fluxes at various energies to investigate the effectiveness of the shielding required for the protection of electronics and sensitive detectors.

Not all the important radiation as far as the design of a space instrument is concerned comes from the Sun. At the very highest energies where individual particles can carry more than 10^{15} electron–volts per nucleon of energy the source of radiation seems to lie outside the solar system and perhaps outside the galaxy itself. But these particles are not just scientific curiosities. They carry such a massive amount of energy individually that collision with the appropriate part of a semiconductor can cause temporary or permanent damage resulting in the corruption of data or the loss of function. Means of designing circuits robust enough to survive such events will be dealt with in Chapter 4 and Fig. 1.7 shows the fluxes incident on equipment in near Earth orbit during solar minimum. Since the damage caused is proportional to the square of the charge carried by the ions, even infrequent collisions with iron ions can have a marked effect on space equipment unless it is properly designed.

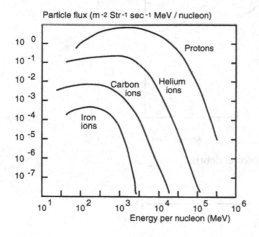

Fig. 1.7. Energy spectrum of galactic nuclei.

1.2.4   *Space debris*

The relative velocity of objects in the solar system can be as high as 70 km/s, which means that even extremely light particles of solid matter can cause enormous damage if they impact on a space instrument or spacecraft. Naturally occurring material, much of it remnants of the proto – solar nebula, is falling onto the Earth by means of gravitational capture at a total rate of thousands of tonnes per year. These

objects range from the scale of dust to that of large meteorites weighing many
kilograms, but with such large encounter velocities possible between objects in
different orbits around the Sun, even dust particles can punch holes in aluminium
structures and destroy functional components in space instruments. Added to the
natural debris in the solar system, and preferentially gathered in low Earth orbit and
geostationary orbit, are the debris left by Mankind's journeys into space. Solid
particles ejected during rocket burns, parts accidently removed from satellites during
injection and remnants from dead and decaying spacecraft are all cluttering up the
space environment and offering hazards to working spacecraft. Some of these pieces
are large enough for defence agencies to track by radar (items with a radar
cross-section greater than 30 cm^2) while others are too small to detect and track. Fig.
1.8 gives the spatial density of items with a cross-section greater than 1 m^2 as a
function of altitude while Fig. 1.9 demonstrates the clustering of such items around the
geostationary orbital altitude where the gradual effect of atmospheric drag will *not*
clear them from orbit as happens in low Earth orbit.

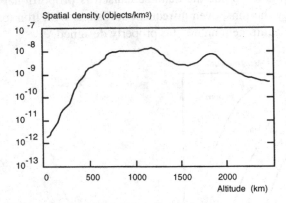

Fig. 1.8.   Spatial density of orbital debris.

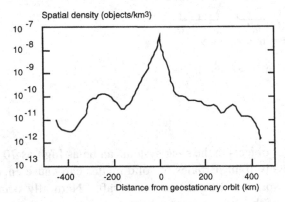

Fig. 1.9.   Spatial density of orbital debris.

1.3   The system design of space instruments

The seven chapters which follow provide some detailed guidance on how to design a scientific instrument for use in space. They address, in turn, mechanical design, thermal design, the design of various electronic functions, mechanism design and finally the establishment of the relevant management systems to enable the whole enterprise to be successfully carried out. These are all complex and specialised issues which require great care and attention if they are to be an effective part of the complete instrument. At the early stages of the evolution of a space project, however, broader issues are more important than the details. If these broader issues are not satisfactorily sorted out then the effort spent on details will be wasted. An overall 'system–level' study is therefore needed to determine the main parameters of the instrument and to establish the scale of the overall project.

1.3.1   *The system design process*

A path through a system design process is given in Fig. 1.10. At the foot of the chart the four main system parameters are shown as the outputs of the study.

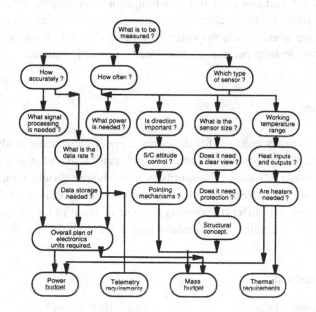

Fig. 1.10. A system design process.

The most important question which initiates the estimates and decisions that determine these outputs, in the case of a scientific instrument, is 'What is to be measured?'. Once this is defined, questions as to the accuracy and frequency of the measurements need to be answered in order to estimate the data rate generated by the instrument. When measurements of specially high accuracy are involved then provision for sophisticated signal processing may be needed in the electronics. The pattern of telemetry access to the instrument will determine whether the data can be read out to the spacecraft data system directly, or whether some provision for data storage is needed. The choice of sensor to carry out the required measurements will determine many of the size parameters associated with the overall design. In particular, the mechanical design, mass budget and probably the thermal requirements are all strongly dependent on the choice of sensor, including any optics which it may need to function adequately. The sensor requirements may also affect the spacecraft pointing requirements in such a way that the overall mission costs will be determined. Following through the questions suggested in the diagram will help to determine what on–board equipment is actually needed and what requirements will be made of the spacecraft

It is very important to carry out at least a preliminary system–level study of the kind outlined above at the earliest stage in a space project. Space opportunities present themselves in various ways and one possibility is that a mission may be available to the experimenter but only if they accept certain size or orbital constraints predetermined by another user. Such opportunities cannot be taken up with any confidence unless the main system parameters are already well determined.

1.3.2   *Some useful facts*

At the level of accuracy required for the initial stages of a systems study there are certain average properties of space equipment which can be useful in scaling the various aspects of the design. None of the numbers listed below should be taken as absolute standards, they are taken from several instruments recently constructed in the UK. Individual instruments will differ by possibly a factor of three in either direction. However, the use of these estimates at the earliest stages will avoid the most serious divergence from reality.

Density of space equipment

Overall instrument density	$150 - 1000 \text{ kg}/\text{m}^3$
Electronics modules	$1000 - 4500 \text{ kg}/\text{m}^3$
Structural elements	$60 - 500 \text{ kg}/\text{m}^3$

Power consumption

Overall instrument	$65 - 300$ watts/m^3
	$0.4 - 3.2$ watts/kg
Electronics modules	$2000 - 4500$ watts/m^3
	$1.0 - 7.0$ watts/kg
Mass memory	$0.2 - 1.0$ 10^{-6} watts/byte
Heat loss	$1.0 - 7.0$ watts/m^2 of MLI

1.3.3   A brief example

In order to make the idea of a preliminary system study more accessible, the following example is described, step by step. A telescope is required to observe distant galaxies and the scientific specification demands a focal length of 4 metres and an entrance aperture of 1 metre diameter. The sensors in the focal plane work at room temperature and read out the x and y coordinates of each detected photon with a precision of 12 bits in each coordinate. The maximum count rate is about 200 per second averaged over the detector and the data can be read out to ground stations once every 24 hours, necessitating on–board storage. From this rather small set of data much can be done in sizing the instrument and quantifying the resources required from the spacecraft.

The overall size of the instrument is already specified by the optical performance necessary to achieve its scientific goals and we may assume a cylinder of diameter about 1.2 metres and 4.5 metres long allowing for additional items outside and behind the optical path. An instrument of this size could have a mass as low as 700 kg if there is little internal structure or as much as 4000 kg if it is densely packed. Since it is an optical payload and there will need to be large volumes occupied by the optical paths, we assume the lower figure until other factors indicate otherwise. The surface area of the structure is 15.8 m^2 and therefore one expects a heat loss of anything up to 110 watts from the surface alone without considering the open aperture.

The photon rate of up to 200 events per second will generate a data rate of about 5000 bits per second, or around $300 - 350$ 16 bit bytes per second. In a day the storage requirements will come to about 30 megabytes without allowance for error correction or missed ground station passes. At this stage in the design a figure of 70 Mbytes might be appropriate, requiring a mass store with power requirements of perhaps 17 watts. The processing of the signals prior to storage will require an electronics module of, say, 12 watts consumption and there may be the need for a similar unit to store commands and control the instrument in operation.

If the telescope can rely entirely on the spacecraft for its pointing and attitude reconstruction then no further allowance for electronics is needed save for thermal control. The loss of thermal energy from the main surface will require the previously calculated power of 110 watts to keep the instrument at an equilibrium temperature, assumed to be 20°C. The open aperture will radiate to cold space and the power loss using Stefan's constant could be as much as 470 watts if no attempt is made to reduce the solid angle of space seen by the components at 20°C. Assuming this can be reduced by a factor of ten then an allowance of 50 watts must be made for heat loss from the optical aperture.

From these input data we can summarise the outline design requirements for accommodating the telescope on the space platform.

Mass	700 kg
Power consumption	210 watts
Size	1.2 m dia
	4.2 m long
Data storage	70 Mbytes

More detailed inputs on the exact performance specification and required functionality would be needed to refine these numbers to the point of agreeing an interface with a spacecraft contractor, but at least the scale of the instrument is becoming clear and boundaries and targets for the design are appearing as a result of these system–level estimates. Chapter 2, which follows, takes up the detailed issues surrounding the mechanical design of the instrument.

2

Mechanical design

For spacecraft and their instruments, the engineering disciplines of mechanical and structural design work together, and both are founded on the study of the mechanics of materials. We create designs, then prove (by calculation and test) that they will work in the environments of rocket launch and flight. Mechanisms, by definition having relatively–moving parts, are not the whole of mechanical design; they are interesting enough to get a later chapter of this book (Chapter 5) to themselves. But any mechanism is itself a structure of some kind, since it sustains loads. We therefore define structures, which are more general than mechanisms, as 'assemblies of materials which sustain loads'. All structures, whether blocks, boxes, beams, shells, frames or trusses of struts are thereby included.

Mechanical design should begin by considering the forces which load the structural parts. The twin objectives are to create a structure which is (i) strong enough not to collapse or break, and (ii) stiffly resistant against deforming too far. The importance of stiffness as a design goal will recur in this chapter. Forces may be static, or dynamic; if dynamic, changing slowly (quasi–static) or rapidly, as when due to vibration and shock. The dynamic forces are dominant in rocket flight, and vibrations are a harsh aspect both of the launch environment and of environmental testing, generating often large dynamic forces. To analyse each mechanical assembly as a structure, we shall need the concept of inertia force to represent the reaction to dynamic acceleration. (Thereby, we can understand that a structure, or any part of it, is in equilibrium between actions and reactions; from the loads we calculate stresses.) Loads vary greatly, from the full thrust of a large rocket motor, via the inertial reactions of many distributed masses, down to the thrust of an attitude–control actuator, and the infinitesimal inertial reaction of each small or remote part. All these loads, as they occur, must be sustained. Further, the materials must sustain the loads without unacceptable elastic deflection, or permanent distortion, or buckling, or fracture.

The need for strength to survive the launch environment is clear enough, but stiffness is a less obvious requirement. Some space (and other) instruments must maintain their dimensions to high accuracy in order to perform well. An optical bench carrying lenses and mirrors at separations precise to a few micrometres is a good example. Perhaps the structure should not displace a component by even a micrometre when gravity forces disappear in orbit. The property of stiffness is the capability to resist deformation (elastic, as we shall see) under the action of force. But other, less demanding, pieces of structure must in any case be stiff, to resist vibration. As already remarked, rocket vibrations are harsh. Stiffness dictates that all naturally resonant frequencies of the structure will be high, and the vibration oscillations, induced by the noisy launch, acceptably small. We shall see that both the choice of materials and also the design decisions for the disposition of structure parts are important for stiffness, whether dynamic or static (in orbit) deformations are in mind.

The aims of structural–mechanical design are therefore summarised as

- adequate strength (to withstand tests and launch)
- adequate stiffness (to minimise deflections, whether launch vibration, or elastic relaxation when 'weightless' in orbit)
- least weight (at allowable cost).

The aims of this chapter are to

- discuss types of structural components and their qualities
- review the basic elements of stress analysis and deformation prediction
- explain spacecraft strength and stiffness requirements, including margins of safety
- introduce elastic instability problems
- outline spaceflight vibration and shock problems, providing some theory and guidelines for their solution
- review spacecraft materials from the structural–mechanical viewpoint
- offer a few points of good design practice.

Beyond design drawings and calculations lie mechanical workshop activities of manufacturing processes, mostly not specific to space instrument making, so omitted here. But assembly, integration and verification (AIV) are managed with special care, as will be clear from later paragraphs.

2.1   Space instrument framework and structure

In a structure such as that found in a spacecraft or launch vehicle we can differentiate between primary and secondary structure. Primary structure is that which carries and distributes principal loads between the sources of thrust and the more concentrated reacting masses, such as fuel tanks and installations of payload and equipment. Secondary structure is what remains, to sustain all the lighter items such as solar cells and thermal blankets. Only the primary structure is essential for a test under static loads. The partition is an engineering judgment, made from considerations of previous experience.

The baseplate, framework or chassis of a small instrument built into a spacecraft may rate as structure in the secondary class. But the distinction becomes artificial for a large instrument such as a space telescope frame or the support for a large antenna. Each component of structure needs to be treated according to its load–sustaining function.

A good piece of structure can be admired for the efficiency (economy) and elegance of its design whether it is regarded as part of an instrument or a spacecraft. Efficiency can be quantified in terms of the strength–weight ratio or the structure–mass proportion. Elegance is a desirable characteristic of functional design. A functional and efficient design may be recognized as elegant through subjective qualities such as simplicity and symmetry. The best form of structure usually depends on the job it has to do, and we shall now discuss some forms of structure.

A good design begins from a clear specification of loads, dimensions, and other physical constraints. Some constraints will be put onto both instruments and spacecraft by the possible need for disconnecting and separating them. We recognise an interface of connecting points or surfaces. There are significant interfaces between a spacecraft and its launch vehicle, as well as between the spacecraft and every piece of equipment built into it. The methods of engineering graphics allow the interface to be delineated and dimensioned. To such an interface the instrument structure connects. A crucial step is to define the constraining loads at each interface. Further below we examine the loads at the interface between spacecraft and launch vehicle, then proceed to trace the reactions from the instrument payload and its parts.

2.1.1   Forms of structure

The traditional container for an instrument is a box, and the tradition continues into space. But there is fierce competition for maximum accommodation for sensors and electronics at minimum expense of parasitic structure, and lessons from aerospace technology apply here. The paragraphs below discuss the design approaches to structures of minimum mass, the principal choice being between frames and shells. Solid, monolithic designs tend to be too strong and stiff, and are therefore unnecessarily heavy. This is true even for optical 'benches' (beams) where an enveloping frame, or stiffened (reinforced) tube, will be lighter for an adequately high natural frequency.

2.1.2   Shell structures

The best way to lighten a 'rather solid' construction is to hollow it out, to leave a more – or – less thin shell. Such a form is likely to have good strength in tension, compression or bending, with elastic stability. For a given system of loads it may be possible to proportion the shell so that stresses approach the maximum strength of the material, thus achieving a minimum weight structure. This ideal, named 'monocoque' from the French for a cockleshell boat, inspires the design of ships, aircraft and space vehicles.

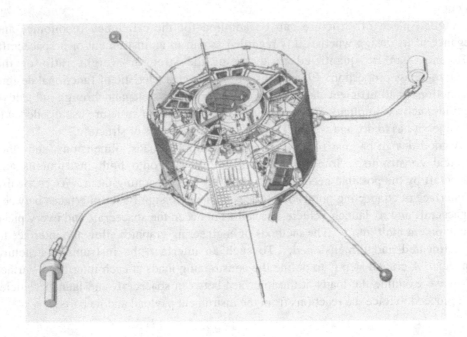

Fig. 2.1. AMPTE UKS (1984), a small geophysics spacecraft, based on
a stiffened conical shell primary structure (Drawing by F. Munger, with
permission. From Ward *et al.*, 1985).

The limitation of the thin shell is its predisposition to elastic instability and failure
by buckling when in compression. Weakness of this kind can be avoided by using
thicker material of lower density, adding stiffening ribs, or introducing honeycomb
sandwich panels (Figs. 2.2 and 2.15). There is a wealth of experience in successful
solutions found in aircraft designs (Niu, [1988]). An example is the tension field
(Wagner) beam, where shear loads are carried by the tensile strength of thin material
which has already buckled elastically under the compressive actions in a perpendicular
direction.

A practical disadvantage of the ideal shell is that it is closed. Structures containing
openings will usually require more material for the same strength and stiffness. There
will be stress concentration around the aperture, requiring careful analysis.

Boxes are shell structures. A weakness to be avoided is inadequate shear
connection of panels at their edges, causing low stiffness in torsion and bending.

Fig. 2.2. Primary and secondary structure. AMPTE UKS static test prototype,
showing primary stiffened cone within lightweight secondary framework,
which carries honeycomb sandwich panels for instruments (MSSL photo).

2.1.3   Frames

Frame structures might seem more appropriate to buildings and bridges (or
helicopters and some aircraft) than to spacecraft. Their superiority to shells is likely to
be found in situations where the optimum shell material ought to have a density lower
than any material actually available, or where general access to the volume inside the
structure is required (as in some racing–car chassis frames, or Serrurier trusses for
space telescopes, including the Hubble Space Telescope frame). Gordon, [1988] states
that the spaceframe is always lighter than the monocoque if torsional loading can be
avoided.

In the concept of the 'just stiff' spaceframe, economy is achieved by using the
minimum necessary (but sufficient) number of compression struts and tensile ties in a
three dimensional assembly. Sometimes with added members, which are redundant in
some loading cases, the spaceframe concept is applied in spacecraft. (It has long been
useful for aircraft engine mountings, large ground–based telescopes, ultra–light
racing cars etc.) Load distribution analysis by computer is almost obligatory,
following initial sizing by simple statics.

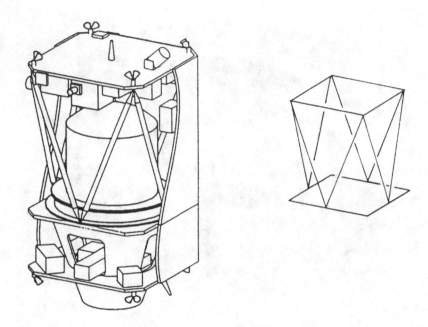

Fig. 2.3.   Frame structure proposed for STEP spacecraft (ESA, [1996]).
Note Serrurier truss of 8 canted struts, inset, springing from a virtually rigid base.

A space frame of the Serrurier truss type, as used also for large telescopes on the ground, is shown in Fig. 2.3.   More elaborate examples are the Hubble Space Telescope and the projected Space Station Columbus structure.

The plane frame is a two dimensional structure, easy to draw and analyse, but limited in its application due to out–of–plane flexibility.   Examples are found among solar arrays, but a honeycomb sandwich panel is usually preferred.   Some spaceframes are simply boxes of plane frames.

2.1.4   Booms

By the deployment in orbit of booms folded or telescoped into the confines of the launch vehicle fairing a larger spacecraft is made possible.   The boom may be a simple tube or a long frame.   Either way it is a long extended structure.   The unfolding of large solar cell arrays from body –stabilised spacecraft illustrates this.   The design of hinges and latches is an aspect of mechanism design (Chapter 5).   Some illustrations of hinges, telescopic booms and self–erecting booms are given in the references, e.g. Sarafin, [1995].   Both hinged and telescopic booms are seen in Fig. 2.1.

The deployment motion is a problem in mechanism kinematics and dynamics (again see Chapter 5). This kind of spacecraft dynamics can be complex when boom inertia is of the same order as the basic craft or the boom stiffness is low.

Gravity gradient tension occurs in long vertical booms. The force, calculable by orbital mechanics, is small. The tension between two 100 kg satellites tethered 60 m apart, the one above the other in orbits of mean radius 7000 km, is only 0.01 N (one gram force).

2.2   Stress analysis: some basic elements

Unless the reader is already familiar with mechanics of materials, the summary below should be followed by study of a detailed text, such as Benham, Crawford and Armstrong, [1996], or Sarafin, [1995]. To begin, we need to define some essential terms.

2.2.1   *Stress, strain, Hooke's Law, and typical materials*

Stress (σ) is load per unit area of cross–section. We distinguish tensile stress from compressive, positive sign for tensile, negative for compressive. Either way, this is a direct stress. Shear stress (paragraph 2.2.2) is found to be a biaxial combination of tension and compression stresses directed at 45° to the shearing force. Units (as for pressure) are newtons / metre ^{2} , i.e. pascals or megapascals; 1 MPa is closely 145 pounds per square inch (145 psi, 0.145 kpsi).

For the three dimensional nature of the most precise statements of a state of stress, and the components of the stress tensor, see the referenced text, or, for rigorous detail, other books on mathematical elasticity.

Strain (ε) is increase of length per unit length, and is a dimensionless ratio, negative if strain is compressive; it becomes an angle (γ) for shear strain. It follows from tendencies of materials to strain near constant volume that a direct strain ε_x in the direction of a direct stress σ_x will imply strain ($-\varepsilon_y$) laterally in the absence of any σ_y. There is a constant, Poisson's ratio (v), such that $\sigma_y = - v\sigma_x$.

Hooke's Law of elasticity is the statement that load is proportional to extension, from which we find that, within a limit of proportionality,

$$\sigma = E \varepsilon$$

where E is Young's elastic modulus, given by the initial slope of the stress–strain curve. E determines the 'stiffness' of the material. If the material is used for a tensile link in a frame, and the link has cross-section area A over length L, the link member's stiffness is defined as

$$\text{stiffness} = \text{extending force/extension} = \sigma A / \varepsilon L = EA / L$$

Stiffness requirements (section 2.4) are distinct from those of strength.

An elastic modulus occurs in a formula for shear stress (τ) related to shear strain (γ), in a way analogous to direct stress and strain, by :

$$\tau = G\gamma$$

and elasticity theory shows that

$$G = E/2(1 + v)$$

The *stress–strain curve*, Fig. 2.4, plotted from a tensile test, is typical for ductile metals, which show permanent plastic distortion beyond the elastic limit, an ultimate tensile stress (UTS) at the maximum strength, fracture (after plastic thinning of the sample's cross-section, known as necking), and often a high (many per cent) permanent elongation at fracture. Brittle materials fracture with little or no yielding or elongation; ceramics and glasses are typical.

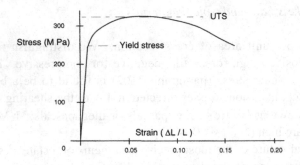

Fig. 2.4.   Stress – strain curve for alloy AlSiMg 6082.

For more discussion of yield stress, which is in practice the allowable stress for ductile materials, see section 2.7.2.

Some properties for five typical spacecraft materials are given in Table 2.1. For more information on materials, see section 2.7.

2.2.2   *Calculation of sections for simple design cases*

The properties of the different materials, shown in Table 2.1, can be used to estimate the minimum cross-sectional area required for various structural elements to stay within acceptable regions of material behaviour, under load conditions (see section 2.3).

Table 2.1. *Properties of typical spacecraft materials*

Material	Spec.	Density $\rho/\mathrm{Mgm^{-3}}$	UTS MPa	Yield stress σ_y / MPa	Elastic modulus E / GPa
Aluminium alloy	2014 T6	2.80	441	386	72
Titanium alloy	6Al 4V	4.43	1103	999	110
Stainless steel austenitic	S 321	7.9	540	188	200
Carbon fibre 61% uni − dir. epoxy	T 300 / 914	1.58	1515	not ductile	132 axial 9 trans.
Polyimide, Vespel	SP 3	1.60	58	N.A.	3

Tension members ('ties'): Anticipating elaboration later, proof−factored limit load, which is the load to apply in a proving test, is here called simply 'proof load'. Then, to avoid yielding under test,
minimum cross−section area required = proof load / yield stress of material.

Compression members ('struts'): Compression yield can be demonstrated but it is rarely published. So long as elastic instability and or plastic buckling (see 2.5.1 below) are not likely, the allowable stress can be approximated as the mean of the tensile yield and ultimate stresses.

Shear: See Fig. 2.5, in which a rivet is a typical member in shear.
Shear stress allowed is approximately $0.5 \times$ (tensile yield stress).
Then single shear area required is   proof load / 0.5 (yield stress).

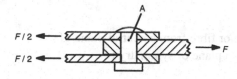

Fig. 2.5. Shear, illustrated by rivet loaded in double shear.

Bending members (or 'beams') : A beam is long, at least compared with transverse dimensions, and generally carries transverse loads. Starting from a diagram of all the external forces, which should be in static equilibrium, a cut is imagined at each section of interest, and then 'free–body' diagrams are drawn to reveal the internal equilibrium of *shear force* (*F*) and *bending moment* (*M*) , as in Fig. 2.6 (a) and (b). At the cut, *M* in newton metres is calculated from

$$M = R \times \text{ distance to its line–of–action}$$

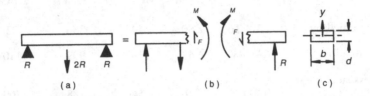

(a)   (b)   (c)

Fig. 2.6. (a) Beam under 3 – point loading; (b) cut to show shear force and bending moment in each of 2 free–body diagrams. End view (c) shows cross–section with neutral plane.

In the central plane of the beam there is no extension of the material and this is called the neutral plane, whereas the lower and upper surfaces of the beam are extended and compressed respectively. In calculating the effect of the geometry of the beam on its stiffness, small elements of material at a distance *y* from the neutral plane are considered. The further these elements are from the neutral plane (the larger the value of *y*) the greater is their contribution to the bending moment because the force acts with a larger lever arm. In addition to this factor, the stress increases with the value of *y* itself, since the extension of the material increases away from the neutral plane. These two factors involving *y* mean that the ability of the beam to resist bending increases as y^2 and so the stiffness can be increased markedly by positioning much of the material at large values of *y*, as in tubes or the skins of honeycomb panels.

It is shown in the referenced texts that, if plane sections remain plane (or nearly so, as in common cases), then

$$\frac{\sigma}{y} = \frac{M}{I} = \frac{E}{R}$$

Here,   *y* is the distance of a particle or fibre from a neutral plane of
       zero bending stress, usually a plane of symmetry for
       symmetrical cross–sections
    I is second moment of area of cross–section given by
         $\int b \, y^2 \, dy$ (see formulae below)
    R is the (local) radius of curvature of bending, varying along
    the beam as *M* varies
    $Z = I/y_{max}$, the 'section modulus', is used to give the simpler statement

$\sigma_{max} = M/Z$ for maximum bending stress.

Symbol y is also used to represent the transverse deflection of the neutral plane as a function of distance x along it.

Then $(1/R) = d^2y/dx^2$ very nearly, for dy/dx is small.

This gives a differential equation which can then be solved for the deflection y in terms of x, with two integrations, and insertion of boundary conditions of slope and deflection.

Torsion: A formula, similar to that for bending, is

$$\frac{\tau}{r} = \frac{T}{J} = G\frac{\theta}{l}$$

where r = radius from axis of torsion
$\quad\quad\quad T$ = twisting torque
$\quad\quad\quad J$ = polar second moment of area of bar (see formulas below)
$\quad\quad\quad \theta$ = angle of twist
$\quad\quad\quad l$ = length of twisted bar

This is used to analyse members such as the rod in a torsion bar

suspension. Torsional stiffness can be express as $\dfrac{T}{\theta} = \dfrac{GJ}{l}$

Table 2.2. *Geometric properties of common sections*

Cross–section	I	Z	J
Bar, rectangular	$bd^3/12$	$bd^2/6$	$bd^3/3$
Bar, circular	$\pi d^4/64$	$\pi d^3/32$	$\pi d^4/64$
Tube, thin wall	$\pi D^3 t/8$	$\pi D^2 t/4$	$\pi D^3 t/4$
Honeycomb plate (sandwich with thin skins)	$bd^2 t/2$	bdt	—

where $b =$ breadth
$\quad\quad\quad d =$ depth or diameter of bar; average depth between skins of honeycomb plate
$\quad\quad\quad D =$ mean diameter of thin tube
$\quad\quad\quad t =$ thickness of thin wall or sandwich plate

Plates : in bending under planar forces, behave much as wide beams, except that lateral strains near the surfaces, due to Poisson's ratio, are resisted (or else an anticlastic curvature is caused). It can be shown that this stiffens the beam. Hence, in deflection calculations, E should be divided by $(1 - v^2)$. Typically $v = 0.3$, and $(1 - v^2) = 1/1.1$. To a first order, stress is unaffected.

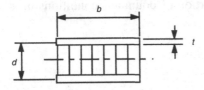

Fig. 2.7.   Section of sandwich panel and skins.

♦ *Example 2.1.   Bending stress in a sandwich panel*

A sandwich panel is to be used as an 'optical bench' for a 6 kg X – ray telescope package.
Design data :   transverse dynamic loads cause a maximum bending moment 400 Nm
Dimensions are : $b = 0.25$ m , $d = 0.019$ m , $t = 0.7$ mm
　　　　　　Calculation : From the table $Z = b\,d\,t$
　　　　　　　　　　Hence $\sigma = M / Z = 120$ MPa

2.2.3   The Finite Element analysis method

The basic formulae above have for long been applied to solve simpler stress analysis problems and to give insight into more complex ones. They are based on the mathematical theory of elasticity. One of the advances made possible by the use of computers has been to encode the elasticity equations for arbitrarily small elements of material, described as two dimensional plates (triangular or rectangular) or three dimensional blocks, then to ensure compatibility of strains at the common surfaces of adjacent elements. With appropriate software, the analysis begins by describing the whole load–bearing object in a mesh of element boundaries, and ends with a comprehensive statement of stresses and distortions. This computing tool is best understood by practising with the software. The principles are further outlined in Benham, Crawford & Armstrong, [1996]. It needs to be emphasised that computer programmes can only analyse a design already supplied to them as input, and the more practicable that initial design is, the fewer iterations are required in the design process. The use of simple conceptual models of the structure, which can be assessed and sized by means of the kind of formulae outlined above, is therefore greatly to be recommended as a start to the design task.

2.3   Loads

2.3.1   Loads on the spacecraft in the launch environment

A spacecraft is a system which, for our purposes, conveniently breaks down into subsystems labelled electrical power, communications, attitude (and orbit) control, command and data handling, thermal environment control, science payload, and structure.

From an early stage of design these subsystems will require allowances of mass and volume. Hence the whole spacecraft will have a size and mass determined by its intended mission, orbit and life. These factors will limit the choice of launching vehicle, and thereafter the launcher capability will constrain the mass and volume of the spacecraft during its development.

The spacecraft will usually separate from its launch vehicle. The separation interface is clearly an important one. The mating parts are likely to be of a standard design to facilitate connection and release. Clampbands with explosive bolts (or nuts) and tapering engagements are usual.

The spacecraft and its payload are accelerated by (1) thrust and vibration transmitted across the launcher interface and (2) acoustic excitation via the atmosphere within the launcher fairing or payload bay doors.

The direct interface loads comprise steady thrust, low–frequency (1–40 Hz) transient accelerations, and random vibration. The accelerations vary as fuels burn and stage succeeds stage (Fig. 2.8(a)). Lateral interface loads result from wind shear and trajectory corrections during ascent. The acoustic noise excites random vibration, particularly of spacecraft panels, directly. This noise originates from air turbulence due to efflux from the rocket engines at lift–off and from the turbulent flow around the whole vehicle, particularly after sonic speed has been passed, so peaks may occur early in flight and later at maximum dynamic pressure. (See section 2.6.4). As propellant is burned, the flight–path acceleration increases and the vehicle compression strain increases; the release of strain energy at engine cut–off excites a significant low–frequency transient vibration (Fig. 2.8(b)). Shock and high–frequency transients are caused at separation by the release of energy, both elastic and chemical, at the firing of pyrotechnic fasteners (Fig. 2.8(c)).

Launch environment data will be found in the launch vehicle manual (e.g. Ariane 4, [1985]). Launch–to–launch thrust variations will have been allowed for, usually on the basis of mean plus two standard deviations. With the mass estimated for a new spacecraft, the preliminary figures for longitudinal interface loads are found, as in the example below. The manual will guide the allowance for amplitude of low–frequency transient acceleration to be added to the steady value achieved at engine cut–off, the product of this quasi–steady acceleration with the spacecraft mass giving a quasi–static axial load to be used in preliminary spacecraft design. Subsequent analysis of dynamic coupled loads, taking into account a provisional

structure design and mass distribution, will permit refinement of this and other
calculations. The result of the calculation is a limit load. The limit loads calculated
for all design cases make up a set of all maximum loads expected. The probability of
these loads actually occurring during the service life of the structure may be quite low.
The loads are frequently expressed in terms of the acceleration producing them rather
than the force in newtons, and the acceleration is specified in units of the gravitational
acceleration, $g = 9.81$ ms^{-2}.

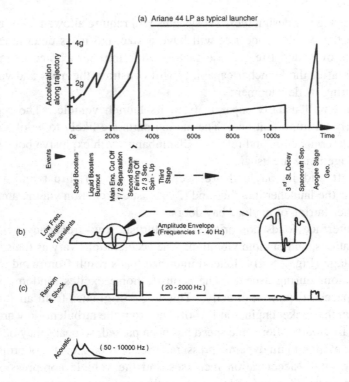

Fig. 2.8. Acceleration (quasi–steady, sensed, along trajectory), in successive
launch phases (Ariane 44 LP); (b) concurrent low–frequency vibration (1–40 Hz);
inset shows possible transient at main engine thrust decay. (c) Trends of
vibroacoustic, random vibration and shock excitations.

Lateral actions must be considered with the dominant longitudinal actions.
Preliminary lateral limit loads are also derived from launcher manual data, and may be
combined with the maximum axial load to create different patterns of loads to compose
design cases. For example, Table 2.3 for Ariane 4 limit loads shows + 1.5 g lateral
acceleration at maximum dynamic pressure (q) combined with a longitudinal of 3 g,
but 7 g at thrust cut–off, combined with lateral + 1 g, is probably a worse case for
most spacecraft. See worked example at section 2.3.5 .

Table 2.3. *Ariane 4 accelerations for calculating quasi–steady limit loads*

Flight event	Acceleration, apparent [*] units of g (9.81 ms^{-2})	
	Longitudinal	Lateral
Maximum dynamic pressure	3.0	± 1.5
Before thrust termination	7.0	± 1.0
During thrust tail – off	– 2.5 †	± 1.0

[*] The apparent acceleration is that sensed by an accelerometer along the longitudinal or lateral axis, and is a component of the vector sum of the true (kinematic) acceleration and the acceleration due to gravity.

† The negative sign denotes a retardation of the spacecraft during a vibration half – cycle in which the launch vehicle is in tension rather than compression.

An allowance for random mechanical or acoustic excitation of spacecraft modes of vibration may reasonably be added, but is a matter for engineering judgement by the spacecraft designer if made in advance of analysis and testing. Provisional estimates can be made by taking vibration acceleration levels lower than the specified test levels – say at two–thirds – and multiplying by an arbitrary magnification factor. This dynamic magnification, covering the probability of some resonance, is commonly given the symbol Q. A reasonable value to use initially is $Q = 4$. See discussion, at section 2.6.6 , which focuses on response to random excitation.

Transit and handling conditions, prior to launch, are typically transient accelerations about 3.5 g , and determine design loads for transit containers. They are unlikely to be critical for the spacecraft and its components.

2.3.2   *Design loads for instruments and equipment*

The structural design for an instrument assembly begins in the same way as for any other spacecraft unit. Data (such as for Ariane, Table 2.3) may be elaborated by the spacecraft contractor to specify accelerations which depend on the subunit centre of gravity position, with acceleration components given in a cartesian or cylindrical coordinate system. Shuttle payloads are often recoverable, so that landing cases must be examined too. If components can occur simultaneously, the superposition principle allows stresses to be calculated for each component separately, then finally added.

The specified accelerations permit the calculation of inertial reactions. The loading actions that cause these reactions are sometimes called the quasi–static limit loads. These forces include those which determine bolt or fastener sizes and the cross–section areas of load–bearing joints.

But we should also consider vibrations at those higher frequencies which excite resonance of an instrument's structure. Immediately many complexities enter the picture. The instrument assembly is a piece of secondary structure. At this stage there is neither a detailed mathematical model of the instrument structure, nor knowledge of how the spacecraft structure will modify the launcher excitations.

In any case, the most exacting conditions imposed are usually the random vibration tests to qualification levels. Because of the uncertainty of flight vibrations, their complexity, and the statistical element in specifying them (see section 2.6.6 below), these tests are likely to be more severe than flight conditions and so may become the dominant design requirements. Hence, a way forward is to interpret the launch agency's random vibration test spectrum (Fig. 2.9) as a conservative envelope of response accelerations, using the formula $3.8 \sqrt{(f Q W)}$. (This is discussed further at section 2.6.6 below). This, on Fig. 2.9, shows that a response acceleration of order 100 g might be expected for the stiffer and smaller subassemblies, which will have the highest natural frequencies (e.g. 700 Hz, with $Q = 25$).

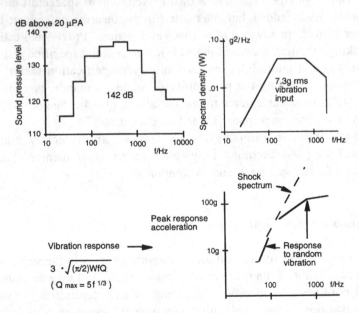

Fig. 2.9. Spectra of Ariane 4 environments for acoustic test (top left) and random vibration tests (top right). The bottom graph compares possible responses, derived according to the formula at bottom left, with the shock environment spectrum. Note that transient responses of 100 g are possible.

Another way is to examine computer output from dynamic load analyses for previous designs. From such studies a broad relation of acceleration to instrument mass can be attempted, as is discussed in the next section.

2.3.3   *Loads from a mass – acceleration curve*

It will be evident from the above that, in many instances, analysis for a projected design will show that dynamic loads are dominant. It may be that a mandatory vibration test to arbitrary levels will cause the highest load, although in the nature of vibration analysis, there will be uncertainties about damping terms and hence response levels. Various predictions of the limit loads may show a large degree of scatter. The uncertainty, expressed as 2 standard deviations, may be as much as the mean value again.

Another expectation is that heavier assemblies will respond less to transients than smaller ones because of their inertia and damping capacity. Since acceleration amplitudes tend to go down as linear dimensions go up, a $(-1/3)$ power law index, as in Fig. 2.10, makes a rough fit.

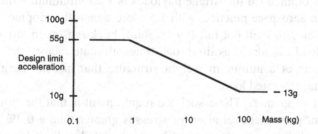

Fig. 2.10. Mass – acceleration curve (as specified for AstroSPAS by MBB).

Both scatter and mass dependance are shown in the work of Trubert and Bamford, [Trubert 1986], who proposed the mass – acceleration curve. From a large number of coupled load dynamic analyses for the Galileo spacecraft, some of them using 9000 finite elements, they graphed peak responses against generalised mass, then fitted a (mean + 2 standard deviations) curve to guide further design. Fig. 2.10 is an example proposed for preliminary design for MBB's Shuttle–launched Astrospas. Such curves, although based on the statistics of earlier design analyses, are somewhat arbitrary, but make useful starting points.

2.3.4   *Pressure loads*

Fluid pressures may arise in special circumstances, as where gases are stored in bottles and associated pipes and valves to supply ionisation detectors. Much lower pressure differentials may occur, transiently, during the venting of internal spaces as the launch vehicle climbs. A design rule for vent area is given under section 2.11 below.

There is biaxial stress in the wall of a cylindrical pressure vessel. Circumferential stress is (pressure) × (diameter) / twice thickness. The similar formula for longitudinal stress is ($p\, d / 4\, t$).

2.3.5   *Strength factors*

The structure must, at least, be adequate to sustain the limit loads (section 2.3.2 above). To ensure an adequate margin of safety, factors are applied to arrive at design loads. In principle, a margin is necessary to cover the possibility that the highest limit load was underestimated, or the stress analysis and materials data not conservative enough.

The ultimate factor recommended for Ariane payloads is 1.25 minimum – this is a typical minimum value in aerospace practice, with 1.5 more generally adopted. The requirement is that the structure shall not fail (by fracture, buckling, or in any other way) under an ultimate load calculated as limit load times ultimate factor. An ideal, perhaps never achieved, is of a minimum weight structure that has no margin of strength beyond the ultimate factored load.

The yield factor is 1.1 or greater. The associated requirement is that the structure shall not show significant permanent set after, or stresses greater than a 0.2% yield strain value during, the application of a proof (test) load calculated as limit load times yield factor.

Pressure vessels may be required by safety authorities to have higher factors.

The term 'reserve factor' is convenient to quantify the results of analysis, and so indicate any excess of strength :

reserve factor = estimated strength / required strength

The ideal minimum weight structure would be so sparely designed that all its reserve factors are unity.

Analysts also use:

margin of safety (MS) = reserve factor (RF) − 1

for which the design goal is, in the same spirit, zero.

♦ *Example 2.2.   Preliminary loads calculation*

A 50kg 'micro' – satellite is to be launched by ASAP, the Ariane Structure for Auxiliary

Payloads. Calculate the limit and factored loads, in the axial case, for the clamp ring.
Arianespace give axial acceleration to be 7 g, i.e. $7 \times 9.81 = 68.7$ ms^{-2}
limit load axially $= 50$ kg $\times 68.7$ ms$^{-2} = 3.43$ kN
Proof load $= 1.1 \times 3.43 = 3.78$ kN
Ultimate factored load $= 1.25 \times 3.43 = 4.29$ kN compression

◆ *Example 2.3.   Attachment loads for preliminary design of a large space instrument*

(i) A telescope assembly budgeted at 45 kg could be launched on a spacecraft for which the transient dynamic events are within the envelope of Fig. 2.10, which passes through 15.5 g at 45 kg.
If its CG is at the centre of 3 attachment bolts equispaced round a large circle, what is the factored tensile load per bolt ? Take yield factor 1.1.

Average proof load / bolt $= 1.1 \times 15.5 \times 9.81$ ms$^{-2} \times 45$ kg / 3 $= 2.5$ kN
(This load is additional to that caused by tightening torque.
A titanium bolt, dia. 5 mm with yield strength 965 MPa can carry
$965 \times 5^2 \times \pi/4 = 19$ kN and has ample reserve.)

(ii) If the telescope assembly includes a filter wheel subunit of 2 kg, and this mounts internally on 3 equispaced axial bolts on a 0.16 m dia. circle, installed 0.1 m axially from the subunit CG, what is the factored tensile load per bolt ?

Mass acceleration envelope gives 43.7 g for 2 kg, acting axially or laterally.
Consider the case of lateral inertia load offset (0.1 m) from the 0.08 m radius circle:
Load / bolt $= 1.1 \times 43.7 \times 9.81$ ms$^{-2} \times 2$ kg (0.1 m / 0.08 m) $= 1.18$ kN
(The same result is obtained for any orientation in axial roll of the equispaced – bolt triangle. It is additional to tightening load, as above.)

2.4   Stiffness

A structure can be light and strong enough yet deform too much; it lacks a measure of rigidity, or, as we prefer to say, stiffness. In the simplest case, a tie member in a frame, its stiffness is quantifiable from the definition of Young's elastic modulus:

$$\text{stiffness} = \text{load / extension} = \sigma A / \varepsilon L = E A / L$$

where σ, ε, A, L are stress, strain, cross–section area and length respectively.
Stiffness is constant and positive, apart from (i) a decline, which is usually negligible, as stress approaches the level of yield stress, and (ii) compression near to any buckling instability (see section 2.5 below).

It is sometimes convenient to discuss elastic flexibility in terms of compliance, the reciprocal of stiffness, i.e. deformation per unit load.

The deformation of a given structure, under a set of loads in static equilibrium, can be calculated by the methods (which include computer codes) of mechanics of materials. Parts will be identified and modelled as bars, beams, plates etc.. Of course, the overall stiffness of a structure derives from a combination of the stiffness of all load–carrying component members. (Where stiffnesses are in parallel, they add; so do compliances in series.)

Mathematical modelling, as a number n of members, of the details of the structure enables stiffness to be represented in a stiffness matrix, a diagonally symmetric array of terms k_{ij} each of which is the rate of force increase at node i per unit displacement at j. With the mass distributed between the nodes, as a column matrix, it enables the $3n$ natural frequencies of vibration to be calculated (see section 2.6).

Hence it is convenient to specify stiffness in terms of an arbitrary minimum natural frequency, perhaps 80 – 120 Hz for a small instrument, 60 Hz for an average subsystem, 24 Hz for a large instrument such as the Hubble Space Telescope, and 20 Hz axially for a whole spacecraft. This, from the interrelation of displacement amplitude, frequency and acceleration, (see section 2.6.1 below) constrains vibration displacement, for example 3 mm for a transient of 43 g at 60 Hz. This leads to a design rule–of–thumb that 'rattle spaces should be at least 3 mm', following from the observation that vibration impacts between inadequately stiff parts generate destructive shocks.

The orbiting 'weightless' structure, because its gravity is no longer opposed by reactions from its supports, will relax to an undeformed shape which (if it is particularly flexible) may be slightly but significantly different from its shape as erected on Earth. Critical optical or other alignments may change. (There is a parallel with ground–based telescopes whose elastic deformation changes with their orientation.) Such problems are solved by elasticity analysis to calculate the deformation due to the Earth gravity weights and reactions.

Elastic instability is a mode of failure of structures of insufficient stiffness which is discussed (at section 2.5) below.

Thermoelastic instabilities have affected some boom structures in space, sometimes enough to upset alignment and pointing accuracy. The transverse position of the end of a long tubular or framework boom is sensitive to gradients of temperature across the boom (due to differential expansion). In turn, these gradients could be influenced by transverse position, affecting solar heat absorption, for some sun angles. Feedback instabilities can be predicted by analysis and cured, at a weight cost, by more stiffness and better heat conduction.

2.5   Elastic instability and buckling

2.5.1   Struts and thin–wall tubes

The design of frames (Fig. 2.3) composed of struts and ties requires an informed decision on how slender a compression strut can be. A slender strut may bow sideways and collapse. Euler's classical analysis (Benham *et al.*, [1996]) gives the critical stress, for the bowing instability of a pin–ended strut, as

$$\sigma_c = \pi^2 E / (\text{slenderness ratio})^2$$

where slenderness ratio is : length L/radius of gyration ρ.

The quantity ρ^2 equals cross–section second moment of area, divided by area.

For a thin–wall tube, $\rho = r/\sqrt{2}$ where r is the tube's mean radius, so that

$$\sigma_c = 0.5 \, \pi^2 E \, r^2 / L^2$$

On a graph of σ_c versus (L / r), as Fig. 2.12, this relation for Euler bowing is a hyperbola. If the strut ends are fixed rather than free to tilt, the effective length is reduced, down to half L, and the critical stress for bowing instability is raised.

If this stress is close to the yield stress, because the material is ductile, the lower tangent modulus should be used instead of E. Alternatively, the Johnson's parabola approximation gives acceptable predictions for slenderness ratios where Euler stress would be above half the yield stress σ_y, with similar results:

$$\sigma_c / \sigma_y = 1 - \frac{\sigma_y \left(L / \rho \right)^2}{4 \, \pi^2 E}$$

However, a thin–wall tube can crinkle and collapse by local buckling, which is a mode of instability quite different from Euler bowing. It can be demonstrated by carefully stepping on the end of a drinks can; just before collapse, a dozen or so dents and bulges appear in a kind of pin–cushion pattern. It depends on thickness ratio, not slenderness, and may appear before the Euler bowing condition is reached. By classical theory for the elastic instability of a thin cylinder of thickness t, there is a critical stress given by

$$E \times \frac{1}{r} \times \frac{1}{\sqrt{3\left(1 - v^2\right)}} = 0.6 \, E \, \frac{t}{r}$$

for Poisson's ratio $v = 0.3$, which is close to the value for most materials.
This upper–bound formula overestimates actual test results (up to $\times 3$), because of defects such as imperfect cylindricity and end constraint effects. Lower bounds are estimated using empirical correlation factors, denoted γ, Fig 2.11.

Data from the NACA Handbook of Structural Stability are reproduced in Agrawal, [1986] and Niu, [1988]. A review by Hanna & Beshara, [1993] is in agreement. The ESA Composites Design Handbook ESA, [1986] adopts the simple coefficient $1/3 \times 0.6 = 0.2$, which is independent of (r/t). Following this lead, we may use, as a conservative approximation to the buckling stress of a thin–walled tube,

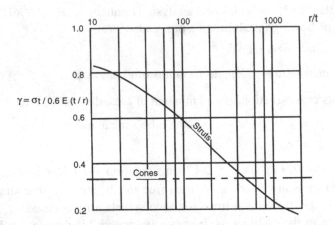

Fig. 2.11. Empirical correlation factors for local buckling of thin–wall tubes, plotted against (radius / thickness) ratio. A recommendation by NASA of $\gamma = 0.33$ for truncated cones matches an ESA, [1986] guide for struts of composite material.

$$\sigma_t = 0.2 \, E \, t/r$$

Hence it can be shown that there is a stress which satisfies the Euler bowing and local buckling relations simultaneously, given by

$$\sigma^3 = \frac{\pi}{20} E^2 \left(\frac{P}{L^2} \right)$$

The stress–like quantity (P/L^2) is the strut loading index. For efficiency we seek an optimum strut, carrying the proof–factored load P over its length L for a minimum value of mass m, given by the identity

$$m = \rho L \times \text{cross–section area} = \rho L \times (P/\sigma)$$

so that we can combine strut index with effective material properties to investigate strut density in the form of the parameter

$$\frac{m}{L^3} = \rho \left(\frac{P}{L^2} \right) \bigg/ \sigma = \left(\frac{P}{L^2} \right) \bigg/ \left(\frac{\sigma}{\rho} \right)$$

where ρ is material density, and (σ/ρ) is specific stress.

Evidently, to achieve lowest weight per unit length, the component must be designed for the highest specific stress. Since, for a given stress, thickness reduces as radius is enlarged, we maximise σ by finding the lowest slenderness ratio which just avoids local buckling. The maximum value would appear to be given by the σ^3 relation derived above. However, an empirical correlation factor is absorbed in this, and its validity is compromised by prescribing simultaneous instability in both bowing and local buckling modes. But a safe optimum design will exist near (if not at) this stress, with practical values for r and t determined accordingly.

For large values of the index (P/L^2), and if a ductile (metal) material is chosen, the optimum stress may be above half the yield stress, and cannot lie on the Euler curve. Bowing instability will occur at a stress better predicted by Johnson's parabola, or Euler's equation written with the tangent (plastic) modulus rather than the elastic (Young's) modulus. A graphical technique which answers this problem is to plot the local buckling stress as a line across the Euler–Johnson graph. By rearranging the formula for σ_t above, the local buckling stress is

$$\sigma_t \;=\; 0.178 \cdot \sqrt{E\,P}\,/\,r \;=\; 0.178\,\sqrt{E\left(\frac{P}{L^2}\right)}\cdot\frac{L}{r}$$

In this expression, σ_t appears as linearly dependent on (L/r) and as a function of the stress–like index (P/L^2). This is illustrated in Fig. 2.12, for the example below.

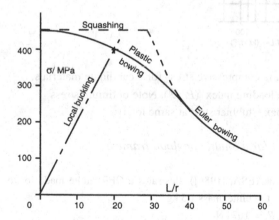

Fig. 2.12. Graph to determine optimum stress in a thin–walled tube in compression, plotted for the data given in Example 2.4.

Note that the crossed point of intersection gives simultaneous optima for compression stress and slenderness ratio.

Hence the intersection of the local buckling line with the Euler–Johnson curve determines the highest practicable stress, the optimum slenderness ratio, radius, thickness and weight. Note that the solution depends on the material properties E and σ_y on the one hand, and on the structure loading index (P/L^2) on the other.

The identity

$$r/t = 2\,\pi \left\{\,\sigma \Big/ \left(\frac{P}{L^2}\right)\cdot\left(\frac{L}{r}\right)^2 \right\}$$

which is provable by expanding the equation for stress, can be used to check the solution.

Figs. 2.13 and 2.14 show the influence of (P/L^2) on the optimum stress and density parameter (m/L^3) for example materials.

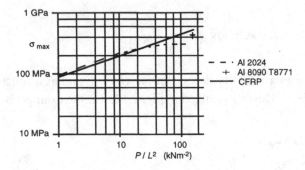

Fig. 2.13.   Strut maximum compressive stress for 3 candidate materials, as a function of structure loading index (P/L^2). Note optimum stress increases with higher index (stubbier strut at same load).

♦ *Example 2.4.   Optimum strut for a Shuttle payload frame*

The Composites Design Handbook (ESA, [1986]) illustrates a CFRP tube made for the frame structure of the SPAS Shuttle pallet satellite; mass 1.5 kg.

Design data are :   factored load P = 167 kN

length L = 990 mm   (between pin ends).

Is a lighter strut possible in a high specific stiffness alloy, aluminium − 2.4% lithium, specification 8090 ?

Material properties are: E = 79 GPa ρ = 2.54 Mgm^{-3}

σ_y = 450 MPa   (in T 8771 damage− tolerant condition).

Since $(P/L^2) = 0.17$ MPa   in this problem we find, for the local buckling stress,
$\sigma_t = 20.6$ MPa $\times (L/r)$

Sketching this on the Johnson—Euler curve drawn (for 8090 properties) as Fig. 2.14, and solving for the intersection with the parabola, we reach a provisional solution with $\sigma = 401$ MPa at $(L/r) = 19.5$ $r = 50.8$ mm and $t = P/2\pi r\sigma = 1.3$ mm.

This tube would have a mass of $2\pi rtL\rho = 1.0$ kg   (plus end fittings).

We conclude that the 8090 alloy strut appears to be lighter and a strong competitor against the composite design as flown.

An even lighter design results from using cross–ply carbon composite with the properties :
82 GPa,   600 MPa   1.49 Mgm^{-3}.

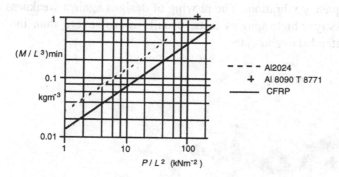

Fig. 2.14. Compression strut minimum specific mass;
variation, with structure loading index, for the candidate materials.

2.5.2   *Thin shells*

The elastic instability of a thin cylinder used in bending may lead to crippling by buckling on the compression side. The effective end load due to the bending moment M on a thin cylinder of radius r is $(2M/r)$, causing circumferential line load ' flux' $(M/\pi r^2)$, and stress $(M/\pi r^2 t)$. Compared with pure end loading, higher stresses may be achieved, since there is a chance that the worst initial imperfection is not in the zone of highest compression. An empirical correlation is given by Niu, [1988]. Refer also to Agrawal, [1986] for panels with attached stiffeners and also honeycomb sandwich panels (Fig. 2.15, overleaf).

2.6   Spacecraft vibration

2.6.1   *Rockets and mechanical vibrations*

We have already noted that dynamic loads, particularly vibration test loads, are critical for the spacecraft instrument designer. Also we have drawn attention to the role of stiffness, in relation to given mass, for promoting acceptably high natural frequencies.

The vibrations found in rocket vehicles during the spacecraft's launch are random, as with other transport vehicles. Mostly they are not the periodic vibrations which occur in rotating machinery, as discussed in introductory vibration theory. Further, they may be very intense, particularly in the vicinity of rocket efflux, which radiates a strong acoustic field. At release and staging there are shocks, which excite transient modes of higher frequency vibration. The proving of designs against weakness under vibration is a necessity which spawns tests, usually more severe than the actual vibrations they are intended to simulate.

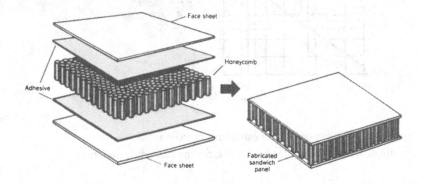

Fig. 2.15. Sandwich panel, showing typical construction. See 2.1.2, 2.5.2.

On the other hand, the rocket–powered launch period is short. The aggregate of the combustion periods of all stages is measured in minutes, and a typical test lasts 2 minutes in each of 3 perpendicular axes. Hence, the failures to be avoided are not in most cases those which occur in other transport machinery after many millions of fatiguing stress cycles, but are effects of high stress found towards the low cycle end of the $\sigma - n$ (or $S - n$ or stress / number of cycles) curve. Nevertheless, the failures are recognisably fatigue phenomena; that is, there are (i) stress–raising features in a failed design, (ii) cracks caused by the high stress, or perhaps imperfect material processing, (iii) crack propagation due to stress cycling and (iv) terminal rapid fracture.

We review vibration theory only to highlight the relevant elements of analysis, supplementing these with remarks about good design practice. Mathematical symbols are summarised at the end of the book.

Useful relations between vibration displacement and amplitude can be derived from consideration of the sinusoidal vibration $x = A \sin \omega t$, whose displacement amplitude is A at frequency f (Hz) with $\omega = 2\pi f$.

Differentiation yields a vibration velocity

$$\dot{x} = A \omega \cos \omega t$$

and vibration acceleration

$$\ddot{x} = -\omega^2 x$$

showing that the peak acceleration and displacement amplitude are related by (frequency)^2. These often useful relations are charted on Fig. 2.16. Note that, in practice, vibration accelerations are given in units of g (9.81 ms^{-2}) rather than directly in ms^{-2}, the SI unit in calculations.

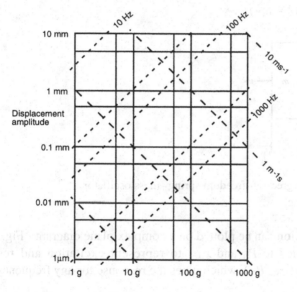

Fig. 2.16. Displacement amplitude versus peak acceleration in units of g (= 9.81 ms - 2); for varying frequency.
Loci of peak vibration velocity also shown.

2.6.2   *The simple spring–mass oscillator*

The equation of motion of a mass m on a light elastic structure (or spring) of stiffness k, subject also to some viscous damping (as in Fig. 2.17) and excited by a sinusoidally varying force of peak value P_0 is

$$m\ddot{x} + c\dot{x} + kx = P_0 \cos \omega t$$

or $\qquad \ddot{x} + 2D\dot{x} + p^2 x = \dfrac{P_0}{m} e^{i\omega t}$ $\qquad\qquad\qquad\qquad$ (1)

where $\quad 2D = c/m, \quad p^2 = k/m$

and the use of $\exp(i\omega t) = (\cos \omega t + i \sin \omega t)$ implies that any terms in the solution which turn out to be labelled by i can be ignored as superfluous. Leaving aside the complementary function part of a complete solution of equation (1), since it represents a damped transient vibration only, we find

$$x = \frac{P_0 \cos(\omega t - \phi)}{m \sqrt{(p^2 - \omega^2)^2 + 4D^2\omega^2}} \qquad\qquad\qquad (2)$$

where $\quad \tan \phi = 2D\omega /(p^2 - \omega^2).$

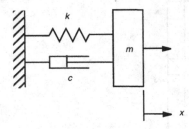

Fig. 2.17.   Single degree–of–freedom spring–mass oscillator, with viscous damping.

This steady – state vibration can be plotted on a complex plane diagram (Fig. 2.18) with rotating phasors, scaled to P_0 and x_0 , to represent excitation and resultant motion. The complex ratio (x_0/P_0), which gives the response for any frequency :

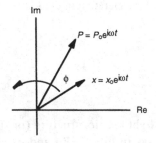

Fig. 2.18.   Complex plane with rotating phasors.

$$\alpha(i\omega) = \frac{1}{m\,(\,p^2 - \omega^2 + 2iD\omega\,)}$$

$$= 1/(\,k - m\omega^2 + ic\omega\,) \qquad (3)$$

This ratio, the receptance, reduces to compliance $(1/k)$ at $\omega = 0$.

It is convenient to express the damping non–dimensionally by Q, the magnification at resonance. This is given by the peak value of the receptance and, assuming small damping, is

$$Q = \sqrt{km}\,/\,c \qquad (4)$$

At this point we recognise that material damping in structures is not actually viscous, as though with a multitude of distributed dashpots. Another model, the hysteretic, implies a dissipation of energy per vibration cycle independent of frequency (but increasing as x_0^2). With the small mathematical restriction that the motion be virtually sinusoidal, the spring term is specified to be complex and denoted

$$k(1+i\eta) \qquad \text{where } \eta \text{ is given by } (1/Q).$$

Clearly this makes damping concomitant with stiffness, which corresponds with the physical idea that both are due to strains at the atomic level. (The assumption that a viscous damping varies inversely with frequency has the same effect.) Hence equation (1) becomes:

$$\ddot{x} + \eta p x + p^2 x = \frac{P_0}{m}\,e^{\,i\omega t} \qquad (5)$$

The solution (2) for the response now is

$$x = \frac{P_0\,\cos(\,\omega t - \phi\,)}{m\,\sqrt{(\,p^2 - \omega^2\,)^2 + \eta^2 p^2 \omega^2}} \qquad (6)$$

with $\tan\phi = \eta\,p\omega/(p^2 - \omega^2)$

The receptance is again (x_0/P_0), now given by

$$\alpha(i\omega) = \frac{1}{m\,(\,p^2 - \omega^2 + i\eta p^2\,)}$$

$$= \frac{1}{k\,(1+i\eta) - m\omega^2} \qquad (7)$$

The magnification at resonance is simply given by the dynamic magnification factor

$$Q = \frac{1}{\eta} \qquad\qquad (8)$$

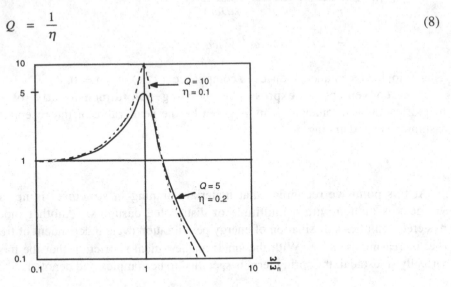

Fig. 2.19. Frequency response of simple oscillator.

The frequency ω_n at resonance, in the usual case where $\eta \ll 1$, is seen from (7) to be virtually at

$$\omega_n = \sqrt{k/m} \qquad (=p)$$

i.e. for a result in hertz $\quad f_n = \dfrac{1}{2\pi}\sqrt{\dfrac{k}{m}} \qquad\qquad (9)$

Non–dimensional graphs of the magnitude of the receptance, expressed as its ratio to the static displacement (P_0/k), are drawn in Fig. 2.19 for $\eta = 0.2$ $(Q = 5)$ and also for $\eta = 0.1$.

2.6.3   Multi–freedom systems

The spring–mass system with but one degree of freedom falls far short of realistic models of actual structures, but its behaviour is nevertheless an idealisation of each actual natural frequency.

The mechanical oscillator model, shown in Fig. 2.17, can be elaborated as a 2–degree of freedom system as shown in Fig. 2.20. Its frequency response curves, one for each coordinate of freedom, show two resonant natural frequencies. Each has a corresponding mode of vibration.

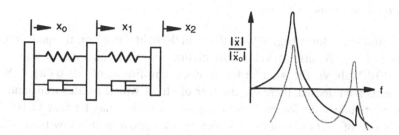

Fig. 2.20. Two degrees of freedom. Model (left) and frequency responses for each coordinate.

The step from one to many freedoms is made by considering the normal modes of a multi–mass system, as discussed in texts on vibration theory (e.g. Newland, [1990]). It suffices here to say that finite–element models can be assembled for N discrete masses, with $N = 10 - 100$ or more; when backed with powerful computing to manage the large matrices these will calculate a large number ($3N$) of eigenvalues and eigenvectors; these are interpreted as natural frequencies, theoretically independent and uncoupled, each with its own distinctive mode. Each vibration mode is a representation of the shape of deformation characteristic for its frequency, and normalised to unity at a given point. When the actual structure is vibration tested, a corresponding set of modes and frequencies should be observable; but the unavoidable errors in choosing parameters for the model will lead to discrepancies, particularly in middle to high frequencies.

Fig. 2.21 illustrates a typical response among the many taken in a typical test.

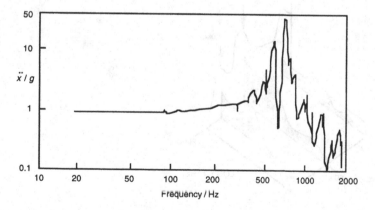

Fig. 2.21. Spectral analysis of response acceleration during instrument vibration test.

2.6.4   Launch excitation of vibration

The first stage burn begins with the vehicle held firmly on the pad. From a few seconds before lift–off, rocket motor efflux generates an intense noise field (possibly 140 dB above the standard reference amplitude of 20 μPa). When hold–down arms are released, there is a transfer of stress which excites transiently the axial modes, particularly the lowest ' pogo' frequency which may be near 15 Hz. The pogo mode is one of vertical motion in which masses above and below the centre of gravity oscillate in opposite directions.

As the vehicle climbs, the sound field is intensified by ground reflection, leading to a maximum random–acoustic excitation. The acoustic peak decays as the vehicle accelerates past unity Mach number in thinner air, but mechanical transmission through the structure will continue.

Until the dynamic pressure ($\frac{1}{2} \rho V^2$) peaks a little further into the flight, the spectral character of the vibration may change as boundary layer turbulence grows. Supersonic shockwave movement with flow separation will add its effects.

At main engine cut–off the vehicle axial modes will again be transiently excited, but at higher frequency (perhaps 30 Hz, due to reduced propellant mass) and there may be some 'chugging' (combustion instability, due to interaction with propellant feed).

The upper stage motors, burning beyond the atmosphere, transmit their random vibrations through vehicle and spacecraft structure.

The evolution and decay of vibration, during a first stage burn, is graphed against time into flight in Fig 2.22. The spectral content of some other samples of flight measurement is shown in Fig 2.23 (from Bendat & Piersol, [1971]).

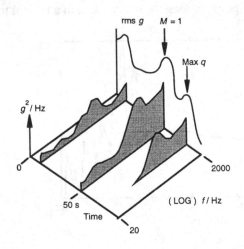

Fig. 2.22.   Vibration evolution recorded in flight of launch vehicle.

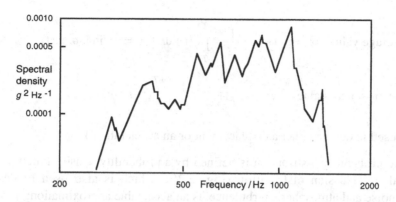

Fig. 2.23.   Spectral analysis of Nimbus spacecraft vibration near time of peak dynamic pressure.

2.6.5   *Random vibration and spectral density*

This section is an overview of the statistical concepts on which the analysis and specification of launch vehicle random vibrations are based. For further detail, see Newland, [1993].

Random accelerations such as might be represented by Fig. 2.24 are stochastic processes; that is, they are not strictly predictable, yet can be determined by probability laws. A probability distribution, deduced from an ensemble of similar records for successive launches, all sampled for the same time into the flight, determines its statistical properties. If any translation of the time origin leaves these

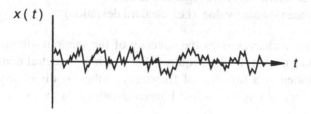

Fig. 2.24.   Random acceleration – time record, as at a point on a spacecraft structure vibrated by rocket combustion or turbulent airflow.

properties unchanged, the process is said to be stationary. The stationary process is called ergodic when ensemble averaging can be replaced by time averaging over a single sample record of substantial duration. The actual vibration environment is observed to change slowly; it is evolutionary. But the stationary and ergodic hypothesis is a good approximation for the purposes of engineering analysis.

Hence:

$$\text{average value} \quad \bar{x} = \lim_{T \to \infty} \frac{1}{2T} \int_{-T}^{+T} x(t)\, dt \quad (= 0 \text{ in lab. test})$$

$$\text{mean--square value} \quad \overline{x^2} = \lim_{T \to \infty} \frac{1}{2T} \int_{-T}^{+T} x^2(t)\, dt$$

(Here x is used to denote either a displacement or an acceleration.)

Now the probability distribution is defined by a probability density function; here, the normal or Gaussian distribution (Fig. 2.25), which is also used to describe electronic noise and atmospheric turbulence, is an acceptable approximation:

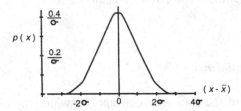

Fig. 2.25. Gaussian probability density distribution.

$$p(x) = \frac{1}{\sigma \sqrt{2\pi}}\, e^{-\frac{1}{2}(x - \bar{x})^2 \sigma^{-2}}$$

where

    $p(x)$ is the probability density

    $\bar{x}$ is the mean acceleration , here regarded as zero

    σ is the root--mean--square value (i.e. standard deviation)

Evidently the rms value σ characterises the severity of the random vibration. But information on frequency content is also required, for which the spectral density must be determined, and plotted as a function of frequency; either by direct analysis (as described below) or by computing an inverse Fourier transform of the autocorrelation function for the data.

Consider a vibration record of finite length T. From the mathematical theory of Fourier Series, it can be established that any physically realisable function of time can be expressed as a series of sine and cosine functions at harmonic multiples of a fundamental frequency $\omega_0 = 2\pi / T$. Thus $x(t)$, however random and variable, can be written as

$$x(t) = \Re \sum_{n=-\infty}^{\infty} C_n \, e^{\, in\omega_0 t} = C_0 + 2 \sum_{n=1}^{\infty} |C_n| \cos(n\omega_0 t - \alpha_n)$$

where

$$C_n = \frac{1}{2T} \int_{-T}^{+T} x(\xi) \, e^{-in\omega\xi} \, d\xi$$

is the complex amplitude and $2T$ is the record length.

The mean–square value is σ^2 or

$$\overline{x^2} = \frac{1}{2T} \int_{-T}^{+T} x^2(t) \, dt = C_0^2 + 2 \sum_{n=1}^{\infty} |C_n|^2$$

indicating that (σ^2) can be synthesised as a sum of components, up to the highest frequency of significance, matching some multiple of the arbitrary fundamental frequency, ω_0. See Fig. 2.26 (a).

The number of components can be increased 2–fold if the loop is lengthened by being spliced with a copy of itself, but new coefficients will be evaluated with about ($1/2$) of their previous values. See Fig. 2.26 (b).

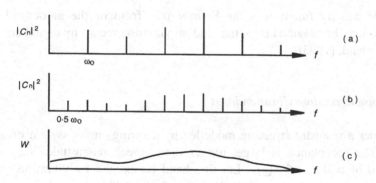

Fig. 2.26. Spectrum analyses of random vibration, as determined from magnetic tape loops or other vibration records.

(a) Record length T, frequencies $\omega_0 = (1/T), \ldots n\omega_0$;
spectral lines are Fourier coefficients, but only for these frequencies.
(b) Period $2T$, frequency samples doubled.
(c) Continuous spectrum, spectral density determined, as if from unlimited samples.

Since σ has a fixed stationary value, we see (in the equation immediately above) that the contribution to the mean value of (x^2) is inversely proportional to spectral line spacing $\Delta f \ (=1/nT)$, and

$$\frac{\overline{\Delta x^2}}{\Delta f} = \frac{2|C_n|^2}{\omega_0} = \frac{2|C_n'|^2}{\frac{1}{2}\omega_0} = W(f)$$

which is called the spectral density; it characterises the spectrum of the random vibration. The area under the spectral density plot gives the value of σ^2, and the ordinate of the plot is the acceleration variance per unit of frequency.

It is often called power spectral density because, being the square of an amplitude, it represents the spectral distribution of the rate at which excitation forces are doing work against dissipative forces. In this context we properly should say 'mean–square acceleration spectral density'. This name should be used if confusion with either random displacement spectral density, or random force spectral density, might occur

To summarise this overview, the spectral analysis of a random record, which could be performed electronically with a spectrum analyser using analogue filters, leads to a plot of spectral density, on which the bandwidth of the filters determines the spectral resolution. The rms acceleration is the square root of the area under the spectral density plot.

The spectral density function is the Fourier transform of the autocorrelation function, and so may be obtained from the random vibration record by digital analysis techniques (Newland, [1993]).

2.6.6   Response to random vibration input

Consider a resonant structure modelled by a spring–mass system of given (f_n) and Q. The receptance is large, higher than   (peak magnitude $/\sqrt{2}$), in a frequency band of width (f_n/Q). Let this band be excited by vibration whose spectral density W is virtually constant in the range $f_n \pm (B/2)$.

Intuitively we expect a mean–square response proportional to both input spectral density and the bandwidth of receptance. Newland, [1993] and others produce the result that, at the input

$$|\ddot{x}|^2 = WB \times \frac{\pi}{2}$$

Hence the rms input acceleration is

$$\sqrt{\frac{\pi}{2}WB} \qquad \text{where} \ \ B = \frac{f_n}{Q}$$

with motion at the resonant mass greater by factor Q.

Thus the rms response is

$$\sigma = Q\sqrt{\frac{\pi}{2} W \frac{f_n}{Q}} = \sqrt{\frac{\pi}{2} f_n Q W}$$

The probability distribution of acceleration data has been supposed Gaussian, implying that the peak accelerations are in a Rayleigh distribution (Fig. 2.27). It follows that if we choose to ignore those high peaks with probability below 1 in 100, we can estimate the highest response acceleration as 3σ, that is

$$3\sqrt{\frac{\pi}{2} f_n Q W} \cong 3.8\sqrt{f_n Q W}$$

(This quantity is sometimes called Miles' criterion.)

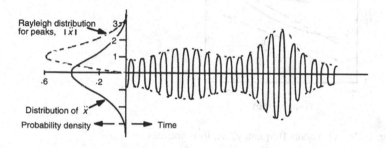

Fig. 2.27. Rayleigh and Gaussian distributions for the narrow band, quasi–resonant response of a structure vibration mode.

Since this result implies knowledge of the bandwidth of the vibration's response, it can be applied to the estimation of acceleration wherever B can be found, as by an appropriate resonance search experiment, and is not restricted to idealised single–degree–of–freedom cases. Fig. 2.28 illustrates how B can be derived from a spectral density plot for use in the response acceleration formula:

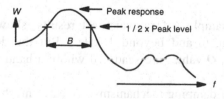

Fig. 2.28. Determination of resonance bandwidth from spectrum of response, in vibration test.

$$ 3.8 \sqrt{f_n Q W} \quad = \quad 3.8 f_n \sqrt{\frac{W}{B}} $$

If an appropriate mass can be estimated – and this requires sound judgement – then this formula can guide the calculation of vibration–induced force and stress levels, always provided that W is given reliably (as it should be in predicting for the effect of a specified test) and that the (f_n , Q) for each resonance are known. Since the latter is usually not the case, the formula is limited to extrapolations from statistics of Q versus f_n , such as Fig. 2.29. Accepting the absence of precision, its value as a guide

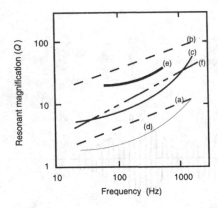

Fig. 2.29. Q versus frequency, various sources :
(a) (b)   R A E ,lower & upper 95% confidence limits ;
(c) North American, mean ;   (d) (e)   D H , lower & upper limits;
(f) $Q = \sqrt{f} / \mathrm{Hz}$.

to design judgement remains. The criterion is, that if the quasi–steady acceleration which the parts are calculated to withstand exceeds $3.8 \sqrt{f_n Q W}$, then the unit is probably strong enough.

2.6.7   *Damping and Q data*

Fig. 2.29, compiled from samples of relevant test results, shows some tendency of Q to increase with f_n up to and beyond 1 kHz. But the dominant observation is that, for any frequency, Q values are scattered within a band nearly a decade wide.

This scattering implies a variety of damping mechanisms. A basic mechanism is Kelvin's thermoelastic effect, whereby a strained material shows a small change of temperature, and energy is radiated away. Then η varies between materials, being lower (10^{-4}) for the stiffer materials, higher (10^{-2} or more) for glassy polymers, and yet higher for rubbery polymers. Metals stressed beyond elastic limits, composite materials, honeycomb panels, and glued structures are lossy. Assemblies made by rivetting or bolting exhibit joint slip with lossy Coulomb friction and lower Q.

An oft–repeated observation in vibration testing is that fully integrated assemblies show lower Q (more damping) than structure test models. Another is that low levels of input excite relatively high Q responses which do not increase proportionately with increased input, i.e. Q values are lower at higher test levels, with more joint slip. In that higher damping reduces stress at resonance this effect is welcome; so long, that is, as (i) joint slip does not allow unacceptable misalignment, and (ii) fretting does not generate particulate contamination.

◆ *Example 2.5.   Acceleration and inertia force resulting from random vibration input*

(i) A filter wheel for a space telescope has been designed as a pierced circular plate, perpendicular to the wheel's shaft. Under vibration the rim oscillates in the direction of the shaft axis, with a dynamic magnification, Q, of 6 ; this occurs at a lowest resonant frequency of 180 Hz.

Estimate the 3–sigma acceleration of the wheel–rim response to a random vibration input with spectral density 0.1 g^2 Hz^{-1} at 180 Hz.

For given data , $\quad |\ddot{x}| = 3.8 (180 \text{ Hz} \times 6 \times 0.1 g^2 \text{ Hz}^{-1})^{0.5}$
$$= 39.5 g$$

(ii) If the filters, their mounts, and their wheel–rim have total mass of 0.3 kg, what is the 3 – sigma inertia force which the shaft bearings must transmit in the axial direction ?

Force $= 0.3 \text{ kg} \times 9.81 \text{ ms}^{-2} \times 39.5 = 116 \text{ newtons}$

2.6.8   Vibration tests

The random excitation vibration test has been widely adopted as the principal test for proof of mechanical design. The only alternative would be a test in an acoustic chamber. A typical test requirement, whether for the whole spacecraft or a subunit, is 2 minutes in each of 3 perpendicular axes to some arbitrarily simple spectrum such as shown in Fig. 2.8(b). The flat peak of the spectrum contrasts starkly with the spikey spectra of records taken in flight. But it will have been evident from the paragraphs above that any Gaussian random test will be unavoidably arbitrary in several respects. The launch environment cannot be predicted accurately, and in the early stages of development of new designs arbitrary overtests are less risky than no tests. Such tests are of much value in exposing weaknesses in design, but they fall short of being simulations of the environment.

The means of exciting a test specimen, by using an electrodynamic vibrator, with spectrum generator, amplifier, acccelerometers, and brackets is outlined on Fig. 2.30 (see also McConnell, [1995]). It is standard equipment in environmental test laboratories, and is usually also capable of performing swept sinewave tests (e.g. Fig. 2.21). Hence the low frequencies, 5 – 100 Hz, are usually excited sinusoidally, as well as by that low–power part of the random spectrum which overlaps into this range.

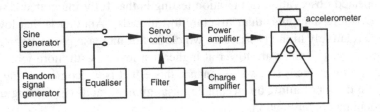

Fig. 2.30. Sine and random vibration test system.

In the management of a development programme the opportunity arises to test the component units and instruments for a spacecraft separately. The natural course is that a research institute or instrument manufacturer will expect to build prototypes independently of the spacecraft constructor. But the separation of parts of an assembly of coupled units will disturb significant dynamic interactions, and it is not practicable to redefine the complex force spectrum for the interface. In theory the test spectrum intended for the whole spacecraft could be modified, by a frequency–dependent transmissibility factor, to reshape it for the subunit; but a reliable computational model is rarely available.

The crude approximation sometimes made is to multiply by a factor (typically 4) right across the frequency range of test. If this is the Q value at some resonant frequency for the subunit, the effect will be right at that frequency, and an overtest at most other frequencies. For early development, the inherent conservatism is laudable. But, for proof–of–design qualification and flight acceptance tests later, less wasteful levels derived from tests of complete structures should be demanded.

Another view of this vexing problem is given by the Bamford curve. Coupled load analysis uses large computer models for the complete launcher / spacecraft payload assembly to calculate the transient responses to real or arbitrary transient launch inputs. Trubert and Bamford (Trubert, [1986]) showed by different coupled load analyses and test specifications how, starting from the relatively severe vibrations demanded for small components, lower vibration levels have been deemed appropriate for more massive assemblies and spacecraft. See Fig. 2.10.

Notching is another practice with the objective of bringing vibration test levels, as monitored by accelerometers, down to match force amplitudes set by coupled–load analysis or strain gauges. When applied, the notching is an amplitude reduction in the vicinity of spacecraft lowest frequencies.

2.6.9   *Vibration measurement and instrumentation*

The recording and analysis of flight and test laboratory vibrations depends on miniature crystal accelerometers. In essence, the accelerometer is a stiff piezoelectric crystal, assembled with a dense, dynamically reactive mass, and screwed or cemented onto the vibrating surface. (The latter might be a launch vehicle interface, a test shaker surface, or the extremity of an instrument on test.) The piezoelectric output is amplified and can be displayed on an oscilloscope or recorded on magnetic tape. Calibration of the system will have established a flat low–pass output from a few hertz to a few kHz, with higher frequency resonances filtered out. Random outputs can be processed by a spectrum analyser to display the excitation in contiguous narrow bands up to 2 kHz, the bands often either one–third octave or 25 Hz wide.

2.6.10   *Acoustic vibrations and testing*

As well as the mechanical vibration paths, the atmosphere transmits rocket acoustic energy. In basic acoustics the sound wave has pressure amplitude and frequency. The sound pressure level (SPL) of the amplitude is, by convention, given as n decibels (dB) according to the formula

$$n = 20 \log_{10}(p/p_0)$$

where p is the rms sound pressure and p_0 is a reference level pressure of 20 mPa rms. Hence, the random noise field of a rocket can be defined by a spectrum of octave–band sound pressure levels, implying an overall SPL, typically about 140 dB; see Fig. 2.9.

Agrawal, [1986] shows that a sound power spectral density (analogous to vibration acceleration spectral density) can be obtained :

$$s = \sqrt{2} \cdot \frac{p^2}{f} = \sqrt{2} \cdot 10^{0.1n} \cdot \frac{p_0{}^2}{f}$$

Techniques of statistical energy analysis (SEA) have shown some promise for computing the dynamic response of complex structures, with hundreds of conceivable vibration modes, to acoustic inputs given for different zones within the launch vehicle. See Elishakoff & Lyon, [1986] and ESA (Structural Acoustics Design Manual, [1987]).

In the light of the description of the launch vibration environment given in section 2.6.4 above, it is evident that a fully realistic simulation would combine acoustic with mechanical excitation, but this is too costly. Practical policy dictates that separate acoustic testing be reserved for large, light panels whose vibrations on a shake test machine would be heavily damped by the ground–level atmosphere. This requires a facility in which reverberant chamber, random siren, and microphones are provided (instead of vibration laboratory, random vibrator and accelerometers).

The design problems are outlined in the ESA Structural Acoustics Design Manual, (ESA, [1987]).

2.6.11   Shock spectra

Spacecraft and their payloads are subject to shocks, when pyrotechnic actuators are fired, as at stage separation or boom deployment. As remarked above, the effects appear as transient vibration responses of resonant modes. Launch manuals refer to a shock spectrum; this is a statement, for a given arbitrary value of Q, of the peak response of any of these resonant modes which occur in the frequency range of the spectrum (e.g. $10^2 - 10^4$ Hz). The response acceleration may be larger than would be expected in a vibration test, although the transient displacement amplitudes are quite small. Shock tests can reproduce the spectra, by calibrated blows, or transient pulse trains on an electromechanical vibration test machine. But experience indicates that the weaknesses are usually revealed in a vibration test. In any case pyrotechnic firings are usually performed in ground tests; this, except for unavoidable gravity, is more realistic than simulation.

2.6.12   Design for vibration

A space instrument must survive the launch environment and therefore it must survive the greater rigours of the vibration tests. The following code of practice is recommended:

i)   Every detail of design should be examined to eliminate stress–raising sharp corners and sudden changes of section.

ii)   Stocky, sturdy constructions are to be sought, well braced and buttressed. Flimsy and flexible parts are suspect. While there will always be some low–frequency modes, the aim should be to get as many as possible above 200 Hz.

iii) Every screw fastening should have a locking feature.

iv)   Materials (with their heat treatments) should be specified with precision so that quality can be assured.

v)   Test brackets and adaptors are worthy of as much careful design attention as the space hardware itself, with the difference that they should be stiff up to 2000 Hz where possible.

2.7   Materials

2.7.1   *Requirements for the launch and space environments*

Space is often regarded as a hostile environment for which only exotic materials will serve. This extreme view carries a grain of truth, but much has been achieved with well–tried good materials and sound engineering. The earlier section on structures and mechanical design showed how engineering analysis focuses on requirements for stiffness and strength. The purposes of this section are to show what the materials which have been used in space have to offer, and how the space environment constrains their choice. A table of selected spaceworthy materials is included (Table 2.5 at end of section).

Materials for space instrumentation, as for the spacecraft and its systems, should be strong, light, stiff, have low volatility, low outgassing and low sublimation in vacuum, have low thermal expansion and be resistant to penetrating cosmic radiation. Resistance to solar UV radiation and to atomic oxygen erosion (in low earth orbit, LEO) is required for materials on the surface of the spacecraft. Corrosion resistance is desirable for ground storage (sometimes for several years). Permanent magnetism is troublesome except where magnetic flux is ferromagnetically contained, as within an electric motor. The stability of dimensions, in the sense of resistance to creep (in response to locked–in preload stresses) at moderate temperatures, will be considered where alignment must be maintained, as with detectors and optics. Special properties such as optical transparency will be required for special purposes. For manned spacecraft, crew safety demands low flammability and toxicity.

At altitudes from 200 to 600 km, O_2 and O are the dominant constituents of the outer atmosphere (Chapter 1.2.1). Since the return of early Shuttle flights it has been clear that, because the highly reactive oxygen arrives at the spacecraft with a relative velocity of the order of 10 km s^{-1}, most exposed polymers are vigorously attacked and eroded. Data on fluence and erosion rates for many representative materials resulted from the long duration exposure in a LEO environment of LDEF, launched in 1984 and recovered in 1990 (Stein & Young, [1992]). There was contamination (a 'nicotine'–like stain) of metal surfaces, possibly from by–products of plastics erosion. The possibility of interactive synergism with UV radiation is a subject of current research.

The ultraviolet part of the solar radiation has long been recognized as damaging to exposed polymer materials including some paint binders, plastic films, and adhesives. UV photons may inflict a 'sun–tan' change of the thermo–optical properties which determine temperatures (Chapter 3), as well as causing embrittlement leading to cracking under thermal cycling.

The Van Allen belts of trapped protons and electrons are part of the spectrum of corpuscular radiation, which includes cosmic particles (See section 1.2.3.) above. This radiation may seriously damage semiconductor materials in detectors and sensitive circuits. But the particles can be absorbed by structural metals, which thereby act as space radiation shields without suffering significant damage. Optical glasses, however, may darken.

Meteoroid impacts may puncture a thermal blanket film but are relatively infrequent. LDEF was found to have collected particles in the size range of 0.1 to 2 mm at the average rate of one per sq. decimetre per year (Stein and Young, [1992]). See also space debris, section 1.2.4 above.

2.7.2 *Mechanical properties*

Stiffness is characterised by the elastic moduli and Young's modulus (E) in particular. The stiffness / density ratio or specific stiffness E/ρ is a useful figure of merit; the fact that it does not vary greatly between steels and the structural alloys of titanium, aluminium and magnesium would mean that, for a given weight, a similar stiffness is obtainable from any of them. But the advantage goes to the lowest density alloys where a large cross–section is required to resist elastic instability under compression. Among metals, the best specific stiffness is offered by beryllium (which, however, is brittle and somewhat toxic). Only high modulus carbon fibre composites can match it.

Strength properties are ultimate tensile stress and yield stress. The 0.2% yield stress is the stress sustained at the onset of plasticity, as indicated by 0.002 strain after unloading. It may be a little higher than the 0.1% yield for which 0.001 is taken as the permanent strain level. The yield stress is sometimes called the proof stress and is of more interest in practical design with ductile materials than the ultimate stress. The strength will depend on materials processing, particularly for strong metallic alloys, on forming operations, plastic working, and heat treatment. Treatments such as precipitation hardening greatly improve strength and are important for beryllium copper and many aluminium and other alloys. Forming and shaping operations are often required before heat treatment. Few materials are isotropic, but variations of strength according to direction in the material can usually be neglected except for composites. At the ultimate stress, a ductile material may be elongating prior to fracture, and the ductility makes the material 'forgiving', in the sense that the yielding allows alternate load paths to take more strain without overall failure. But a brittle material will fracture, by propagation of a crack. Materials which are strong and also resistant to cracking (often because they are ductile) are said to be tough.

Toughness can be quantified as a critical stress intensity factor, a material property which can be determined by experiment, and is related to ability to absorb energy during fracture. The result is usually quoted as the plane–strain fracture toughness. Fracture mechanics analysis can then assess the susceptibility of a loaded structure to possible cracks. The most immediately applicable use of toughness data, where available, is for calculating the critical crack length $(K_{Ic} / \pi \, \sigma_y)^2$, where K_{Ic} is the fracture toughness, and σ_y the yield stress. In situations where it cannot be guaranteed that there is no crack above the critical length, brittle fracture is predicted; otherwise, a high stress will cause yield before fracture. The critical length is in the order of a millimetre for many metals, composites and polymers, but only some micrometres for ceramics which are otherwise strong. See Fig. 2.31.

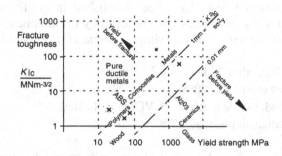

Fig. 2.31. Chart of fracture toughness versus yield strength. The sloping lines represent critical crack lengths.

Resistance to fatigue, which is the growth and propagation of cracks under oscillatory stress, is a desirable rather than a compelling requirement. Although the launch vibration environment may be critical if stresses are high (particularly due to stress–raising design features), the number of stress cycles is rarely high enough to require examination of a material's $\sigma - n$ curve. This would not be the case, however, for a component undergoing many stress cycles in a long orbital life. An example would be a flexure in a scanning mechanism, or material exposed to frequent thermal cycling.

A phenomenon affecting the strength and vibration resistance of some metals is stress corrosion cracking. This is a more subtle effect than rusting and surface oxidation; the latter are avoidable by surface treatments and careful storage. The corrosion is localised, with surface pit formation and intergranular effects, the atmospheric attack being triggered and sustained by stress in the affected metal. The stress may be due to the preloading of spring elements, or the unintentional result of differential contraction after welding, heavy machining, or other fabrication processes. Dunn, [1997] gives lists of SCC resistant steels and aluminium, nickel and copper alloys.

For springs of all kinds, the material should be strong enough to show a high strain (σ_y / E) before yield. Copper beryllium, given in Table 2.5, is an example. Not all such materials are also resistant to stress corrosion cracking, but many are.

2.7.3   Outgassing

Space vacuum exposes the tendency of materials to create a rarefied atmosphere, to sublimate, to outgas and release adsorbed gases. Even slight mass loss is undesirable, as is any propensity for evolved gases to condense on and contaminate optical and thermal surfaces. Contaminants on such surfaces may be polymerised by UV radiation and so change critical surface properties. Outgassing rates decrease with time in vacuum; for some materials they have been published in the literature of vacuum science. But measurement labour has been reduced by the adoption, both by NASA and ESA, of a standard test with relatively simple criteria of acceptability :

i)   Total mass loss (TML), after 24 hours at 125 °C
        and 0.13 mPa $(10^{-6}$ torr) :   < 1.0%
ii)   Collected volatile condensable material (CVCM), collected
        for 24 hours on an adjacent plate, at 25 °C :   < 0.1%

The criteria are obviously quite arbitrary; tighter limits might be sought for materials in the proximity of sensitive optics.

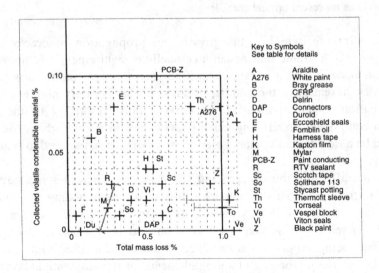

Fig. 2.32.   Graph of CVCM test results for spacecraft materials which
show significant outgassing. From Table 2.4.

The graph (Fig. 2.32) and table herewith show that, whereas metals and ceramics are generally acceptable, there is wide variability among polymers. (Compare the negligible outgassing of PTFE with the on–limits results for polyurethane paint specimens). There is also a fair degree of consistency between tests in separate laboratories on similar materials. Also note that probably most of a measured mass loss will have been uncontaminating water vapour, so TML outside 1% is less deplorable than CVCM outside 0.1%. Further note that percentage figures alone ought not to be invoked to exclude an otherwise useful outgassing material of which only a relatively small mass will serve the purpose in view. There is a rough correlation implying that 0.1% TML results from 1 μgs^{-1}m^{-2} outgassing as measured after 10 hours in vacuum.

Table 2.4. *Typical spacecraft materials , whose outgassing criteria are plotted on Fig. 2.32*

Listed in order of decreasing condensed contaminants (CVCM).

Material type	Name	Use	Key	TML (%)	CVCM (%)
Silicone	PCB–Z	White matt paint, conductive		0.6	0.10
Urethane	A 276	White paint		0.99	0.08
	Thermofit RT 876	Wire, insulation sleeve	Th	0.8	0.08
Silicone rubber	Eccoshield SV–R	Conductive seals	E	0.3	0.08
Epoxy	Araldite AV 100 / HV 100	Adhesive	A	1.1	0.07
Grease	Braycote 602	Lubricant	B	0.15	0.06
Epoxy	Stycast 1090 / 9	Potting foam	St	0.55	0.04
PETP	Gude Space DPTH	Harness tape	H	0.5	0.04
Urethane	Aeroglaze Z 306	Black paint	Z	0.92	0.03
Polyester	Scotch 850 Silver	Thermal tape	Sc	0.6	0.03
Epoxy	Scotchweld 1838	Adhesive	Sw	0.65	0.03
Silicone	RTV 566	Sealant, adhesive	R	0.27	0.03
Polyimide	Kapton H	(Thermal) film	K	1.03	0.02
Fluorocarbon	Viton B910	Rubber seals	Vi	0.5	0.02
Polyacetyl	Delrin 550	Plastic parts	D	0.39	0.02
Urethane	Torrseal	Sealing resin	To	1.0	0.015
PETP	Mylar A	(Thermal) film	M	0.25	0.015
Epoxy–Carbon	Cycon C89 / HM–S (40/60)	Structure composite	C	0.6	0.01
Urethane	Solithane 113	Potting resin	So	0.37	0.01
Fluoralkylether	Fomblin Z 25	Lubricating oil	F	0.06	0.01
Polyimide	Vespel SP–3	Machined insulators	Ve	1.08	0
DAP		Connector bodies		0.44	0
PTFE/glass/MoS$_2$	Duroid 5813	Bearings, composite	Du	0.08	0
Glass, woven	Betacloth	Thermal blanket	W	0.03	0

The materials selected by designers are usually checked against space agency criteria by product assurance personnel. During thermal vacuum testing, contaminants have been measured using quartz crystal microbalances.

In laboratory ultra–high vacuum practice bake–out at up to 300°C has long been standard practice. A temperature as high as this is often unacceptable for space systems because :

i)   it precludes the use of many useful polymers;
ii)   it weakens precipitation–hardened aluminium alloys, by
      coarsening the precipitates;
iii) creep strain rates, harmless at lower temperatures, may be
      dangerously accelerated.

But bake–out to (say) 70°C, maintained for several days, ought to be equally effective as a once–for–all measure to drive off volatiles before they can contaminate in orbit. (On the supposition that release of adsorbed molecules requires an activation energy, then the equation of Arrhenius applies and absolute temperature can be exchanged for duration on a power law basis.)

2.7.4   Thermal properties

Thermal conductivity plays a major role in heat balance, the subject of a later chapter. The structural metals conduct heat well, and their performance can only be bettered by heat pipes, which in any case are metallic containers for a 2–phase fluid. Ceramics, and polymers to a lesser degree, are thermal insulators. Heat capacity is the other thermal property of interest. Thermal diffusivity is the parameter $(K/C\rho)$, formed from conductivity K, heat capacity C, and density ρ, with units (m ^{2} s $^{-1}$), and is significant in some transient heat conduction calculations.

Heat conduction across joints and interfaces is sensitive to surface finish, as indeed are the thermo–optical properties (Chapter 3).

Table 2.5. *Some selected materials used in space instruments*

Specification	Density ρ/Mgm⁻³	UTS MPa	Yield σy/MPa	El. modulus E/GPa	Spec. mod E/ρ	CTE μm/mK	Thermal capacity J kg⁻¹K⁻¹	Thermal conductivity Wm⁻¹K⁻¹
Metal alloys								
Al 2014-T4	2.80	441	386	72	26	23.5	920	201
Al 5083-0	2.7	275	125	70	26	23.8		120
Al 6061-T6	2.71	310	120 +	69	27	23.8	960	171
Al 6082-T6	2.71	320	240	69	27	24	920	201
Cu3Be B101	8.24	1540	1080	128	16	17	420	107
Ti6Al4V	4.43	1103	999	110	25	8.8	523	23
Steel S321	7.9	540	188	200	25	16	500	15.9
Composites								
CF/epoxy T300/914	1.58	1515	not ductile	132 along 9 across	84			
Vespel SP3 polyimide glass	1.60	58	N.A.	3	2			
Ceramics								
Deranox 97.5% alumina	3.79	206 makers figure		338	89	8		9.3
Macor MGC	2.52	103 rupture modulus		64	25			
Glasses								
BK7	2.51	50 ±?	N.A.	81	32	7.1	858	1.11
Zerodur	2.53	10 - 80	N.A.	91	36	0.12	812	1.65
Polymer								
PTFE	2.2	17		0.34	0.15	100	1050	0.25

2.7.5   Selection of materials

Within the broad divisions of metals, ceramics, polymers and composites we identify (following ESA, [1990]) eight or nine classes for practical use. Within these classes are found less than one hundred materials judged fit for the requirements outlined above. Their proven reliability does not mean, however, that they are perfect; the strength and other properties of some typical materials are shown in Table 2.5. Launching agencies properly call for reviews of materials by specialists, and these are occasions for assessing the risks of promising but unproven materials.

2.7.6   Metals

The basic materials for structures are light alloys of aluminium. The aluminium–silicon alloys 6061 and 6082 perform well for machined parts. The aluminium–copper alloy 2014 is stronger but less resistant to corrosion, except in the Alclad sheet form. There is limited use of titanium for its lower thermal expansion, high–temperature strength and high thermal conductivity; also magnesium alloys and beryllium for lightness. Magnesium offers vibration damping. Screw fasteners are frequently austenitic stainless steel (e.g. 321), which is largely (unlike martensitic steels) non–magnetic; they are sometimes of titanium alloy. Plumbing is effected with aluminium alloys, stainless steels and titanium alloy. Circuit wiring employs copper and solder alloys. Plating for electronics, corrosion protection and thermal control is done with copper, silver, gold, aluminium, and nickel. Magnetic alloys of nickel, cobalt and rare–earth metals are required for motors and actuators. Memory–metal alloys (of nickel and titanium) can be designed into electro–thermal actuators.

Fabrication employs all the methods of manufacturing technology, i.e. bolting, riveting, welding, brazing, soldering and adhesive bonding. Aerospace industry standards ensure reliability.

It is important to note that the strength of a metal depends on the material's history of heat treatment (e.g. precipitation hardening, annealing) and working (extruding, forging). A full specification defines these aspects as well as the alloying composition.

Corrosion can occur any time up to launch. Electrolytic couples (ESA, [1990]) must be avoided or inhibited by insulation. Plating or painting are optional protections. Conversion coatings, such as chromate–phosphate (Alocrom) on alloys of aluminium, are quite adequate for clean–room storage. Anodising leaves an electrically non–conducting surface and is not preferred, but is unchanged by atomic oxygen exposure in orbit. Magnesium is at the bottom of the electrochemical potentials table and very prone to corrosion; chromating is virtually obligatory.

Cadmium, zinc and tin (in some solders) are to be avoided on account of high vapour pressure; they tend to form whisker crystals which may cause electric breakdown. Here is a reason, additional to their ferromagnetism, for avoiding cadmium–plated commercial steel bolts.   .

Metals are better than polymers, but worse than ceramics, in resisting temperatures. The aluminium alloys begin to lose strength at 150°C.

The propensity of clean, uncoated metal surfaces to cold weld if pressed and moved against each other under vacuum is a serious concern, but is avoidable by tribological design and surface treatment. See lubricants, section 2.7.13.

2.7.7   *Plastic films*

Plastic film material is incorporated in electronic components to perform as insulator, dielectric or printed wiring base. But the more exposed application is in multilayer thermal insulation, assembled from aluminised polyimide (Kapton) or Mylar. Flexible solar reflectors are made from metallised (Ag, Al) film. Gas–inflatable structures have been made from polyethylene–terephthalate film.

The polymers in common use are polyolefins, polyesters, fluorinated ethylene propylene, polyimide, polycarbonates and polyacetals. Assembly is by sewing (e.g. with PETP Dacron thread), adhesives, and adhesive tape.

The film may charge up electrically then breakdown destructively if exposed to the plasma in the spacecraft's orbiting atmosphere without having been electrically grounded.

The outgassing of plasticisers, if present from manufacture, may lead to stiffening of the film material. Thermal blankets must be vented.

Radiation deforms, embrittles and discolours films. On LDEF, 5.8 years exposure to atomic oxygen eroded silverised PETP blankets as much as 30 μm; it appears that AO attack was faster after UV had caused radiation damage. Meteoroids delaminated the metallising for about 1mm around the impact point, reducing thermal insulation marginally.

2.7.8   *Adhesives*

Modern adhesives are versatile, but no single glue will meet all bonding needs. Structural adhesives include modified epoxy film used (by specialists) to assemble honeycomb panels. Adhesives include epoxies, polyurethane, silicones and also cyanoacrylates. Insulators and glasses can be joined. Screws may be locked with epoxy or thread–lock resin. Conductive epoxy can ground a metallised blanket. Adhesive tapes are commonly used for thermal control, and a typical material is an aluminised polyester tape carrying acrylic pressure–sensitive adhesive.

Surfaces to be bonded require pre–treatment (abrading, cleaning, perhaps etching or priming). It is important to use an adhesive as the manufacturer intended. The control of humidity during curing can be important for the strength of the bond and outgassing in vacuum.

2.7.9   Paints

Thermal control is rarely effected without use of paint, but there are problems with application, outgassing, and deterioration in the space environment. A properly adhering coat of paint is an effective corrosion inhibitor on magnesium alloy.

Paints consist of pigments in a binder, which may be epoxy, acrylic, silicone, polyurethane or silicate. The pigments are particles such as carbon black, aluminium powder, or titanium dioxide, The paint usually needs a solvent, for application by spray or brush. Some paints are electrically conducting where spacecraft charging might be a problem. On aluminium alloys, conversion coating pre-treatment is advised (and is often an adequate finish on its own).

Degradation in orbit by UV and atomic oxygen particularly affects white paint. The LDEF flight showed that PCB-Z, a silicone electrically conductive matt white paint, was stable in LEO. Meanwhile, A 276 deteriorated to a degree which depended on the incidence of the oxygen fluence, because the surfaces darkened under UV exposure but were eroded by atomic oxygen.

2.7.10   Rubbers

We refer here to those rubbery polymers or elastomers from which seals and gaskets are extruded or moulded to their final shape, often an O-ring. A typical O-seal material is fluorocarbon rubber. A metal-filled silicone rubber, for conductive seals, has had some space use. UV generally hardens rubbers. Fluorinated rubbers and polyurethanes withstand ionising radiation better than commoner rubbers.

2.7.11   Composite materials

The carbon fibre composite materials draw attention by virtue of outstanding specific stiffness and high strength. In addition, the low thermal expansion offers good dimensional stability for optical assemblies such as telescopes and radiometers. A disadvantage not found with metals is some small degree of moisture absorption of epoxy resin matrix material, threatening dimensional stability (i) during ground testing if humidity is not controlled, and (ii) for the early orbits outgassing period.

Also, the composite is orthotropic rather than isotropic, especially if the fibres are unidirectional. The anisotropic properties are an essential consideration in design.

ESA has published a Design Manual, (ESA PSS–01–701, [1990].

The aromatic polyamide fibre Kevlar 49 shows excellent specific strength, and is listed by NASA, [1990] as used for structural members and pressure vessels.

The honeycomb sandwich panel is a widely useful structural component, frequently comprising two thin aluminium alloy sheet or biaxial carbon fibre laminate skins, spaced apart by core material of alloy foil formed into hexagonal cells, the whole being adhesive bonded.

2.7.12   *Ceramics and glasses*

Space instrumentation has included a few modern ceramics, such as high alumina ceramics and machinable glass ceramic; optical glasses should be included. (But see Chapter 6.) The concern is their brittleness with respect to the launch environment. However, they are stiff and sometimes strong, albeit with wide scatter of strength.. Fig. 2.33 illustrates a 95% alumina ceramic with a Weibull modulus of 2.5 (where 50 or more would be typical for the narrower scatter of a metal. See, for example, Weidmann *et al.*,[1990]). The ceramics make useful electrical and thermal insulators.

High alumina ceramic (such as Deranox 975) can be specially processed (by slip casting or hot isostatic pressing, then fired) to form components of the required shape. Precision is good enough for compatibility with machined metal parts. The materials called glass–ceramics are (monocrystalline) glasses transformed (devitrified) to the crystalline state by high–temperature heat treatment. Examples are CerVit, and Macor, which is machinable in the laboratory workshop. Optical solar reflector (OSR) tiles are fused silica glass wafers, polished and silvered on the back surface, to act as second–surface mirrors.

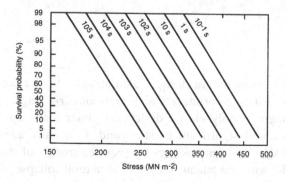

Fig. 2.33.   Strength / probability / time diagram for 95% alumina ceramic.
Plotted on Weibull probability axes, for various load durations. Davidge, [1979].

Foamed silica tiles have been used for protection against re–entry aerodynamic heating (on the Shuttle Orbiter and in European tests for similar vehicles). They illustrate the high–temperature resistance of ceramics in general. But not all ceramics have the thermal shock resistance of silicon carbide, silicon nitride, and CerVit.

2.7.13   Lubricant materials

Oils have uses limited by the necessity to seal them from space vacuum, unless of exceptionally low vapour pressure. The perfluoroalkylether oil Fomblin Z25 is reckoned to have a vapour pressure below 5×10^{-12} torr (1 nPa). The grease Braycote 602 is based on this oil, with PTFE thickener and added molybdenum disulphide, MoS_2. Barriers against oil migration are advised. Silicone oils are also very prone to creep–spreading, and are out of favour.

It is courting seizure to move surfaces over one another in vacuum without lubricant. The alternative to oil is the solid lubricant MoS_2, or a self–lubricating solid such as PTFE, a polyimide (e.g. Vespel), or the PTFE / glass / MoS_2 composite Duroid. Thus ball–bearing assemblies (which, for non–space application, would be oil lubricated) can be fitted with Duroid ball cages. A PTFE film is transferred to balls and rings during the necessary running–in, and thereafter good lives in space vacuum are achieved. Some instrument–size bearings are readily available with this construction.

Very thin coatings of MoS_2 or lead can be applied by ion or RF sputtering, by specialist institutes.

Gold is a special case of a soft metal solid lubricant ; the plating with gold of brass pins in electrical connectors protects them from corrosion before launch and facilitates demating and reliable reconnection, although remating in space vacuum may be required on manned missions.

2.8   Tests of structures

Tests greatly increase confidence in design calculations. It is a common practice to perform static load testing of primary structure at an early stage. Proof testing, to demonstrate that there is only elastic deflection under loads up to the yield–factored design load, without significant plastic yield, is usually adequate. Testing to ultimate–factored design load demonstrates the required margin of strength, but risks the occurrence of yield, with permanent 'set', or of structural collapse. Test to destruction is rarely specified, but may occur due to local miscalculation or accidental overload. Modes of failure comprise tensile fracture, unacceptable plastic yield, elastic instability, and buckling (under compression or shear).

In the past centrifuges have been built to test strength under launch accelerations, but fell out of use because of airdrag and size limits. The same purpose can be served by using a vibration test machine set in sinusoidal motion, to the required acceleration amplitude, at a frequency well below the lowest natural frequency of the test item.

It has already been remarked that the most exacting tests of a structure may be the random vibration tests to qualification levels, which are planned to be more severe than flight acceptance tests or actual launch conditions. An acoustic chamber may be used to excite similar vibrations (but in three axes of motion simultaneously) of a complete spacecraft fully integrated with its payload.

2.9   Low–temperature structures for cryogenic conditions

Because of the great improvements in signal–noise ratio obtained from silicon and other detectors, many instruments have been built whose parts, optics and mechanisms (but not all their electronics) operate at cryogenic temperatures, such as that of liquid helium II at 2.1 K. The special problems that arise are:

i) Differential shrinkage of dissimilar materials, leading to stress or change of pre–stress on cool–down. The contraction for aluminium alloy from 22°C / 295 K to 4 K is about 0.2% ; other materials are similar. Dimensions resulting at the cryogenic temperature from manufacture at 295 K must fit well, without unacceptable thermal stress. But the manufactured dimensions must be capable of being assembled, and probably tested too, at room temperature. This is likely to influence the design by constraining dimensions, limits and fits. It may require compliance (i.e. deliberate reduction of stiffness) to accommodate changes on cooling. Incidentally, cooling should be slow enough to avoid high values of transient temperature stresses. See example below.

ii) Change of material properties, such as increase of elastic modulus, brittleness and loss of ductility (but with increase of fracture strength). See Barron, [1985].

iii) The difficulty of performing tests, particularly vibration tests, at the cryogenic temperature.

♦ *Example 2.6.   Differential shrinkage*

The magnets for the etalon drive solenoids of a Fabry–Perot interferometer are held between nickel–alloy endcaps and secured to an aluminium–alloy chassis plate by a stainless steel screw through all four parts.

To find the increase in the screw's tensile stress after cool–down from 295 K to 4 K.

Although the aluminium shrinks more than the steel, the magnetic materials shrink less, as the following data for integrated shrinkage from 295 to 4 K shows:

Al alloy	6061	4.15×10^{-3}
Ni alloy	Permendur 24	$1.8 \ \times 10^{-3}$
Samarium cobalt	SaCo 5	$1.2 \ \times 10^{-3}$
Austenitic stainless steel	321	$3.0 \ \times 10^{-3}$

Dimensions (mm) are:
   Length of screw, from under head to mid–length of engaged thread: 9

Thickness:	plate	2
endcaps	2 + 1 = 3	
magnet	5	

Hence thermal interference, offset by tensile straining, is
   $9 \times 3.0 - 2 \times 4.15 - 3 \times 1.8 - 5 \times 1.2 \ = \ 7.3 \ \mu m$
Neglecting the slight compression of the magnet parts, average screw strain is
   0.0073 mm / 9 mm = 0.0008.
Hence screw stress due to cool–down is
   0.0008×230 GPa = 187 MPa,   where the elastic modulus is the value at 4 K

2.10   Mass properties

   Once detailed working drawings exist, good mass estimates can be made by calculating the sum of the (volume × density) products for constituent parts. There is an art in adjusting calculable block sizes so that trivial contributions of material under fillet radii can be ignored. However, mass guesstimates are needed long before sets of drawings have been completed. In this situation, knowledge of a similar instrument is useful. In the earliest stages of a project, we must adopt some rule of thumb, such as 'structure may be 20 (± 10) % of the whole system weight'. A contingency, of perhaps 15 % of each part, should prudently be added. Later it will be pruned away as actual weights are progressively measured.

   Systematic logging of masses and centre of gravity coordinates will be a part of good design management, using a computer. The reckoning of contributions to moments of inertia and products of inertia finds a natural place here.

2.11   Structure details and common design practice

   Before summarising the many considerations above, it is appropriate to offer some points of design practice which are widely advised for spaceworthy structures.

   The implementation of different structures for space flight has much in common with aircraft practice, from which it evolved. Detail design work of good quality is worthy of study, with drawings where available. Fig. 2.34 is a cross–section of a plasma detection instrument.

For frame structures, the tie members will be tubular with end fittings rivetted, welded or (for composites) glued on; for example titanium alloy fittings bonded to carbon fibre composite tubes.

Rivetting has not been superseded, even by advances in welding, and provides an energy–absorbing joint for vibration absorption, especially for plates in single shear, as in many box constructions.

Bolted joints tend to be heavier, but allow dismantling at interfaces. Lightweight nuts can be procured, and shakeproof stiffnuts are obligatory in those laboratories which work to the highest standards.

The range of modern adhesives includes epoxies and others which combine high strength with acceptably low outgassing properties (see section 2.7.8)

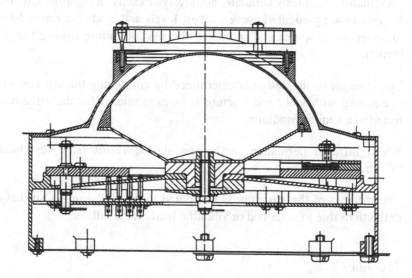

Fig. 2.34. Cross–section drawing of plasma detector   (MSSL Cluster Mission).

A mark of much successful and elegant vibration–resistant design is the generous provision of corner radii, to minimise stress concentration and so discourage fatigue cracking.

The venting of trapped air must be provided for by appropriate apertures and drillings. A guide for the vent cross–section is 10 (mm)² per litre.

2.12   Structure analysis and mechanical design: a postscript on procedure

2.12.1   Themes

At the end of this long chapter we can recapitulate certain themes for composing a space instrument design:

- The created design is some kind of a structure; that is, an assembly of materials, carefully selected, whose purpose is sustaining loads.

- The critical loads are generally dynamic, induced most often by the qualification vibration tests.

- The qualification tests simulate, not always exactly, a rigorous environment. Vibrations are particularly severe. Test levels and loads are raised by factors to cover uncertainties and minimise the risk of failure during the ensuing launch.

- The *strength* of the design is determined by computing the *stresses* in it and comparing with yield and fracture stresses tabulated for the same materials tested in a similar condition.

- No significant permanent deflection should result from test loads and vibrations.

- The *stiffness* of the structure is computed to be high enough so that elastic deflections due to sustained or vibrating loads are small.

2.12.2   Procedure

The flow chart, Fig. 2.35, plots a typical series / parallel progress from mission concept and instrument specification into instrument structure design. Perhaps with a backward glance at earlier designs, if any, an outline sketch will have emerged. As in other branches of engineering, progress to detailed drawings and manufacture can proceed by the following steps :

- Envision the size and shape of parts and the construction of assemblies, choosing form of structure (box or frame etc.). To avoid ambiguity and misunderstanding, nothing betters orthographic 3–view drawings, whether computer–created with CAD software or pencilled on a drawing board. Solid–modeller graphics software, and card or wood mock-ups, are powerful aids.

- Estimate mass breakdown.

- Establish the factored loads, and trace them through the main structural parts, checking that each part is in equilibrium. If launch vehicle unspecified, work with some typical but conservative data.

- With assumed sizes and cross–sections estimate stresses, and compare with strength of proposed material. Preliminary designs are usually analysed by approximate, even 'back–of–envelope', hand calculations.

- Revise parts that appear too highly stressed, or unnecessarily bulky and heavy.

- Estimate deflections, and lowest natural frequencies; invoke any stiffness, stability, or frequency requirement to establish if the evolving design is driven by stiffness rather than strength; revise as necessary.

At this stage the design may be sufficiently mature to allow the following activities to proceed in parallel :

- Finite element modelling for stresses, deflections, and natural frequencies.

- Detailed manufacturing drawings, with planning of logical manufacture and assembly.

Both the above are interactive, with iterations until the design, as defined by the dimensions on the drawings of significant parts, converges on the design requirements, and the drawings are issued for manufacture.

- As suggested near the bottom of the flow–chart, the manufacture and assembly of parts leads to trial fits with other assemblies, and so to early test. Where tests uncover errors in design, modifications must be made, with a fresh cycle of make–and–test.

- The modifications are recorded on the drawings, and the analysis updated.

- A final FE model, not necessarily of great complexity, but representing the essential dynamics of the instrument, may finally be composed, for incorporation in a total spacecraft–on–launch–vehicle dynamic model. This will permit the launch organisation to run a Coupled Dynamic Load Analysis, with the purpose of verifying all the structure analyses in the light of the best knowledge of launch excitations.

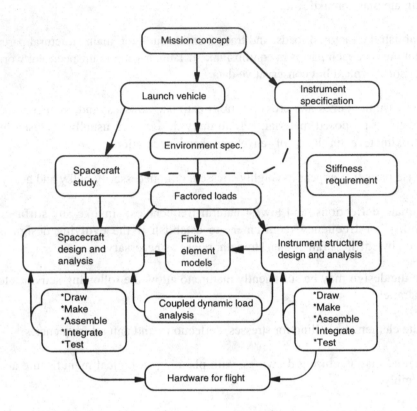

Fig. 2.35.   Flow chart, showing some typical steps and iterations
through structure analysis, from concept to hardware.

3

================

Thermal design

3.1   General background

3.1.1   Preamble

The temperature of laboratories in which space experiments are assembled, calibrated and tested is nominally 20°C (293 K) and it is thus not surprising that in general this is a most desirable operating temperature for that same equipment in space. There is nothing unique about this temperature. It is, within a relatively small band, a typical temperature that is experienced anywhere on the Earth's surface and, as fossil records show, has remained remarkably stable over billions of years.

Interestingly an Earth satellite is in a similar thermal environment, modified of course by the presence of the Earth. It is instructive therefore to consider what each of these thermal environments are and in what subtle ways they differ.

3.1.2   The temperature of the Earth

Essentially the Earth's surface temperature of about 290K results from the fact that the Earth orbits the Sun, which has a luminosity of 3.9×10^{26} W, and is at a mean distance of 1 Astronomical Unit (AU) from it, that is 1.5×10^{11} m . The emitted solar power crosses the surfaces of a succession of concentric, imaginary spheres centered on the Sun. The sphere which intercepts the Earth has a radius equal to the Astronomical Unit, so it is a simple calculation to show that the energy flux density at the distance of the Earth is 1.37 kWm^{-2}. Thus the power equivalent to a one bar electric fire is received across every square metre of the Earth's projected area. All the power received by the Earth is re −radiated back into space; if this were not so then a net amount of heat would either be gained by or lost by the Earth and thus its temperature would change accordingly until a steady state was reached. The mean temperature of the Earth is a measure of that balance between the rate at which energy is received and subsequently lost. It is assumed here that the Earth does not dissipate any stored energy. This is not entirely true since the core of the Earth is very hot and some heat is lost to the surface. More important, and certainly more topical, is that energy originating from the Sun and stored in the form of fossil fuels is now being burned at an ever increasing rate. Globally the direct effect is still negligible although the indirect effect through the release of by-products into the Earth's atmosphere, the so called greenhouse effect, is of considerable significance. As an aside however, it is interesting to note that the large amounts of heat lost from buildings in large cities due to poor thermal insulation is also significant to the extent that it does have an impact on the local weather. It was stated earlier that the Earth's temperature resulted from the amount of radiation received from the Sun and which is subsequently re-radiated. At night-time the dark side of the Earth does not receive its 1.37 kW m^{-2} although it is still able to radiate energy away. But night-time temperatures are not significantly different from daytime temperatures. This is because of the influence of the convective transport properties of the Earth's atmosphere in sharing out the Solar heat received on the dayside through atmospheric winds and oceanic currents. The Moon however does not have an atmosphere and so its dayside and nightside temperatures differ very significantly.

That the mean temperature of the Earth is about 290K can be demonstrated in a number of ways though two are of particular relevance. Both involve the application of Stefan's Law.

$$P = \sigma A T^4 \qquad\qquad\qquad\qquad\qquad (1)$$

where P is the power radiated
A is the effective radiating area
T is the temperature of the radiating surface
σ is Stefan's constant, 5.7×10^{-8} Wm^{-2}K^{-4}

The concept of a blackbody and blackbody radiation is an interesting and difficult one and the reader is referred to any standard physics text. For our purposes it suffices that a blackbody is a perfect emitter and a perfect absorber of radiation at all wavelengths. This condition is only approximated to in nature and in general bodies are not black over the full spectral range. At a given wavelength however the absorptivity and emissivity will be the same and will be some fraction of that which would be appropriate for a true blackbody. Thus the absorption and emission coefficients are scaled by the absorptivity and emissivity of the surface, these being numbers less than or equal to unity.

Assuming the Earth to be a perfect, spherical blackbody of radius R_E at some temperature T_E ,

$$\text{Power received} = \pi R_E^2 \times 1.37 \times 10^3 \ \text{ W}$$

$$\text{Power radiated} = 4\pi R_E^2 \sigma T_E^4 \ \text{ W}$$

Hence, under steady – state conditions

$$T_E = \left[\left(1.37 \times 10^3\right) / \left(4 \times 5.7 \times 10^{-8}\right)\right]^{\frac{1}{4}} = 278K$$

This result is applicable to a 'black' sphere of any size.

Another way to approach this same problem is to pose the question 'why is the temperature at the Earth's surface not the same as that of the Sun's surface (~ 6000K) since, under equilibrium conditions, temperatures equalise ?'. This would be the case if the Earth were completely surrounded by Suns, that is to say, if the Earth were in an oven the walls of which were at a temperature of 6000K. Thus in some way the Earth's temperature is scaled from 6000K by the fraction of the sky which the Sun occupies. Stefan's Law dictates the scaling to be by the temperature to the fourth power. The fraction of the sky occupied by the Sun is given by the ratio of the solid angle subtended at the Earth by the Sun compared with that subtended by the sky. The solid angle is defined as the area of the object normal to the line of sight divided by the square of the distance to it. Thus

$$\Omega = \frac{A}{d^2} \text{ st}$$

More readily understood perhaps is that the solid angle of an object as viewed is its projected area expressed not in terms of the normal metric area but by its angular area, that is to say to characterise the dimensions of the object in angular distances where the angles are expressed in radians (not degrees) and then to use conventional geometric formulae. To illustrate both these, using the former to calculate the solid angle subtended by the sky and thinking of the sky as being represented by the surface of a sphere of radius r gives

$$\Omega_{SKY} = \frac{4\pi r^2}{r^2} = 4\pi \text{ st}$$

and using the latter to calculate the solid angle of the Sun given that the *angular* diameter of the Sun is 1 / 2 degree

$$\Omega_{SUN} = \pi \, r_s^2$$

where r_θ is the *angular* radius in *radians*.

Thus

$$\Omega_{SUN} = \pi \left[\frac{\frac{1}{2}}{2} \times \frac{2\pi}{360}\right]^2 = 6.0 \times 10^{-5} \text{ st}$$

and therefore

$$\left[\frac{T_E}{T_S}\right]^{\frac{1}{4}} = \frac{\Omega_{SUN}}{\Omega_{SKY}} \qquad \therefore T_E = 6000\left[\frac{6.0\times10^{-5}}{4\pi}\right]^{\frac{1}{4}} = 280K$$

What this result exemplifies is that the Earth receives power in the form of radiation at a temperature of 6000K from a small part of the sky and re–radiates that power to the whole sky at a temperature of about 280K.

Clearly what has been shown above is a simplification. For example the Earth is not a perfect absorber since it has an albedo of about 0.3, that is to say about 30% of the incident power is reflected back into space without being absorbed. Neither is the Earth a perfect radiator. However it can be seen that the resultant temperature as calculated by the first method is unchanged provided that the Earth is as imperfect an absorber as it is an imperfect radiator, that is if the absorptivity and emissivity were not unity but none–the–less identical.

What we can see from the above expression is that the temperature of the *absorbed* radiation is 6000K (i.e. in the visible) and that of the *emitted* radiation is about 300K (i.e. in the infrared). If we now take the more reasonable view that the Earth is not a perfect blackbody our assumption of equal absorptivity and emissivity is no longer valid because of the fact that the associated wavelengths of the absorbed and emitted radiations are significantly different. This can readily be taken into account by the introduction of the coefficients of absorptivity and emissivity (α and ε respectively) into the equations we generated earlier for the power received and emitted by the Earth. The result of this is to give a steady – state temperature for the Earth of

$$T_E = 278\left(\frac{\alpha}{\varepsilon}\right)^{\frac{1}{4}} \; K$$

that is, the temperature is simply modified by the ratio α/ε. Bearing in mind that α and ε can each have values ranging from 0 to 1 the effect of the absorptivity and emissivity can be considerable (see section 3.5.2). In the literature α is conventionally associated with absorption (emission) at optical wavelengths and ε with emissivity (absorption) at infrared wavelengths. We should note again that when referring to the exchange of radiation at a *single* (or *similar*) temperature the ratio will have a value close to 1. For example this will be the case for any internal surface of a body which can be considered to be isothermal.

Were the calculation to be performed more rigorously taking into account the accepted values of absorptivity and emissivity the resultant temperature would be rather less than 280K. This would be the effective temperature at the top of the Earth's atmosphere. The temperature at the Earth's surface is higher than this because the atmosphere acts as a blanket and thus provides some thermal impedance to the flow of energy across which a temperature difference can be established. This is particularly noticeable if clouds are present at night – time or if there are clear skies. This principle is of importance and will be taken up later.

3.1.3   *The temperature of satellites*

As was stated earlier, this result is independent of the size of the object and so equally well applies to any spherical satellite in orbit round the Sun at a distance of 1 AU provided there are no internal heat dissipations. One obvious difference however is that a man – made satellite has no atmosphere and hence heat transport will be by radiative transfer and conduction only, not by convection. As will be seen later however thermal insulating blankets are used extensively in satellites and here there are some similarities in principle with the Earth's atmosphere. In general however satellites are not placed in orbit round the Sun but round the Earth either in so called near – Earth orbits or in highly elliptical orbits which can have apogees of order 10^5 km. The presence of the Earth, particularly in the former case, does have a significant effect on the heat balance of these satellites and hence on their mean steady – state temperatures. This effect is twofold. First, as already mentioned, the Earth is not a true blackbody and reflects about 30% of the incident sunlight back into space and this visible radiation which corresponds to a temperature of order 6000K, can also be intercepted by a satellite. Secondly the Earth re – radiates, at IR wavelengths, the absorbed solar radiation. This IR radiation (at a temperature of order 300K) can also be intercepted by the satellite. Thus in very general terms although satellites can spend some time of their orbit in the shadow of the Earth, a blackbody satellite would come to a steady – state temperature in excess of the equivalent steady – state temperature of the Earth. Hence one of the main problems of thermal design is to keep satellites cool rather than to heat them up. This is an oversimplification since satellites themselves cannot be treated as blackbodies but the scenario is a valid one.

One of the tricks of thermal design is to take advantage of the non – blackbody properties of thermal surfaces. We have already shown how the ratio α/ε can be used to control the mean temperature of a body exposed to sunlight. This obviously becomes more complicated in the case of a subsystem forming part of some complex configuration but the principles are the same. Designing for an acceptable mean temperature is therefore not very difficult since it results from balancing powers. However, this is not the real issue because of the presence of thermal gradients within a spacecraft. Heat input to the system is usually asymmetric and cannot be transferred from one part of the system to another in an ideal way. Thus the main problem of thermal design is one of adjusting the temperature gradients that exist within spacecraft in order to bring the temperatures of the different subsystems into acceptable ranges.

3.1.4   *Thermal modelling*

In order to do this in detail it is necessary to develop a thermal mathematical model on a computer which involves setting up and solving a large number of equations describing the radiative and conductive heat flows between the different elements of the model. The results are only as good as the model and the skill lies therefore in establishing an accurate and representative structure for the model. As this is a time consuming process and cannot therefore be iterated readily, it is usual first to establish a somewhat simpler approximate model in order to develop the design concept so that different ideas can be tried out relatively rapidly. Hence thermal design incorporates three main stages of modelling; the rather basic, simple calculations to determine the overall heat balance, the conceptual mathematical model to give an approximate idea of the thermal properties of the system and to identify the important components of the design, and finally the detailed model to predict the likely temperature distributions within the system.

The purpose of this chapter is to discuss the basic equations describing the transport of heat and how these are used in setting up a model, to describe how a conceptual and / or a detailed model can be constructed, to discuss the range of thermal materials and components available for both a passive and an active thermal design, and to discuss some of the relative merits of those conceptual designs. It is not the purpose of the chapter to give an exhaustive data base for thermal materials nor to describe the details and relative merits of different software systems used currently in thermal design. It is intended to adopt a physical, rather than an engineering, approach so that the physical principles underlying thermal design can be emphasized.

3.2   Heat, temperature and blackbody radiation

Heat and temperature form part of our everyday experience: we can feel warmth and our nervous system will react involuntarily when we touch something very hot or very cold. The physical meaning of heat and temperature however are difficult concepts to understand. It is quite possible to construct thermal models and to perform thermal analyses without an appreciation of what is meant by heat and temperature. However an insight into these concepts leads to a much fuller understanding of the physics of what is happening. This is of particular importance since we are increasingly having to accept the results given by computer modelling and find it difficult to question them. Although some of the following discussion may seem rather academic and somewhat obscure it does lead us to an understanding of 'emissivity' and 'absorptivity' which is particularly useful, especially in the context of multiple reflections. This is very important in thermal systems in which the radiative coupling of heat is the dominant heat exchange mechanism.

There is an obvious connection between heat and temperature; when a substance absorbs heat it gets hotter, that is its temperature increases, and when it loses heat its temperature falls. Thus heat is a form of energy and, since heat can be transported then heat must also be some form of energy that can also be transported whether by conduction through a solid, by radiation through a vacuum or by convection through a gas or liquid. In addition to a substance being able to gain or lose energy, it can also store energy and this stored heat is known as its internal energy. It is a general principle that energy can neither be gained nor lost by a body unless there is an available energy state to which that body can rise or to which it can fall. Thus the lowest energy state for this internal energy is that which defines the absolute zero point of Kelvin temperature scale; a body at this temperature can lose no further heat.

A reasonable question therefore is 'when a body absorbs heat, how much does its temperature rise?'. For a given amount of heat and for a given mass of a body, the temperature rise is determined by the specific heat of that body. This is analogous to the electric charge Q that can be stored on a capacitor of capacitance C; namely $Q = CV$ where V is the voltage. Thus Q is analogous with heat, V with temperature and C with the specific heat (sometimes called the specific heat capacity). Thus if a body has a high specific heat then the temperature rise will be small and if it has a low specific heat then the temperature rise will be large. Alternatively, a body with a high specific heat is efficient in storing its heat, that is it can store heat with only a small increase in its temperature. These effects are described by the relationship

$$U = c\,m\,T$$

where U is the stored energy,
        m is the mass of the body,
        c is its specific heat ($Jkg^{-1}K^{-1}$) and
        T is the temperature in degrees kelvin (K).

Thus when T is zero it has lost all its stored energy.

We can see from this that the unit of heat is the joule which is the same unit as that of energy and, for that matter, as work.

If we put two bodies at different temperatures in an enclosure and leave them they will come to the same intermediate temperature after some period irrespective of the substance comprising the bodies. This is accomplished by there being a net flow of energy from the hotter body to the cooler body until the temperatures are equalised. Hence we refer to a flow of heat (or to an energy flux if we are talking in terms of a flow per unit area). If this flow rate is high then equalisation will be achieved quickly, if not it will take longer. The units for heat flow is joules/second which is power. Thus we can readily see that power is the rate at which energy is being produced (or the rate of doing work). We talk here of temperatures equalising, that is, of the system coming to some thermal equilibrium. When the temperatures are the same there is no net flow of energy. This can equally be said of two different locations within the same body; hence when there are no temperature gradients within a body it is isothermal, in a state of thermal equilibrium and no heat flows. It is now at a single temperature throughout. But what is it that physically describes and quantifies that temperature?

Table 3.1. *The specific heat capacities of some common substances*

Aluminium alloy	920
Copper	420
Stainless steel	500
Titanium	523
Ceramic	800
Concrete	3350
PTFE	1050
Nylon	1680
Carbon fibre	840
Water	4190
Air @ STP	993
Helium	5250

Units are in $Jkg^{-1}K^{-1}$

Consider the following experimental setup shown in Fig. 3.1. A large number of identical billiard balls are fired at a given velocity at a hole in the side of a large box so that they enter the box. We assume the absence of gravity, that the box remains stationary, and that all collisions between the billiard balls and each other and with the box walls are perfectly elastic. We then wait for a billiard ball to emerge from the hole, measure its velocity, and register the square of its velocity on a histogram.

When we had measured the emergent velocity of a statistically significant number of balls (assuming this number also to be negligibly small compared with the total number in the box) we would then find that there was now a distribution of velocities rather than the single velocity with which the balls entered the box. The features of this distribution is that there are no balls with zero velocity (clearly otherwise they would not emerge from the box), that a few would have very high velocities but that most would have some intermediate velocity. Thus the histogram, which is a measure of the energy of the balls, becomes a continuous curve which goes through the origin, peaks at intermediate energies and is asymptotic with the energy axis.

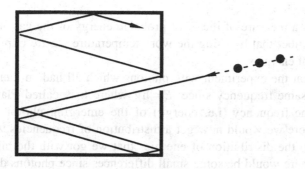

Fig. 3.1.   The box cavity.

This curve, shown in Fig. 3.2, is called a Maxwell–Boltzmann distribution and shows how the original input of energy has now been re–distributed between the billiard balls as a result of collisions (which in fact have to involve three or more bodies).

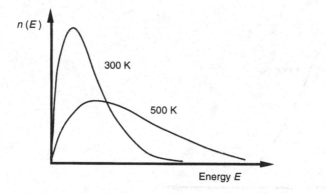

Fig. 3.2.   The Maxwell–Boltzmann distribution.

We would get the same result if we performed the experiment with gas atoms. Indeed the same curve can be determined from a statistical analysis and the curve can thus be mathematically described in terms of a variable which is called the temperature. By differentiating the expression, the energy corresponding to the location of the peak can also be defined in terms of the temperature, thus

$$\frac{1}{2}m\langle v^2 \rangle = kT$$

where $\langle v^2 \rangle$ is the mean–square velocity and k is a constant called Boltzmann's constant. This mean velocity would, in fact, be identical to the single velocity of the gas atoms as they entered the box.

Thus the temperature is a measure of the most probable energy of the gas atoms, but it is important to remember that by using the word temperature we are implicitly assuming this distribution of energies.

If we were now to repeat the experiment with photons which all had an identical energy (that is, all at the same frequency since $E = h\nu$ where h is called Planck's constant) and measured the frequency (i.e. energy) of the emerging photons and plotted a histogram as before we would now get a distribution of frequencies which would look very similar to the distribution of energies that we got with the billiard balls and the gas atoms. There would be some small differences since photons do not behave as atoms, for example they do not undergo elastic collisions although they can exchange energy with the walls of the box, but the curve would exhibit a peak as before with a most probable frequency. Again, this result can be reproduced from a statistical analysis, the differences arising as a result of the different way a photon is described as compared with a particle. This distribution so obtained is the Blackbody curve (or the Planck Function) and is again described by the single parameter, the temperature. This spectrum is shown in Fig. 3.3.

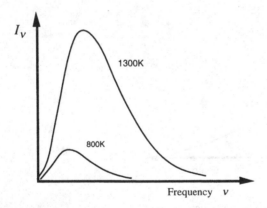

Fig. 3.3. The blackbody spectrum.

Thus we can ascribe the property of temperature to radiation in a similar way in which we ascribed it to matter. Indeed, by differentiating the Blackbody function we find that the value of the most probable frequency as identified by the location of the peak is given by

$$h\langle \nu \rangle = kT$$

This expression is very similar in form to that for the gas. Noting that the frequency and wavelength of light are related to the velocity of light by the expression $\lambda \nu = c$, then we can see that

$\lambda T = \text{const}$

This is the well known Wien's Law. The value of the constant is 0.03 mK. Thus it is that blackbodies at temperatures of say 6000K have a peak emission in the visible part of the electromagnetic spectrum at about 500 nm as is the case for the Sun, and that for blackbodies at temperatures of say 300K the peak emission is in the infrared at about 10 μ m as will be the case for say a wall in a heated house.

The photons in our box do not come to equilibrium through interactions with each other but through interactions with the atoms in the walls of the box, that is through the interactions of radiation with matter. Under equilibrium conditions there is perfect thermal contact between the radiation and the matter and they are both at the same temperature. A photon arriving at the hole in the box travelling in an outward direction is certain to pass through the hole. Thus we can describe the *hole* as a perfect blackbody radiator provided that the inside of the cavity is at a uniform temperature. Similarly we can see that a photon arriving at the hole travelling inwards is certain to pass through the hole and so the hole is also a perfect absorber. Thus in absorption the hole will appear black, not only at visible wavelengths but at all other wavelengths also and we can call it a perfect blackbody.

It is a general rule that at a given frequency of radiation the emissivity of a *body* is identical with its absorptivity but the value of both is less than 1 and is not the same at all frequencies. The *hole* in our box however will have an emissivity and absorptivity of 1 and will not differentiate between photons of different frequency. Hence the *hole* as a perfect radiator is only something that can be approximated to by matter in the Universe as we shall see.

The example of the box that we have considered is the well known Blackbody Cavity with its Cavity radiation. The power being emitted from the cavity is given by Stefan's Law (see equation (1) in section 3.1.2) where the temperature is that of the walls of the cavity and the area is the area of the *hole*.

The emissivity of a body is a measure of the probability of a photon at its surface actually being emitted (and likewise its absorptivity). Hence for a blackbody surface a photon incident at the surface from the outside will certainly be absorbed and a photon arriving from the inside is certain of leaving the surface. The opposite of a perfect blackbody is a perfect reflector and it can be seen from this that there can be no exchange of energy and hence no thermal contact between radiation and a perfect reflector. In reality there is no such thing as a perfect reflector or perfect radiator and bodies have emissivities and absorptivities that are neither 1 nor 0 but somewhere in between. But if we view photons as particles of electromagnetic radiation then they have to be either reflected or absorbed, there can be no in between. However a way in which we can understand the absorptivity of a substance being, say 0.7, is that if 1000 photons were incident on its surface, 700 of the photons would be absorbed and the other 300 would be reflected or scattered and would travel onward until they met some other surface.

If that other surface was some other part of the inside wall of the cavity we were discussing earlier then, of those 300 reflected photons, 210 would be absorbed and the remaining 90 would be further reflected. It is easy to show the number of interactions, n, between the photons and the surface required to result in the absorption of 99% of the original number is given by

$$n = \frac{-2}{\log(1-\varepsilon)}$$

where ε is the emissivity.

Thus in the case of the example above, 10 photons will remain after 4 interactions and only 1 photon after 6 interactions. Bearing in mind that photons travel at the speed of light, 6 interactions can occur in a very short time span. That is to say thermal contact is good even under these conditions and equilibrium will be established quickly. We can see from this that the only case where blackbody radiation would not be emitted from a cavity is if the cavity walls were perfectly reflecting which they cannot be. Thus it is interesting to note that although the walls of the cavity are not a perfect emitter or absorber, by making a small hole in the walls, the hole in principle can become a perfect emitter and absorber.

It is a perfectly reasonable question to ask how large a hole can be tolerated. In principle, if the cavity is to be in thermal equilibrium there can be no net thermal gradients and therefore there can be no net loss of energy and thus the hole has to be infinitely small. If this were the case the only way in which we could observe the blackbody radiation would be by our having to be inside a closed cavity.

What happens if we were to make the hole somewhat larger? Under these circumstances there would be some parts of the cavity walls at which the hole would subtend some solid angle and for these parts no energy would be received within this solid angle although energy would be radiated by these parts into this solid angle. Hence there would be some areas which would radiate more energy than they would receive and would thus cool down. This establishes temperature gradients in the cavity walls and so the walls could no longer be deemed to be at a uniform temperature. The spectrum of the emerging radiation from the hole would not be that of a blackbody because it would be a composite of blackbody spectra corresponding to the small temperature band represented by the temperature gradients. The magnitude of the gradients would depend on the magnitude of the solid angles subtended by the hole compared with 4π. If these angles are small then a uniform temperature will approximate and the emerging radiation will have the same temperature as the cavity. Alternatively we can say that we require the emergent photon flux to be a small fraction of the random photon flux or that the area of the hole be small compared with the total area of the cavity walls.

Under circumstances where the hole is not small comparatively, energy will be lost from the system as a result of the temperature gradients being set up. Provided that the lost energy can be replaced from some energy source then the system will not cool down but a thermal balance will be established whereby the energy supplied exactly matches the energy being lost and the temperature gradients will be stable; that is a steady state will prevail in which steady – state temperatures persist.

Besides establishing certain principles, the above discussion introduces the role of the thermal properties of surfaces and highlights the importance of multiple reflections in the case of non–perfect absorbers. This is a topic to which we shall return later and which we will see is of particular significance in thermal systems which are controlled by radiative coupling.

Summarising the above we see that temperatures can be ascribed to both matter and radiation. Photons do not exchange energy with each other but will only exchange energy with matter. Although blackbody radiation could be described as radiation that is in equilibrium with itself this is rather an academic definition. Matter is a necessary component in the processing of photons via emission and absorption interactions and so it is more useful to define blackbody radiation as that radiation which results when radiation is in thermal equilibrium with matter which is itself in thermal equilibrium. Ordinary materials are not perfect blackbodies but approximate to blackbodies only over a limited range of frequencies, although the cavity does begin to approach the ideal blackbody. (We shall see later how the emissivity of the inside surface of a cavity produces a larger effective emissivity which we can ascribe to the cavity aperture.) Even though complete thermal equilibrium is therefore not encountered, a thermal balance can be established in a system such that steady state temperatures prevail.

We now need to look more quantitatively at the conductive and radiative heat flows to the elements of a thermal system.

3.3   Energy transport mechanisms

3.3.1   General

We have seen how a thermal component can come to a state of thermal balance provided that the total amount of heat gained by the component is equal to the amount of heat lost per unit time interval. In this respect we have discussed in general terms the role of absorption and emission of radiation, that is, aspects of the radiative coupling of heat to a component. There are however other mechanisms which are also important in the transfer of heat from one component to another which we will briefly discuss here before returning to a quantitative treatment of them.

When radiation is incident on a surface there are three things which will occur to a greater or lesser extent. The radiation can be reflected, transmitted or absorbed as is shown in the Fig. 3.4.

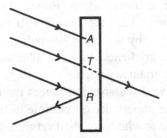

Fig. 3.4. Absorption, transmission and reflection.

Reflection and transmission can both be further described as being specular or diffuse; hence, for example, diffuse reflection is equivalent to scattering.

If we denote the processes of reflection, transmission and absorption by the symbols R, T, and A the efficiency of each of which can have a numerical value in the range 0 to 1 then, by conservation principles,

$$R + T + A = 1$$

If a component is opaque, as will the case be for a metal, then

$$T = 0 \quad \text{and thus} \quad R = 1 - A$$

It follows that, as $\alpha = \varepsilon$ for a given frequency,

$$R = 1 - \alpha \quad \text{and} \quad R = 1 - \varepsilon$$

where α and ε are the absorptivity and emissivity respectively at the given frequency.

If a component is transparent, as might be the case for glass, then

$$R = 0 \quad \text{and thus} \quad T = 1 - A$$

and similarly, for a given frequency,

$$T = 1 - \alpha \quad \text{and} \quad T = 1 - \varepsilon$$

Clearly it can be seen that

if $R = 1$ then $\alpha = 0$ and $\varepsilon = 0$

and if $T = 1$ then $\alpha = 0$ and $\varepsilon = 0$

In each of these last two limiting cases, no energy is absorbed by the component. However, these mechanisms may not be ignored because we still have to consider where any reflected or transmitted energy subsequently goes. Reflection and transmission do have to be taken into account and can be very significant. In general however, transmission is not an important factor in the thermal design of spacecraft. As an aside we can note that transmission is an important mechanism in the thermal design of buildings because glass is used extensively. Glass is an interesting material since it is transparent at visible wavelengths but opaque at infrared wavelengths. That is to say it is a poor emitter and absorber of visible light but a good absorber and emitter of heat; hence the effectiveness of the greenhouse. This underlies the point that although substances have the same emissivity and absorptivity at one wavelength it does not follow that they will have that same emissivity and absorptivity at another unless, of course, they are a good blackbody.

The other major mechanism for energy transport is conduction. The ability for a component to conduct heat through itself depends, as we shall see, on the material properties of the component and its shape. Metals are good conductors of heat and non–metals are good thermal insulators. This is because the atomic processes responsible for the conduction of heat are also responsible for the conduction of electricity. Table 3.2 shows some examples of the thermal and electrical conductivities for some common substances. In general the conductivity of a component is isotropic. However this is not necessarily the case for composite materials such as carbon fibre laminates and due account has to be taken in any thermal model of these types of material.

Table 3.2. *Conductivities of some common space materials*

	(a)	(b)
Aluminium alloy	201	377
Copper	385	588
Gold	300	417
Stainless steel	0 16	10
Titanium	0 23	19
PTFE	0.3	-
Nylon	0.3	-
Carbon fibre	1.8	-

(a) Thermal conductivities ($WK^{-1}m^{-1}$)

(b) Electrical conductivities ($\Omega^{-1}m^{-1} \times 10^{10}$)

We have considered the conductivity of bulk materials, but spacecraft comprise many such components fastened together in some way whether bolted, riveted or with epoxy. It is necessary therefore to treat the interface between thermal components as a separate thermal component itself; indeed it is often found that it is the interfaces that can dominate the conductive response of a thermal network.

3.3.2   Conductive coupling of heat

The flow of heat in a conductor is analogous to the flow of water down a pipe or the flow of charge down a wire (an electric current). In the case of heat it is the temperature difference which drives the flow (whereas it is a pressure difference and a voltage difference respectively for the other situations).

Consider some material of length d and cross–sectional area A as shown in Fig. 3.5.

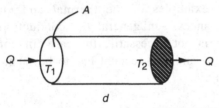

Fig. 3.5.   Conductive heat flow in a cylinder.

The heat flow, Q, is proportional to the temperature difference $T_1 - T_2$ along the length, is proportional to A and inversely proportional to d. The proportionality constant, λ, is the coefficient of thermal conductivity in units of $Wm^{-1}.K^{-1}$. Thus

$$Q = \lambda \frac{A}{d}(T_1 - T_2)$$

Hence we can talk of a thermal conductance which is a measure of the ability of the medium to conduct heat or, conversely, of a thermal resistance which impedes the flow of heat.

The conductance, k, is given by

$$k = \frac{\lambda A}{d}$$

and the impedance, R, is given by

$$R = k^{-1} = \frac{d}{\lambda A}$$

As an exercise the reader might like to show that the radial conductance of a cylinder of inner radius r_1, wall thickness $r_2 - r_1$ and length L is

$$k_c = \frac{2\pi\lambda L}{\ln\left(\dfrac{r_2}{r_1}\right)}$$

and similarly that the radial conductance of a hollow sphere of inner radius r_1 and wall thickness $r_2 - r_1$ is

$$k_s = \frac{4\pi\lambda}{\left(\dfrac{1}{r_1} - \dfrac{1}{r_2}\right)}$$

It is easy to see in both the above that for large r_1 and r_2 and small $r_2 - r_1$ the conductance reduces to that given in the first equation.

The above examples are for simple shapes; real configurations are not normally as simple. In such circumstances it is usually possible to break down the configuration into a series of simpler configurations for which the above equations are valid. Additive rules can then be applied to determine the overall conductance. Thus for series elements of conductance k_1, k_2, k_3 etc. the overall conductance k is given by

$$\frac{1}{k} = \frac{1}{k_1} + \frac{1}{k_2} + \frac{1}{k_3} + \cdots\cdots$$

and for parallel elements the conductance is given by

$$k = k_1 + k_2 + k_3 + \cdots\cdots$$

It is not always necessary to have to calculate the overall conductance. If the element network is part of a configuration for which a mathematical model is being generated by a computer then the elements can simply be described as separate nodes and the program will take care of the computation as we shall see later.
Let us take a simple example. Consider two rods of equal length and equal circular cross – sectional area. The conductance of each is

$$k_1 = k_2 = \frac{\lambda A}{d}.$$

Therefore if they were perfectly joined end to end then the combined conductance is

$$k_{12} = \frac{\lambda A}{2d}$$

which is the same as if we had simply considered a rod of twice the length.

Consider further that the rods are of aluminium alloy 10 cm in length and 1 cm^2 cross-sectional area. It is easily shown that the temperature difference between the ends has to be 5K to drive 1 W along each rod or 10K if the rods are joined end to end.

Again we have taken two simple, identical shapes. They could have been different shapes but we could still have applied the additive rule provided the elements were describing the same piece of material, that is, the elements were perfectly joined. There are situations in reality where this is not the case; the elements may be clamped or joined together with epoxy and so the join is not perfect.

Consider the situation where one rod is placed vertically and the other rod is balanced on top of it end to end. The minimum number of physical contact points between the two rods at their interface is 3 and considering the strength of the alloy, the sum of those 3 contacting areas need only be negligible for the lower rod to support the weight of the upper rod. That is to say, at the interface the effective cross-sectional area is considerably reduced and thus a large temperature difference would be required to conductively drive heat through the interface. Clearly the contact area depends on the surface roughness at the interface; for smooth surfaces the temperature drop will be less and for rough surfaces more.

Let us consider a surface roughness of 1 μm which is fairly typical of the best finish that can be conveniently achieved by so called rough turning. Thus we might suppose that the gap between the two surfaces is 2 μm if the rods are contacting at 3 pairs of peaks. There will be radiative coupling across the interface and, if the rods are in an Earth environment, then there will be air in the gap which will be capable of conducting heat. It is informative to estimate the respective temperature differentials required to drive the 1 W of power in the example that we are considering.

For radiative coupling we can assume the surfaces are thermally black. Since the power coupled is also dependent on the absolute temperature, consider the temperature at one interface to be 290 K. The power radiated by this surface can be calculated from Stefans Law and is only 40 mW. That is to say a temperature difference of 290 K is required to transport 40 mW across the gap and that a greater differential would be required for 1W although such a solution would not be possible in our example without raising the temperature of one rod to over 600 K.

For conduction by air at NTP we can consider a slab of air of circular cross-section of 1 cm^2 and a thickness of 2 μm. The conductance of this slab is easily calculated to be 1.2 WK^{-1} and so, for a power of 1W a temperature differential of 0.8 K is required. (This value is probably an upper limit since the mean free path of air molecules at NTP is of the same order as the size of the gap in this example so there will be relatively few collisions between air molecules in the gap.) In any event, the temperature drop across the gap is small compared with the temperature drop between the ends of the rods.

In space, any residual gas in the gap would have leaked away. The example does serve to show, however, that some medium which 'flows' and can effectively fill the gap to the full – cross section will greatly enhance the conductance. For example, if we select some low vapour pressure grease which may have a thermal conductivity of say 0.1 W K^{-1} then the temperature drop is only 0.2 K. Thus, even though the filler does not have a large thermal conductivity it can have a high conductance by virtue of it comprising a very thin layer. Similar values are appropriate for epoxy resins.

There are circumstances where it may not be feasible or desirable to use fillers to bridge gaps between interfacing materials and the faces have to be dry. Under these circumstances the conductance will depend on the surface finish and on the clamping force. As a practical guide the values shown in the table can be used. These values are for normally machined components. If higher conductances are required the surfaces would have to be polished to give a surface roughness of 0.1 µm or better. Under these circumstances and a heavily clamped join, an interface conductance of a few 1000 Wm^{-2} K^{-1} can be achieved.

Sliding fit	20 Wm^{-2}K^{-1}
Moderately clamped	200 Wm^{-2}K^{-1}
Heavily clamped	1000 Wm^{-2}K^{-1}

Returning to our example, if the two rods of cross – sectional area 1 cm^2 were to be heavily clamped the conductance would be 0.1 WK^{-1} resulting in a drop of 10K across the gap.

From a thermal modelling aspect, the two rods and their interface can be considered as three elements in series with conductances of 0.2, 0.1 and 0.2 WK^{-1} respectively. This gives an overall conductance of 0.05 WK^{-1}. Thus a total temperature difference of 20K will be required to drive 1 W through the system as we have seen above.

In the discussion in the preceding paragraphs we have assumed the coefficients of thermal conductivity to be independent of temperature. This is not true and, in general, the conductivity decreases with decreasing temperature as can be seen in Fig. 3.6. Although the sensitivity with temperature does not generally have to be taken into account iteratively in the computer model, in critical areas it is appropriate to use conductivity values applicable for the relevant operating temperature.

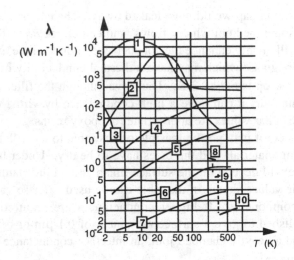

Fig. 3.6. Thermal conductivities of some substances as a function of temperature :
1. Copper, 2. Sapphire, 3. Helium (solid), 4. Aluminium alloy, 5. Stainless steel,
6. Glass, 7. Helium (gas), 8. Ice, 9. Water, 10. Steam.

3.3.3   Radiative coupling of heat

General considerations

    If a surface radiates say W watts isotropically and some other perfectly
absorbing surface (which is at absolute zero) is of such an area and at such a distance
that it intercepts one fifth of the radiation, then $W/5$ watts will be radiatively
transported from one surface to the other. Since no surface can be at absolute zero (not
even deep space) the above is an oversimplification though to think of it like this
serves to illustrate the point.

    Therefore a calculation of the radiative coupling between two surfaces requires a
number of different elements to be determined. These are:

1. The amount of heat radiated per unit solid angle in a given direction
    (which depends on the temperature, emissivity and area of the emitting
    surface).

2. The solid angle of the receiving surface as seen by the emitting surface
    (which depends on the distance between the two surfaces and the projected
    area of the absorbing surface normal to the line joining the surfaces, i.e. its
    cross–sectional area).

3. The absorptivity of the receiving surface.

This is shown schematically in Fig. 3.7. (a).

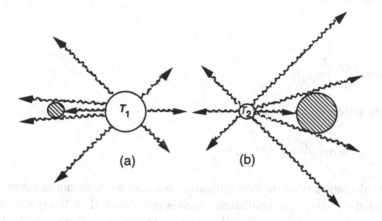

Fig. 3.7. Radiative coupling between two spheres.

Since both surfaces will be at non–zero temperatures then the above applies equally to both surfaces because they are both able to radiate and receive heat (see Fig. 3.7 (b)). If the surfaces are at different temperatures then a determination of the above heat exchanges will show that there will be a net flow of radiative heat from the hotter surface to the cooler surface. This at first glance may seem surprising. For example we could envisage a situation where the cooler body was of large area and high surface emissivity and the hotter body of small area and low surface emissivity and we could then fine tune these parameters such that the flow of heat from the cooler to hotter body exceeded that in the reverse direction, i.e. we could force the heat up the temperature gradient. That this is not so can be readily seen from the following example.

Consider the heat exchange between two spheres separated by a distance d and whose physical parameters are as shown in the Fig. 3.8. We can calculate the heat exchange in the stated direction by following the prescription 1 to 3.

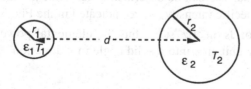

Fig. 3.8. Radiative heat exchange between two spheres.

Heat exchange from 1 to 2

$$H_{1-2} = \sigma A_1 \varepsilon_1 T_1^4 \times \frac{\Omega_{1-2}}{4\pi} \times \varepsilon_2 = \sigma A_1 \varepsilon_1 T_1^4 \times \frac{\pi r_2^2}{4\pi d^2} \times \varepsilon_2 = \sigma \pi \left(\frac{r_1 r_2}{d} \right)^2 \varepsilon_1 \varepsilon_2 T_1^4$$

Similarly

$$H_{2-1} = \sigma \pi \left(\frac{r_1 r_2}{d} \right)^2 \varepsilon_1 \varepsilon_2 T_2^4$$

Hence the net heat flow is

$$H_{1-2} - H_{2-1} = \sigma \pi \left(\frac{r_1 r_2}{d} \right)^2 \varepsilon_1 \varepsilon_2 \left(T_1^4 - T_2^4 \right)$$

Again this calculation is an oversimplification because we have not accounted for the factor $(1-\varepsilon)$ of the intercepted radiation which is not absorbed. Clearly this has to go somewhere and part of it may well return to the surface from which it was originally emitted etc. This is something to which we will return but we will assume here that this component is very small. We can see from the above that the direction of net heat flow is solely determined by the temperatures and indeed there is a positive heat flow from the hotter to cooler body, and we cannot contrive to cause heat to flow uphill. Further we see that, in the absence of any other heat source, this exchange continues until both surfaces come to the same temperature and the laws of thermodynamics are preserved. In our applications however other sources and sinks are usually present and the temperatures are indeed different. In this respect the heat exchange is characterised by the factor $\varepsilon_1 \varepsilon_2$ and it is useful to think of this factor as some effective emissivity associated with the heat transfer between these components. We shall return to this later.

We shall now address more fully and in more detail the three elements mentioned earlier.

The emitted radiation

The radiation emitted by a surface can be described in terms of the radiation field which is characterised by the specific intensity, I_ν, as indicated in the Fig. 3.9.

The specific intensity has units of $Js^{-1}m^{-2}Hz^{-1}st^{-1}$. Thus the elemental power in a given frequency interval emitted per unit area into a solid angle in a direction θ with respect to the surface plane is

$$dP_\nu = I_\nu \, dA \cos\theta \, d\nu \, d\Omega$$

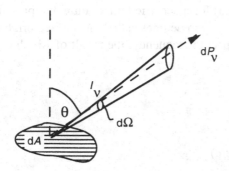

Fig. 3.9. The specific intensity of radiation.

We can see from this that to derive the total power emitted from a surface we require ultimately to perform a triple integral by summing over all frequencies, all solid angles and the whole emitting area.

$$P = \iiint_{\nu\Omega A} I_\nu \; d\nu \cos\theta \; d\Omega \; dA$$

We can illustrate how this is performed by deriving Stefan's Law quoted in section 3.1.2 (equation (1)).

When radiation is in equilibrium with matter the specific intensity becomes identical with a characteristic of the matter which is called its Source function, S_ν, the nature of which is determined by the properties of the atoms comprising it and the physical state of the matter. When the matter itself is in equilibrium, the source function is identical with the Plank function, B_ν, sometimes called the Blackbody function. Hence in equilibrium

$$I_\nu = S_\nu = B_\nu.$$

In general such equilibrium is not possible, but the best approximations to it can be achieved in systems where the density of atoms is high and where the temperature of the atoms is low. We can certainly make such an approximation in the case of solids (as compared with gases) although we shall need to qualify this. The advantage of making this approximation is that the Planck function is only a function of temperature as we saw in section 3.2. Thus

$$dP_\nu = B_\nu \cos\theta \; d\nu \; d\Omega \; dA$$

The Planck function in its frequency form is given by

$$B_\nu = \frac{2h\nu^3}{c^2}\left(e^{(h\nu/kT)} - 1\right)^{-1}$$

and the area under the curve (see Fig. 3.3) for a specific temperature T represents the luminosity per unit area, per unit solid angle (sometimes called the surface brightness). This area is simply the integral with respect to frequency, the result of which is given as

$$B = \int_\nu B_\nu d\nu = \frac{ac}{4\pi} T^4 = \frac{\sigma}{\pi} T^4$$

where   c is the velocity of light
and   a is the radiation constant ($= 4\sigma/c$)
where σ is Stefan's constant.

Hence the total power can now be obtained by performing the double integral over solid angle and area.

$$P = \frac{\sigma T^4}{\pi} \int_A \int_\Omega \cos\theta \; d\Omega \; dA$$

To integrate over the solid angle, consider the power emitted at angle θ into the elemental angle $d\theta$ in the two dimensional case as shown in the Fig. 3.10.

We can see that the elemental solid angle into which the radiation is emitted in the three dimensional case is the elemental surface area of the hemisphere subtended by dA namely $2\pi r \sin\theta \; r \, d\theta$, divided by the square of the distance between dA and the hemisphere. Therefore to sum over all such solid angles in this case reduces to an integration over θ.

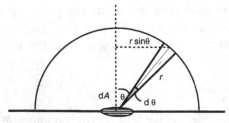

Fig. 3.10.   Power radiated into a hemisphere.

Thus

$$P = \frac{\sigma T^4}{\pi} \int_A dA \int_\theta \cos\theta \frac{2\pi \sin\theta \; r \; d\theta}{r^2} = \sigma T^4 \int_A dA \int_0^{\pi/2} \sin2\theta \; d\theta = \sigma T^4 \int_A dA$$

We see here that the solid angle integral has a value of π rather than the expected 2π. This is because of the inclusion within the integral of the term $\cos\theta$ which takes into account the projection of the emitting area in the direction considered. Without this factor the integral would have a numerical value of 2π which is the maximum solid angle available to any *point* on a plane surface.

Finally we note that we can construct any surface from such elemental areas dA each radiating into solid angle 2π.

Thus

$$P = \sigma \, A \, T^4$$

where A is the surface area. This is the familiar Stefan's Law.

We have assumed that the emitting surface is a perfect absorber/emitter. To a first approximation solids emit blackbody spectra but not necessarily with the full intensity described by the blackbody function. The emission is scaled by the emissivity, ε, which takes a value in the range 0 to 1. Hence as applied to solid surfaces Stefan's Law becomes

$$P = \sigma \, \varepsilon \, A \, T^4$$

It should be reiterated here that in thermal design applications ε describes the emissivity and absorptivity at infrared wavelengths (i.e. radiation at temperatures of a few hundred degrees) and α describes the emissivity and absorptivity at optical wavelengths (i.e. radiation at temperatures of a few thousand degrees).

The example we have just considered gives an expression for the power emitted by a body of emissivity ε which will be the same as that received from it by a perfectly absorbing surface if that surface totally encloses the emitting body. Generally this is not the case and we are more interested in the power coupled to a surface which subtends only some fraction of the maximum solid angle.

Solid angles, view factors and shape factors

These terms all apply to essentially the same thing, that is the extent to which one surface is visible to another surface. The parameters are all dimensionless (although solid angles have *units* of steradians).

A solid angle is a three dimensional angle and characterises the angle at the apex of a cone subtended by a two dimensional surface (which in turn can be the projected cross – sectional area of a three dimensional body). For example, the Sun and the Moon each have an angular diameter of 1/ 2 degree and subtend the same solid angle at the Earth; hence by this accident of nature, eclipses occur whereby the bright photosphere of the Sun is exactly blocked out thus making it possible for observations to be made of the less bright upper solar atmosphere. The projected surface areas of the two bodies are very different but so are the distances as is shown schematically in Fig. 3.11.

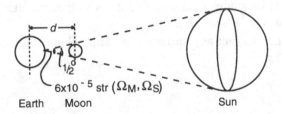

Fig. 3.11. The solid angle subtended by the Moon and the Sun at the Earth.

The solid angle subtended by each body at the Earth is 6.0×10^{-5} steradians and we showed earlier that this is simply derived by expressing the apparent surface area of the discs as an angular area with the angles expressed in radians rather than degrees (see section 3.1.2).

Let us consider just the Moon and the Earth and derive an expression for that power emitted by the Earth, L_E, which is intercepted by the Moon. We can take the Earth as being a perfect emitter at 280K (as calculated earlier) and thus it will emit radiation over the whole frequency band given by the Planck function $B(280)$, (i.e. $B(T)$ with $T = 280$K).

Thus the intercepted power is

$$P = L_E \frac{\Omega_M}{4\pi} = \pi B(280)\, 4\pi R_E^2 \, \frac{\Omega_M}{4\pi} = B(280)\, A_E \, \Omega_M$$

where A_E is the cross–sectional area of the Earth and Ω_M is the solid angle of the Moon subtended at the Earth.

Hence

$$P = B(280)\, A_E \, \frac{A_M}{d^2}$$

where A_M is the cross sectional–area of the Moon.

Therefore

$$P = B(280)\, \frac{A_E}{d^2}\, A_M \;=\; B(280)\, \Omega_E \, A_M$$

where Ω_E is the solid angle of the Earth as seen from the Moon.

Thus we see that

$$A_E \; \Omega_M = A_M \; \Omega_E$$

Since the Planck function is only dependent on temperature and the effective temperature of the Earth is the same from whatever direction it is observed, we see that the incident power can equally be calculated by someone standing on the Earth *or* the Moon using only 'local' knowledge.

More generally for elemental areas the reciprocity relationship can be expressed as

$$dA_1 \; \cos\theta_1 \; d\Omega_{21} = dA_2 \; \cos\theta_2 \; d\Omega_{12}$$

as is shown in Fig. 3.12.

We can define the view factor, dF_{12}, between surfaces dA_1 and dA_2 as simply the fraction of radiant energy emitted by dA_1 which is intercepted by (although not absorbed by) dA_2. Hence, from our earlier example with the Earth–Moon system, the view factor is

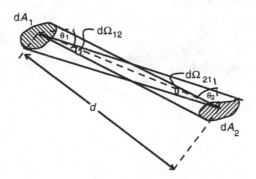

Fig. 3.12. View factors between two elemental areas.

$$F_{EM} = \frac{P}{L} = \frac{\Omega}{4\pi}$$

This indicates a simple relationship between the view factor and the solid angle but a little bit of thought shows that the relationship depends on the geometry of the configuration. In the above example the flux at distance d is independent of θ because of the spherical geometry as seen in the Fig. 3.13 (a).

If the Earth sphere is replaced by a disc, then it is easy to see that the flux at distance d *is* a function of θ as shown in (b). In the plane of the paper it is zero for $\theta = 0$ and is a maximum at $\theta = 90$. This is because the projected area of the emitting surface comes into play in quantifying the flux emitted in any direction.

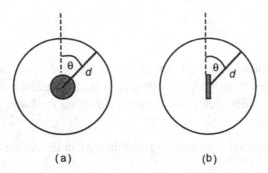

Fig. 3.13.   Solid angles subtended by (a) a sphere and (b) a disc.

Hence in general we cannot simply obtain the view factor by dividing the solid angle of the receiving surface by the total solid angle into which the emitting surface radiates (i.e. 2π or 4π).

To see what happens in the general case we refer to Fig. 3.12. setting the intensity of the radiation leaving dA_1 in direction θ_1 as $I(\theta_1)$.

The radiant power reaching dA_2 is

$$dP_{12} = dA_1\cos\theta_1 I(\theta_1)d\Omega_{12} = dA_1\cos\theta_1 I(\theta_1)\frac{dA_2\cos_2}{d^2}$$

The total power radiated by dA_1 in all directions is

$$L = dA_1 \int_\Omega \cos\theta_1 I(\theta_1)d\Omega$$

Thus the view factor dF_{12} is

$$dF_{12} = \frac{dP_{12}}{L} = \frac{\cos\theta_1 I(\theta_1)dA_2\cos\theta_2}{d^2\int_\Omega \cos\theta_1 I(\theta_1)d\Omega}$$

If we can assume that the intensity is the same in all directions, that is to say if the surface looks to have the same surface brightness irrespective of the angle at which we view it, then

$$dF_{12} = \frac{\cos\theta_1\cos\theta_2 dA_2}{d^2\int_\Omega \cos\theta_1 d\Omega}$$

But

$$\int_{\Omega} \cos\theta_1 d\Omega = \pi$$

and therefore

$$dF_{12} = \frac{\cos\theta_1 \cos\theta_2 dA_2}{\pi d^2}$$

Here we have assumed that dA_1, dA_2 are infinitesimally small, plane surfaces. In order to apply this to real surfaces which are finite and, in general, curved we have to integrate over all dA_2 for any given dA_1 in A_1, and then repeat for all such dA_1 because the view factors will all be slightly different. Thus the total view factor F_{12} can be expressed formally as

$$F_{12} = \frac{1}{A_1} \int_{A_1} \int_{A_2} \frac{\cos\theta_1 \cos\theta_2 dA_1 dA_2}{\pi d^2}$$

In the derivation of the view factor expression we required I to be independent of θ_1 thus greatly simplifying the analysis. This view factor is called the diffuse view factor and we can see that, because of this simplification, the view factor now only depends on the surface geometry and the configuration.

By expressing the above as

$$A_1 F_{12} = \frac{1}{\pi} \int_{A_1} \int_{A_2} \frac{\cos\theta_1 \cos\theta_2 dA_1 dA_2}{d^2}$$

we can see that the interchange of the subscripts 1 and 2 on the RHS does not lead to any change. Thus

$$A_1 F_{12} = A_2 F_{21}$$

which is a simple form of the reciprocity relationship discussed earlier.

Using the general formula it is simple to show, for two plane surfaces of equal area separated by a distance much greater than the characteristic scale length of the surfaces, that for the configurations (a), (b), (c) shown in Fig. 3.14 the view factors are respectively Ω/π, $\Omega/2\pi$, and 0 where Ω is the solid angle of the surface in configuration (a).

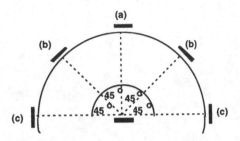

Fig. 3.14. View factor dependence on aspect angle.

Taking case (a), we can now calculate the view factor between two discs each of radius r with their planes parallel to each other and separated by distance h as shown in Fig. 3.15.(a).

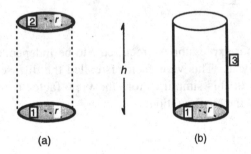

(a)   (b)

Fig. 3.15. View factors between (a) the end surfaces
of a right cylinder and (b) one end surface and the
interior cylindrical surface.

The solid angle

$$\Omega_{12} \approx \frac{\pi r^2}{h^2} = \pi \left(\frac{r}{h}\right)^2$$

and thus

$$F_{12} \approx \frac{\Omega_{12}}{\pi} = \left(\frac{r}{h}\right)^2$$

The above is a very simple geometry. To calculate the view factor between the end surface of a right cylinder and the inside curved surface of the cylinder (see Fig.

3.15 (b)) would be more complicated. However it happens that in fact we can simply write down the answer; it is

$$F_{12} \approx \frac{\Omega_{12}}{\pi} = \left(\frac{r}{h}\right)^2$$

This results because the three surfaces 1,2,3 form a complete enclosure for which the total view factor must be unity. This exemplifies an important law which states that the sum of the view factors of a convex surface (i) with, say, N other surfaces (j) which enclose it is 1, i.e.

$$\sum_{j=1}^{N} F_{ij} = 1$$

Note that i can form one of N if we so choose.

This law and the reciprocity law are very useful in calculating view factors. In general, the calculation of view factors is about the most difficult aspect of thermal modelling even for relatively simple geometries and configurations. For this reason analytical solutions have been derived and exist for the most common configurations as formulae or as graphs. Two examples are shown in Figs. 3.16 and 3.17.

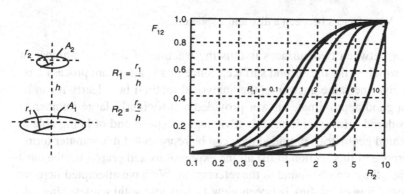

$$F_{12} = \frac{1}{2}\left[x - \sqrt{x^2 - 4\left(\frac{R_2}{R_1}\right)^2}\right] \qquad \text{where} \qquad x = 1 + \frac{1 + R_2^2}{R_1^2}$$

Fig. 3.16.   View factors between two, normal discs.

The art of determining view factors in any mechanical design is one of representing the geometry as an approximation/combination of geometries for which the look–up view factors exist. This can still be a major task but with care reasonably precise view factors can be obtained. Another method is a numerical method which combines Monte Carlo and Ray Tracing techniques. That is, having set up a geometric computer model of the structure, view factors are determined by considering what happens to each 'model photon' which is emitted from any part of a given surface in a random direction.

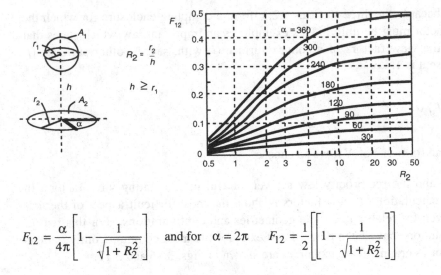

$$F_{12} = \frac{\alpha}{4\pi}\left[1 - \frac{1}{\sqrt{1+R_2^2}}\right] \quad \text{and for} \quad \alpha = 2\pi \quad F_{12} = \frac{1}{2}\left[\left[1 - \frac{1}{\sqrt{1+R_2^2}}\right]\right]$$

Fig. 3.17. View factors between a disc and a sphere.

The view factor between two surfaces is thus the fraction of the total number of emitted 'photons' which strike the second surface. Since this is a random process it is not guaranteed to give the same answer as the analytical method but clearly it can be expected to give a good approximate answer provided a sufficiently large number of 'photons' are considered. However it has the advantage of speed and of being able to cope with complicated geometries provided they can be represented in computer form.

It is not the purpose of this section to list all the expressions and graphs for the most common geometries; these can be found in the references. We have attempted here to establish an understanding of the link between view factors and solid angles; after all, the approximate determination of the latter is easier than the former and can be used if nothing else is to hand. However, before leaving this topic, let us just return to the two geometries in Figs. 3.16. and 3.17 since these are geometries that we have been using as examples throughout.

Firstly for the two parallel coaxial discs if we take $R_1 = R_2 = 0.2$ say then

$$x = 27 \quad \text{and} \quad F_{12} = 0.037$$

whereas from the solid angle approach

$$F_{12} = \frac{\Omega}{\pi} = \frac{\pi(R_1^2)}{\pi} = 0.040$$

Secondly for the sphere to planar case if we assume R_2 to be small then for $\alpha = 2\pi$,

$$F_{12} = \frac{1}{2}\left(1 - \left(1 + R_2^2\right)^{-\frac{1}{2}}\right) = \frac{1}{2}\left(1 - \left(1 - \frac{1}{2}R_2^2 + L\right)\right)$$

$$\therefore F_{12} \approx \frac{1}{2}\left(\frac{1}{2}R_2^2\right) = \frac{r^2}{4h^2} = \frac{1}{4\pi} \cdot \frac{\pi r^2}{h^2}$$

$$\therefore F_{12} \approx \frac{\Omega}{4\pi}$$

which is identical to the result we got from a simple consideration of the Earth – Sun system.

Now that the view factors have been defined we can express our previous equations of radiation exchange between two spheres in terms of the view factors.

Thus

$$H_{1-2} = \sigma A_1 F_{12} \varepsilon_1 \varepsilon_2 T_1^4 \quad \text{and} \quad H_{2-1} = \sigma A_2 F_{21} \varepsilon_2 \varepsilon_1 T_2^4$$

But $A_1 F_{12} = A_2 F_{21}$

$$\therefore H_{1-2} - H_{2-1} = \sigma A_1 F_{12} \varepsilon_1 \varepsilon_2 (T_1^4 - T_2^4)$$

Again, this expression is only approximate since no account has been taken of that radiation which is not absorbed if ε_1 or ε_2 are not unity. As before, a discussion of this will be deferred.

Absorptivity (emissivity)

The absorption, emission and reflection of radiation at a surface comprises some complicated physics and, at its basic level, is concerned with the interaction between photons and atoms. Such interactions can be described macroscopically in terms of the coefficients of absorption and emission which in turn are related to atomic constants and are dependent as well on the physical state of the matter. They are not simple parameters. They can vary with photon frequency, with direction relative to the surface and with the physical nature of the surface itself. However, it is interesting to note that for a blackbody the Planck function, B, is simply the ratio of the emission coefficient to the absorption coefficient at that frequency.

These coefficients have dimensions; in fact they have different dimensions as is exemplified by the fact that B itself has dimensions. However, the terms emissivity and absorptivity are dimensionless and are simply a measure of how efficient an absorber or emitter is compared with a blackbody. Hence they are numbers which lie in the range 0 to 1 and, for a true blackbody are both unity. Otherwise a surface is described as being 'grey'.

In the thermal literature there are a number of differently defined emissivities / absorptivities / reflectivities. However they fall mainly into four categories; monochromatic directional (sometimes called directional spectral), directional, monochromatic hemispherical and hemispherical in recognition of the spectral and directional dependencies of these parameters. The monochromatic directional is that at a single frequency and in a given direction; the directional is that summed over all frequencies. Likewise the hemispherical is averaged over all directions. The directional emissivity depends on the composition and state of a surface (for example polished, oxide, grooved, diffuse etc.) and an idea as to how these can vary for different surfaces is shown in Fig. 3.18.

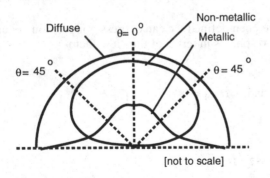

Fig. 3.18   Directional emittance for different surfaces.

We can see from the diagram that the directional emittance for diffuse surfaces is constant and is thus the same as the normal emittance ($\theta = 0$). Even though no diffuse surface can be perfectly so, in general this is a reasonable approximation to make. For other types of surface the directional emittance is approximately constant except at small angles with the surface. The hemispherical emittance is an average over all angles and therefore does not differ significantly from the normal emittance. Since the data on the directional variations are sparse it is generally accepted that either the normal emittance (absorptance) or the hemispherical emittance (absorptance) can be adopted for most practical purposes.

It should be remembered that the hemispherical emittance and/or normal emittance are averaged over the whole frequency range, that is to say they are derived from the directional emittance weighted by the blackbody spectral power and are therefore functions of temperature. This is why at optical temperatures the emissivity is very different from that at infrared temperatures and why the former is referred to in the thermal literature as the absorptivity to emphasise this fact. Even over a fairly limited temperature range (say 100K to 1000K) the emissivity can vary in not an insignificant way but again the data are far from complete.

Hence some care has to be exercised in choosing an appropriate emissivity or absorptivity in thermal design work. In general it is safe to assume that surfaces are grey, diffuse and uniform and in the case of emitting surfaces that room temperature values of the emissivity are appropriate. However in specific applications it is as well to examine carefully all such assumptions.

We have seen in our earlier expressions for the radiation exchange between two spheres that the product $\varepsilon_1\varepsilon_2$ represents some form of effective emissivity though we have been careful to point out that no account has been taken of the non–absorbed components ($1-\varepsilon_1$) and ($1-\varepsilon_2$). Clearly these components have to go somewhere and indeed some of each can be returned to its source thus affecting the heat exchange which itself can be interpreted as a modification of the emissivity. Provided that this returned radiation is only a small part of the emitted radiation then $\varepsilon_1\varepsilon_2$ does represent an effective emissivity. But there are circumstances in which this requirement is not satisfied.

This opens up the more general, important topic of diffusely scattered radiation which we need to explore more fully. As we shall see, this leads to the interesting concept of the indirect radiation link,whether this be back to the emitting surface itself or to a third surface which is not in the direct line of sight of the emitting surface. Such considerations can affect either the effective emissivity between two surfaces or the emissivity of a single surface.

As a way of introducing this let us return to the situation of two spheres placed at a distance d apart as shown in Fig. 3.19 (a).

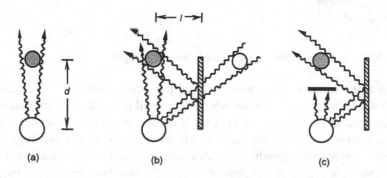

Fig. 3.19. View factors between two spheres; (a) direct, (b) direct and reflected, and (c) reflected.

Let us now place a perfectly reflecting, plane surface parallel to and at a distance l from the line joining the centre of the spheres as shown in (b). For the direct rays the solid angle of the hatched sphere subtended at the unhatched sphere is A/d^2 whereas for the indirect rays this solid angle is $A/(4l^2 + d^2)$. By placing the mirror at $l = d/2$ we can contrive that the latter is half the former. That is, by inserting the mirror the heat exchange between the spheres is increased by 50%. Since there is essentially only a single beam path for this to occur the result is independent of the area of the mirror provided it covers the beam.

If we now place a perfectly absorbing screen at 0K in the direct beam (see (c)) a coupling still occurs via the indirect route and the heat exchange is now half that for configuration (a). Since we have chosen a perfectly reflecting surface the situation we have considered is one of specular reflection. No surface is a perfect reflector and scattering is always present to a greater or lesser extent. But what if the surface was a perfect, diffuse scatterer, that is to say a surface for which the unabsorbed radiation was scattered uniformly in any direction as shown in Fig. 3.20 ?

We see that there is still an indirect coupling which, for the 45 degree path we considered earlier, is reduced from the specular reflection exchange by some factor, since the radiation which was destined for the sphere alone is now shared over a solid angle of 2π. However, unlike the perfectly reflecting case there are now multiple indirect coupling paths and the total exchange now depends also on the area of the scattering surface.

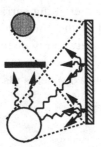

Fig. 3.20. Indirect, diffuse view factor between two spheres.

The fact that the scattering is omnidirectional permits for some of the radiation to be scattered back to the emitting surface which again will affect the heat exchange.

Clearly the situation is becoming increasingly complicated and we can begin to see that the calculation of indirect view factors, whether they be of the self type or the non−self type, is becoming difficult. This is exacerbated with the introduction of more surfaces. If however we limit the number of surfaces to two and contrive, through the configuration, that heat is exchanged only between the two surfaces, then the situation does permit for an analytical solution and allows for the determination of an effective emissivity to describe the coupling. That this can be done results from the fact that any non–absorbed radiation is returned to the other surface. This configuration is that of 'enclosed' surfaces. Thus under equilibrium conditions there are only four heat fluxes we need consider these being I_{ij} for $i, j = 1, 2$ where I_{ij} is defined as the flux incident on j originating from i. Fig. 3.21 shows what happens to those scattered components of I_{ij} which are not absorbed.

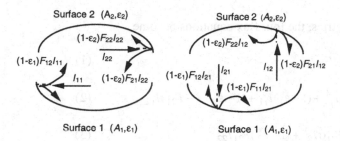

Fig. 3.21. The radiative exchange of power for an enclosed geometry.

By adding the emitted components to the scattered components we can write down the heat flux equation for each I_{ij} as follows.

$$I_{11} = \varepsilon_1 A_1 F_{11} \sigma T_1^4 + (1 - \varepsilon_1) F_{11} I_{11} + (1 - \varepsilon_2) F_{21} I_{12}$$

$$I_{12} = \varepsilon_1 A_1 F_{12} \sigma T_1^4 + (1 - \varepsilon_1) F_{12} I_{11} + (1 - \varepsilon_2) F_{22} I_{12}$$

$$I_{21} = \varepsilon_2 A_2 F_{21} \sigma T_2^4 + (1 - \varepsilon_1) F_{11} I_{21} + (1 - \varepsilon_2) F_{21} I_{22}$$

$$I_{22} = \varepsilon_2 A_2 F_{22} \sigma T_2^4 + (1 - \varepsilon_1) F_{12} I_{21} + (1 - \varepsilon_2) F_{22} I_{22}$$

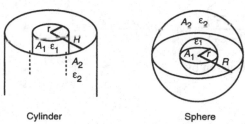

Cylinder   Sphere

Fig. 3.22. Two infinite parallel cylinders and two concentric spheres.

There are three simple geometries that can be considered; two infinite plane parallel surfaces, two infinite coaxial cylinders, and two concentric spheres. Since the latter two can be modified to encompass the first and can also be treated in a common manner it is these which we will take as our examples (see Fig. 3.22).

The only differences between the cylinder and the sphere is the view factor F_{21} which is r/R and $(r/R)^2$ respectively for the cylinder and the sphere. Otherwise for both geometries

$$F_{11} = 0, \ F_{12} = 1, \ F_{22} = 1 - F_{21} \text{ and } A_2 F_{21} = A_1$$

Thus, for both geometries, the four flux equations become

$$I_{11} = (1 - \varepsilon_2) F_{21} I_{12} \tag{1}$$

$$I_{12} = \varepsilon_1 A_1 \sigma T_1^4 + (1 - \varepsilon_1) I_{11} + (1 - \varepsilon_2)(1 - F_{21}) I_{12} \tag{2}$$

$$I_{21} = \varepsilon_2 A_2 F_{21} \sigma T_2^4 + (1 - \varepsilon_2) F_{21} I_{22} \tag{3}$$

$$I_{22} = \varepsilon_2 A_2 (1 - F_{21}) \sigma T_2^4 + (1 - \varepsilon_1) I_{21} + (1 - \varepsilon_2)(1 - F_{21}) I_{22} \tag{4}$$

Eliminating I_{11} from equations (1) and (2) gives

$$I_{12} = \frac{\varepsilon_1 A_1 \sigma T_1^4}{\varepsilon_2 + \varepsilon_1 F_{21}(1 - \varepsilon_2)}$$

Eliminating I_{22} from equations (3) and (4) gives

$$I_{21} = \frac{\varepsilon_2 A_1 \sigma T_2^4}{\varepsilon_2 + \varepsilon_1 F_{21}(1 - \varepsilon_2)}$$

Hence the power exchange is

$$\varepsilon_2 I_{12} - \varepsilon_1 I_{21} = \frac{\varepsilon_1 \varepsilon_2 A_1 \sigma (T_1^4 - T_2^4)}{\varepsilon_2 + \varepsilon_1 F_{21}(1 - \varepsilon_2)} = \varepsilon_{\text{eff}} A_1 \sigma (T_1^4 - T_2^4)$$

where the effective emissivity ε_{eff} is given by

$$\frac{1}{\varepsilon_{\text{eff}}} = \frac{1}{\varepsilon_1} + \frac{(1 - \varepsilon_2)}{\varepsilon_2} F_{21}$$

Thus for infinite cylinders

$$\frac{1}{\varepsilon_{\text{eff}}} = \frac{1}{\varepsilon_1} + \frac{(1 - \varepsilon_2)}{\varepsilon_2} \left(\frac{r}{R} \right)$$

and for concentric spheres

$$\frac{1}{\varepsilon_{\text{eff}}} = \frac{1}{\varepsilon_1} + \frac{(1-\varepsilon_2)}{\varepsilon_2}\left(\frac{r}{R}\right)^2$$

By setting $r = R$ we see that for infinite, parallel planes

$$\frac{1}{\varepsilon_{\text{eff}}} = \frac{1}{\varepsilon_1} + \frac{1}{\varepsilon_2} - 1$$

Following on from our previous discussion, we can now see that the effect of diffuse scattering on the effective emissivity $\varepsilon_1\varepsilon_2$ between two surfaces is to modify it by the factor

$$\frac{1}{(\varepsilon_2 + \varepsilon_1 F_{12}(1-\varepsilon_2))}$$

Since

$$\varepsilon_2 + \varepsilon_1 F_{12}(1-\varepsilon_2) \le 1$$

then $\varepsilon_1\varepsilon_2$ is increased by this factor and hence $\varepsilon_1\varepsilon_2$ can be taken generally as being a lower limit for any effective emissivity.

The other interesting property which follows from this is that for the cylindrical and spherical geometry if $r << R$ then $\varepsilon_{\text{eff}} = \varepsilon_1$. Hence the power exchange is determined solely by the emissivity of the enclosed surface and independent of the emissivity of the enclosing surface. We get the same result if we put $\varepsilon_2 = 1$. This is the situation where a surface has a large view factor with itself with the result that its effective emissivity gets closer and closer to unity. However since the surface is at a single temperature this does not mean that the heat exchange is increasing; the heat exchange with itself is zero irrespective of the emissivity.

Let us now just consider the spheres (still $r << R$) and move the inner sphere to the inside surface of the outer sphere such that, to all extents and purposes, surface 1 now forms a very small part of surface 2. We can see that the effective emissivity of surface 2 as seen by surface 1 is still very close to unity. Finally if we now make surface 1 a hole in surface 2 then the radiation going through the hole will appear to be coming from a surface with an effective emissivity of unity, that is the emerging radiation will be blackbody radiation. We also see that the sphere with the small hole in it appears as a cavity emitting cavity radiation.

This brings us back full cycle to our discussion of the meaning of blackbody radiation with which we began the last section.

In the above we eliminated the reduced surface 1 and replaced it with an aperture. Hence in effect we have constructed a cavity which could, for other geometries, also be a hole or a groove as shown in the Fig. 3.23. What we are saying is that the apertures of such systems have larger effective emissivities than the respective emissivities of the enclosing surfaces.

Fig. 3.23.   Different grooves or cavities.

The magnitude of the effective emissivity of the apertures depend on the surface emissivity and on the number of multiple scatterings that can be generated by means of the geometries. Clearly as drawn in the Fig. 3.23, for a given surface emissivity the effective emissivity of the apertures increases from right to left.

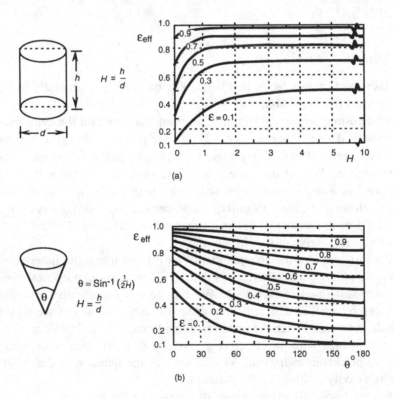

Fig. 3.24.   Effective emissivities for (a) cylindrical cavity, (b) conical cavity.

This can be shown quantitatively. The graphs in Fig. 3.24 show the numerically calculated apparent emissivities for a circular cylindrical cavity and an axi–symmetric conical cavity each with grey, diffuse surfaces for different aspect ratios and for different surface emissivities. For the conic, $\theta = \sin^{-1}(1/2H)$ and so in order to compare the two geometries let us take $H (= h/d) = 1 (-> \theta = 30)$ and $\varepsilon = 0.4$. Thus the effective emissivity for the conic increases by about 50% ($\varepsilon_{eff} = 0.66$) and that for the cylinder by almost 100% ($\varepsilon_{eff} = 0.74$). We note from the graphs that in either case an effective emissivity of unity is never reached. This is because these geometries have open ends. If we were to gradually close off the apertures to form a true cavity as shown on the left of Fig. 3.23, then indeed the emissivity would asymptotically approach unity.

It should be remembered that the effective emissivity is that of the aperture and that the area appropriate to this emissivity is that of the aperture, not the area of the internal, physical surfaces.

3.4   Thermal balance

In the last section we showed quantitatively how heat is transported conductively and radiatively as a result of temperature differences. The equations which respectively describe these are as follows

$$H_C = k_C (T_1 - T_2)$$

$$H_R = \sigma A \varepsilon_{eff} (T_1^4 - T_2^4)$$

It is both instructive and useful to modify the radiative equation to be of the same form as the conductive equation so that we can compare like with like. This can be done through the identity

$$(T_1^4 - T_2^4) = (T_1 - T_2)(T_1^3 + T_1^2 T_2 + T_1 T_2^2 + T_2^3)$$

Hence we see for small temperature differences ($T_1 \approx T_2$) that

$$T_1^4 - T_2^4 \approx 4 T_1^3 (T_1 \quad T_2)$$

and for large temperature differences ($T_1 \gg T_2$) that

$$T_1^4 - T_2^4 \approx T_1^3 (T_1 - T_2)$$

As approximations, the first can be used for internal radiative links and the latter for external links.

Functionally, we can now write the two equations in terms of the temperature difference ΔT, where $\Delta T = (T_1 - T_2)$, as

$$H_C = k_C \Delta T \qquad \text{where} \qquad k_C = \frac{\lambda A}{d}$$

$$H_R = k_R \Delta T \qquad \text{where} \qquad k_R = \sigma A \varepsilon_{\text{eff}} T^3$$

We now have two equations of the same form which show that the heat flow is proportional to the temperature difference with a proportionality constant which is the effective conductivity. This underlines the similarity of the transportation of heat by the motion of particles in a metal or gas and by photons which can be considered as particles.

In any system therefore the temperatures will adjust themselves to transport exactly the heat from one part of the system to another. For the system as a whole if the heat leaving the system does not match the heat dissipated in the system then it will heat up or cool down. Similarly for the heat coming to and leaving any point in the system. Thus in a state of thermal balance, if there is no internal dissipation the gradients are zero, and if there is an internal source of heat then a stable temperature regime is established (we call this a 'steady state' to differentiate it from the 'transient state' in which temperatures are still coming to equilibrium).

With a knowledge of the heat dissipations and the effective conductivities of the elements of a system we are now in a position to use the above in order to calculate what the temperatures in a system are. We will consider two configurations. The first is a simple system in which the elements are only coupled radiatively. The second is rather more realistic which we will initially analyse as a conductive system and then see the effects of adding in the radiative exchanges. In both we shall use the equations that we have discussed but in the second example we shall also be tackling the problems at different levels of accuracy in order that we get some feel for the relative importance and sensitivities of the different approaches.

The first configuration which we are going to investigate is that of a series of thermal shields which are coupled only radiatively. We will take this progressively and start by considering just a single, plane surface in which 22 W is internally dissipated per unit area and which has a view factor to cold space of 1. The emissivity of the surface is 1. For simplicity we assume space to be a plane parallel surface with an emissivity of 1 and at a temperature of 3K. This is shown schematically in Fig. 3.25.

The heat entering each unit area of the surface element (S) is 22 W of internal dissipation and (3^4) W received from cold space. The heat leaving each element is proportional to (T_S^4). Any imbalance of these results in a temperature change in the element, the magnitude of which is dependent on its mass and its thermal specific heat.

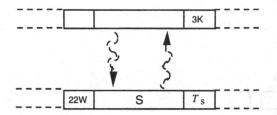

Fig. 3.25.   No thermal shields.

Hence we can write down the heat equation for the material element as

$$mc\frac{dT_s}{dt} = -\sigma T_s^4 + \sigma(3^4) + 22 = -\sigma T_s^4 + 4.6 \times 10^{-6} + 22$$

where m is the mass (per unit area in this case) of element S, c is its specific heat and t is time. Heat flow imbalance thus results in $dT/dt \neq 0$ and no heat flow imbalance results in a steady–state case where $dT/dt = 0$. In the latter case we can solve the equation simply in order to give a value of T_S as

$$T_S = 140.2 \text{ K}$$

We see that the element receives very little heat from space. Indeed, if the amount received were to be say 1% of the internal dissipation (i.e. 0.2 W) then 'space' would have to be at a temperature of 44 K This kind of consideration becomes important when we are considering thermal balance testing in a ground–based thermal chamber where space is typically simulated by thermal shrouds filled with liquid nitrogen at a temperature of 77 K.

Let us now magically suspend a thermal shield of emissivity 1 between the two 'surfaces' (Fig. 3.26) so that there is no conducting material between the three elements thus preserving the radiative only nature of the couplings.

Under steady–state conditions the heat equations for elements 1 and S are respectively

$$0 = -\sigma T_1^4 + \sigma(3^4) - \sigma T_1^4 + \sigma T_S^4$$

$$0 = -\sigma T_S^4 + \sigma T_1^4 + 22$$

Adding gives

$$0 = -\sigma T_1^4 + \sigma(3^4) + 22$$

and hence $T_1 = 140.2 \text{ K}$

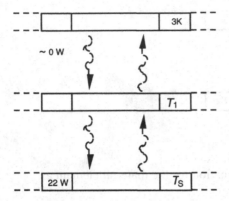

Fig. 3.26.   One thermal shield.

So the temperature of the shield is now the same as that of the material surface in the previous example. This we could have deduced without recourse to equations simply because we know that the outside surface has to radiate the 22 W dissipated in the system. Putting the value of T_1 back into one of the equations gives

$$T_S = 166.7 \text{ K}$$

Thus the net effect of introducing the shield is to raise the temperature of the inner surface as might be expected, in this case by about 26 K.

Adding another thermal shield of emissivity 1 leads to the result shown in Fig. 3.27.

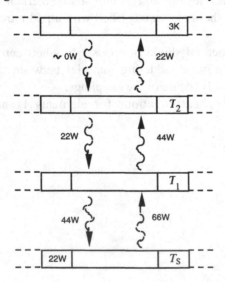

Fig. 3.27.   Two thermal shields.

We see that each shield radiates the same power for each of its two surfaces by virtue of its single temperature. We also see that the *differential* heat flow between layers remains constant and has to match the internal dissipation. In effect each layer adjusts its temperature to ensure this net heat flow and can only do this by increasing the temperature of each successive layer. By inspection it is thus very simple to draw the case for say 6 thermal shields, insert the power exchange between each shield and then calculate the temperature of each shield.

Similarly we can write down an expression for the *n*th shield as

$$T_n = \left(\frac{n \times 22}{\sigma}\right)^{\frac{1}{4}} = 140.2 \ n^{\frac{1}{4}} \ \text{K}$$

and the temperature of the inner surface as

$$T_S = 140.2 \ (n+1)^{\frac{1}{4}} \ \text{K}$$

Thus the inner surface temperature increases as we add in each successive shield; this is shown graphically in Fig. 3.28.

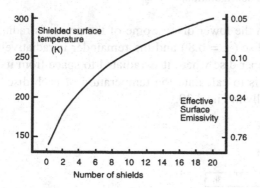

Fig. 3.28. The effect of shields on temperature and effective surface emissivity.

The effect of the shields is to offer an impedance to the flow of heat such that a larger temperature gradient has to be established across the system as a whole in order to maintain that heat flow at the same level. The net heat flow from the inner surface (as for any of the shields) is still the same and equals 22 W, but now the temperature of the surface is increased. (This may be difficult to grasp intuitively, but we did see that the effective conductivity for radiative transfer is a function of temperature, that is the impedance to radiative heat flow increases as the third power of the mean temperature. In our example the net heat flow is constant so the temperature difference between successive shield increases as we go into the stack and the temperature increases; hence the reason for the shape of the graph).

Because the heat loss from the inner surface is unchanged but now occurs at a higher surface temperature we can say that the surface has a lower effective emissivity than its material emissivity of 1. The value of the effective emissivity decreases with increasing numbers of layers as is shown by the right hand axis of Fig. 3.28. This is a topic which we will return to later when we discuss multilayer blankets.

The second configuration that we are going to investigate comprises two thick discs separated by a thin–walled cylinder, the elements being of aluminium alloy. The configuration and its dimensions are shown in Fig. 3.29.

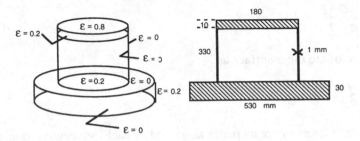

Fig. 3.29. A simple thermal configuration.

10 W of power is dissipated in the lower disc. Some of the power is radiated to cold space from the edge of this disc ($\varepsilon = 0.8$) and the remainder is radiatively and conductively transported to the upper disc where it is radiated to space from its front surface ($\varepsilon = 0.2$). The exercise is to calculate the temperature of each disc. This configuration is shown schematically in Fig. 3.30.

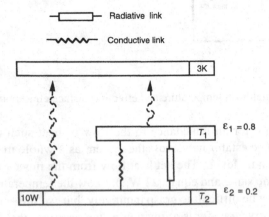

Fig. 3.30. A simple model of thermal nodes and links.

We now calculate the radiative areas and the links.

The radiating areas are straightforwardly

$$A_1 = 2.5 \times 10^{-2} \ \text{m}^2, \quad A_2 = 5.0 \times 10^{-2} \ \text{m}^2$$

The conductive link is calculated as

$$k_{12} = \frac{\lambda A}{d} = \frac{180(\pi \times 0.18 \times 10^{-3})}{0.33} = 0.3 \ \text{W/K}$$

The radiative link arises because the inside surface of the lower disc can see the inside surface of the upper disc; both have non–zero emissivities. Because the inside surface of the cylinder is perfectly reflecting the view factor between the disc surfaces is 1. Since this is now equivalent to two infinitely plane parallel surfaces the effective emissivity of the coupling is given by

$$\frac{1}{\varepsilon_{\text{eff}}} = \frac{1}{0.8} + \frac{1}{0.2} - 1$$

$$\therefore \varepsilon_{\text{eff}} = 0.19$$

Hence, provided that the difference between T_1 and T_2 is not too great we can express the radiative link as

$$I_{12} = A \varepsilon_{\text{eff}} \sigma \bar{T}^3$$

where $\bar{T}$ is the mean temperature.

Since the radiative link will always tend to reduce ΔT, the maximum ΔT will be given by the situation in which all the 10 W flows through the walls of the cylinder, i.e.

$$\Delta T_{\text{max}} = \frac{10}{0.3} \approx 33\text{K}$$

The mean temperature, $\bar{T}$, is simply given by Stefan's Law

$$10 = (A_1 \varepsilon_1 + A_2 \varepsilon_2) \sigma \bar{T}^4$$

which gives $\bar{T} = 276\text{K}$

and we see that ΔT_{max} is only of order 10% of $\bar{T}$.

Putting this value into the expression for l_{12} gives

$$l_{12} = 0.023 \ \text{W/K}$$

Thus, in this case, l_{12} is less than 10% of k_{12} and it can be ignored to a first approximation. Hence we have a conductive model, that is one in which all the internal links are conductive, and the thermal system we have to analyse is that shown in Fig. 3.31.

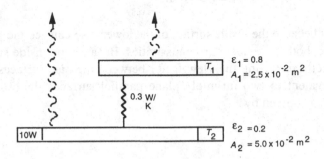

Fig. 3.31. A simplified conductive model.

Having calculated $\bar{T}$ and knowing the radiating areas we can get a good idea of what the heat flows are. Applying Stefans Law to each of the radiating areas we see that surfaces 1 and 2 respectively radiate about 6.7 W and 3.3 W and, since the 6.7 W is conducted through the cylinder it generates a temperature difference of ($6.7/0.3$) K to do this, that is about 22 K (i.e. somewhat less than ΔT_{max} of 33 K). Expressing $\bar{T}$ as the arithmetic mean of T_1 and T_2 i.e. ($T_1 + T_2$)/2 and the temperature difference ($T_2 - T_1$) we can calculate the temperatures approximately, i.e.

$$T_1 = 267 \ \text{K} \quad, T_2 = 287 \ \text{K}$$

Thus the heat flow diagram looks approximately like that shown in Fig. 3.32.

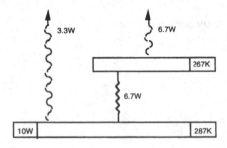

Fig. 3.32. Coupled powers for simplified conductive model.

The powers now radiated are not compatible with the temperatures. Using the calculated values of T_1 and T_2 the powers are modified as

$$6.7 \text{ W} \Rightarrow 5.7 \text{ W}$$
$$3.3 \text{ W} \Rightarrow 3.9 \text{ W}$$

$$\overline{\phantom{10.0 \text{ W}}} \quad \overline{\phantom{9.6 \text{ W}}}$$
$$10.0 \text{ W} \quad 9.6 \text{ W}$$

In order to get the correct total dissipation power which is inviolate, if we make the adjustment

$$5.7 \text{ W} \Rightarrow 5.9 \text{ W}$$
$$3.9 \text{ W} \Rightarrow 4.1 \text{ W}$$

$$\overline{\phantom{9.6 \text{ W}}} \quad \overline{\phantom{10.0 \text{ W}}}$$
$$9.6 \text{ W} \quad 10.0 \text{ W}$$

then T_1 and T_2 are recalculated as

$$T_1 = 268 \text{ K} \quad \text{and} \quad T_2 = 291 \text{ K}$$

Again this is not totally consistent, but we can see now how we might set up some iterative procedure to arrive at an increasingly accurate solution.

The reason for getting an approximate answer by this means is to demonstrate two things. Firstly how, by making some approximations, we can very simply arrive at a solution and secondly how some kind of iteration can lead us to an improved solution.

What we will now do is to adopt a more rigorous approach and set up the heat equations for the elements in the system and see where that leads. The general parameters that we need are shown in Fig. 3.33.

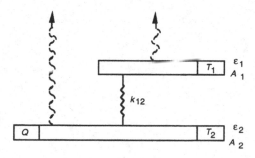

Fig. 3.33. Generalised, simple, conductive model.

Under – steady state conditions the heat balance equations for the 'nodes' 1 and 2 are

$$0 = -A_1 \varepsilon_1 \sigma T_1^4 + k_{12}(T_2 - T_1)$$

$$0 = -A_2\varepsilon_2\sigma T_2^4 + k_{12}(T_2 - T_1) + Q$$

These equations are non–linear, but we can linearise them in T^4 by using the approximation discussed at the beginning of this section

$$k_{12}(T_2 - T_1) = \frac{k_{12}(T_2^4 - T_1^4)}{4\overline{T}^3}$$

Thus

$$0 = -(a_1 + a_2)T_1^4 + a_2 T_2^4$$

$$0 = a_2 T_1^4 - (a_2 + a_3)T_2^4 + Q$$

where

$$a_1 = A_1\varepsilon_1\sigma \qquad a_2 = \frac{2k_{12}}{(T_{1i} + T_{2i})^3} \qquad a_3 = A_2\varepsilon_2\sigma$$

and T_{1i} and T_{2i} are approximate values of T_1 and T_2 .

These two simultaneous linear equations in T^4 can easily be solved for T_1 and T_2 in terms of T_{1i} and T_{2i}.

This can also be done readily by matrix algebra. In matrix notation the two equations become the single equation

$$A\begin{bmatrix} T_1^4 \\ T_2^4 \end{bmatrix} = \begin{bmatrix} 0 \\ -Q \end{bmatrix}$$

where A is the 2x2 matrix

$$\begin{bmatrix} -(a_1 + a_2) & a_2 \\ a_2 & -(a_2 + a_3) \end{bmatrix}$$

This has the solution

$$\begin{bmatrix} T_1^4 \\ T_2^4 \end{bmatrix} = A^{-1}\begin{bmatrix} 0 \\ -Q \end{bmatrix}$$

where $\qquad A^{-1} = \dfrac{C^{T}}{|A|}$

[A^{-1} is the adjunct of A, $|A|$ is the determinant of A, C^T is the transpose of C, and C is the matrix of the cofactors of A].

Hence

$$|A| = (a_1 + a_2)(a_2 + a_3) - a_2^2$$

$$C = \begin{bmatrix} -(a_2 + a_3) & -a_2 \\ -a_2 & -(a_1 + a_2) \end{bmatrix} = C^T$$

then

$$\begin{bmatrix} T_1^4 \\ T_2^4 \end{bmatrix} = \frac{1}{|A|} C^T \begin{bmatrix} 0 \\ Q \end{bmatrix} = \frac{1}{|A|} \begin{bmatrix} 0 & a_2 Q \\ 0 & (a_1 + a_2)Q \end{bmatrix}$$

and hence

$$T_1 = \left(\frac{a_2 Q}{(a_1 + a_2)(a_2 + a_3) - a_2^2} \right)^{\frac{1}{4}} \quad \text{and} \quad T_2 = \left(\frac{(a_1 + a_2)Q}{(a_1 + a_2)(a_2 + a_3) - a_2^2} \right)^{\frac{1}{4}}$$

These temperatures are not unique since each expression contains the constant a_2 which is a function of temperature. However, they do provide for an iterative procedure whereby we set T_{1i} and T_{2i} to some initial arbitrary value in order to calculate a value for a_2 and thus T_1 and T_2 which we then adopt as improved values for T_{1i} and T_{2i}. We then repeat the cycle.

We can illustrate this by applying this to our previous example.

Table 3.3 shows three such sets of iterations each starting from a different level of ignorance of T_{1i} and T_{2i}. At this level of accuracy the iterations appear to have converged once $a_2 = 3.44$. Clearly the better our original guess the fewer the number of iterations required. We see that for the third set, for which our starting value of T_1 and T_2 was that calculated for $\bar{T}$, we require only two iterations. Hence, for this conductive model the consistent steady – state solution is as shown in Fig. 3.34. This is to be compared with our previous, approximate solutions.

It is interesting to note that had we linearised the equations in T rather than T^4 the expressions for T_1 and T_2 are the same as those for T_1^4 and $\cdot T_2^4$ but with the coefficients as

$$a_1 = A_1 \varepsilon_1 \sigma T_{1i}^3 \qquad a_2 = k_{12} \qquad a_3 = A_2 \varepsilon_2 \sigma T_{2i}^3$$

However, this time the iteration becomes unstable and does not converge to a solution. The reader may wish to try this out as an exercise and ponder why this might be so.

Table 3.3. *Results of computer modelling with different initial temperature conditions*

$$a_1 = A_1 \varepsilon_1 \sigma T_{1i}^3 \qquad a_2 = k_{12} \qquad a_3 = A_2 \varepsilon_2 \sigma T_{2i}^3$$

T_1 (K)	T_2 (K)	a_1 $\times 10^{-9}$	a_2 $\times 10^{-9}$	a_3 $\times 10^{-9}$	Net heat in node (W) 1	2	Net heat imbalance (W)
500	500	1.14	0.6	0.57	-71.3	-25.6	-96.9
245	319	1.14	3.34	0.57	18.1	6.3	24.4
275	297	1.14	3.21	0.57	8.0E-2	-1.0	-0.9
268.9	290.1	1.14	3.44	0.57	0.4	-0.4	0.0
269.3	289.3	1.14	3.44	0.57	4.0E-3	7.3E-3	1.1E-2
300	300	1.14	2.78	0.57	-9.2	-14.6	-23.8
268	292	1.14	3.42	0.57	-4.6	-1.3	-5.9
269.4	289.5	1.14	3.44	0.57	2.5E-2	-3.4E-2	-9.0E-3
269.3	289.3	1.14	3.44	0.57	4.0E-3	7.3E-3	1.1E-2
276	276	1.14	3.57	0.57	-6.6	6.7	0.1
269.2	288.5	1.14	3.46	0.57	-0.2	0.3	0.1
269.4	289.3	1.14	3.44	0.57	3.0E-2	4.0E-2	7.0E-2
269.3	289.3	1.14	3.44	0.57	4.0E-3	7.3E-3	1.1E-2

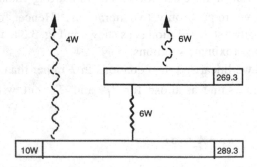

Fig. 3.34. Coupled powers from computer modelling.

Finally, we now have all the tools in place to include the radiative link which we previously removed thus giving us a fully representative model. The system of 'nodes' and 'links' is shown in Fig. 3.35. The radiative link simply introduces an additional term into each of the heat equations.

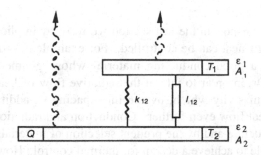

Fig. 3.35. A generalised, simple model with conductive and radiative links.

Thus

$$0 = -A_1\varepsilon_1\sigma T_1^4 + k_{12}(T_2 - T_1) + A_1\varepsilon_{\text{eff}}\sigma(T_2^4 - T_1^4)$$

$$0 = -A_2\varepsilon_2\sigma T_2^4 + k_{12}(T_2 - T_1) + A_1\varepsilon_{\text{eff}}\sigma(T_2^4 - T_1^4) + Q$$

Setting $a_4 = A_1\varepsilon_{\text{eff}}\sigma$ and linearising in T^4 as before we get

$$0 = -(a_1 + a_2 + a_4)T_1^4 + (a_2 + a_4)T_2^4$$

$$0 = -(a_2 + a_4)T_1^4 - (a_2 + a_3 + a_4)T_2^4 + Q$$

which is the same as before except now $a_2 + a_4$ replaces a_2

Hence

$$T_1 = \left(\frac{(a_2 + a_4)Q}{(a_1 + a_2 + a_4)(a_2 + a_3 + a_4) - (a_2 + a_4)^2}\right)^{\frac{1}{4}}$$

$$T_2 = \left(\frac{(a_1 + a_2 + a_4)Q}{(a_1 + a_2 + a_4)(a_2 + a_3 + a_4) - (a_2 + a_4)^2}\right)^{\frac{1}{4}}$$

In our particular example $a_4 \approx 0.03 \times 10^{-9}$ compared with $a_2 \approx 10^{-9}$ and so makes little difference to T_1 and T_2 as we expected.

We can begin to see now what would happen if we increased the number of elements or nodes. If we had n nodes we would have n temperatures and n heat equations which can then, in principle, be solved to give each temperature. Thus we would end up with a more detailed mathematical model. We shall return to this topic when we come to discuss thermal mathematical models (TMMs).

3.5   Thermal control elements

3.5.1   Introduction

In our discussion of heat transport in the last section we were by implication discussing ways in which the flow of heat can be controlled. For example in order to conduct heat readily we require a good conducting material whose geometry is designed to be 'short' and 'fat'. Or, in order to inhibit the radiative flow of heat we require a surface finish of low emissivity which, by the interspacing of additional radiation shields will reduce the heat flow even further. Conduction and radiation are always present in any hardware structure and by the prudent selection of materials and thermal finishes it is usually possible to achieve a design for thermal control. However this might not always be either sufficient or convenient bearing in mind other constraints such as mass and space and a thermal design will then require to make use of other thermal control elements.

There is quite a large range of thermal control devices which can be used depending on the general application and space system. What follows here is an account, though by no means exhaustive, of thermal control elements which could be appropriate for small instruments. They are normally classified as passive or active; here passive is taken to be where the thermal properties remain constant and do not change either autonomously or as a result of active thermal control.

3.5.2   Passive elements

Thermal control surfaces

As we have seen, one of the most effective ways to control radiative exchange is through the emissivity and absorptivity of surfaces. This is especially so because the value of α/ε is not by definition unity for a given surface and can take on a wide range of values for different surfaces. Hence surfaces with high α/ε ratio can produce a net heating effect and surfaces with a low ratio a net cooling effect, depending of course on the external environment. It is useful therefore to categorise surfaces in this way as is shown in Fig. 3.36. Also shown is the equilibrium temperature of a sphere with a given α/ε ratio at the distance of the Earth from the Sun.

It is possible to obtain a surface finish by mechanically treating the surface itself. For example the surface can be polished to make it shiny or can be bead/sand blasted to give it a dull, matt finish. The surface can also be chemically treated or layers of a different metal added via the various plating processes. Alternatively different surfaces can be obtained by the bonding of a layer which has different thermal properties. Paints (black, white, etc.) are used extensively in this way as are the flexible, self adhesive tapes (aluminium, kapton etc.). Also used, though not as extensively, are Second Surface Mirrors (SSM) which are mechanically similar to solar cells and which are epoxyed to the surface.

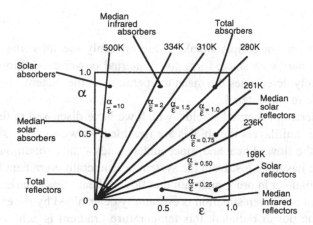

Fig. 3.36. Categorisation of thermal surfaces.

The range of thermal control surfaces available is summarised in Fig. 3.37.

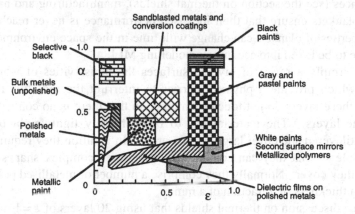

Fig. 3.37. Thermal control surfaces.

The use of SSMs underlies a consideration which has to be taken into account for any coating, and that is the mechanical stability of the bonding agent. The temperature range that such adhesives might be subjected to may cause the bond to deteriorate and hence render unsuitable that particular coating.

A further consideration for any surface finish is that of contamination. This is important for low emissivity surfaces since contamination tends to make them more emitting. Surfaces can be cleaned on the ground and kept clean by an appropriate contamination control plan but in orbit control it is far more difficult. Of particular concern is cryo–deposited water which condenses out onto surfaces which are cold and which can give rise to significant increases in emissivity.

Multilayer insulation (MLI)

Multilayer insulation, or simply blankets, are arguably the most important single element of thermal hardware since, for a given internal power dissipation, it is the blanket which ultimately determines the mean temperature of the system. We will therefore discuss this topic in some detail.

Blankets comprise a series of thermal shields, which we have discussed in the last section, but in the case of multilayers the shields are flexible. As we saw, the shields provide an impedance to the flow of heat and, since under steady – state conditions, the internal power dissipation has to be lost from the system, a temperature gradient has to be established across the blanket in order maintain that flow of heat. Since the external temperature of a blanket in any given situation is essentially determined by the external surface emissivity and the power radiated, this temperature gradient is achieved by increasing the temperature of the internal surface of the blanket. Thus good insulation blankets produce high internal temperatures whereas poor insulation blankets do not.

Although it is readily shown that in theory blankets should provide very large thermal impedances (see the section on thermal shields), manufacturing and handling problems with blankets ensure that their idealised performance is never reached. In addition the properties of blankets can change with time in the space environment and such factors have to be taken into account in modelling MLI.

A blanket is simply a series of layered surfaces the emissivities of which are controlled and which provide a physical barrier to interrupt the radiative flow of energy. Ideally these layers do not touch each other so that there is no conduction of heat between the layers. The material has to be extremely lightweight because blankets are multilayered and cover large surface areas. In addition they require to be reasonably flexible so that they can contour the sometimes complex shapes of the subsystems that they cover. Normally they comprise a number of metallised polymeric films each with a thickness of a fraction of a mm.

We saw in our discussion on thermal shields that using 20 layers of $\varepsilon = 1$ material, the effective emissivity of a surface could be reduced from 1 to 0.048. The same effect could obviously be achieved by replacing the blanket with a thermal control surface with emissivity of 0.048 which is quite feasible and would be far less costly in money and mass. However, it is clear that a much lower effective emissivity would result if instead of choosing black surfaces, reflecting surfaces with low emissivities were chosen. For example, with emissivities of 0.1, the formula suggests that an effective emissivity of 0.005 should be achieved for this 20 layer blanket. In practice this is not so. Blankets cannot be made which are plane and infinite, whose layers do not touch and which require no support. However it is relatively straightforward to manufacture blankets having an effective emissivity of 0.01 which is better by a factor of a few than that which can be achieved with a single thermal control surface.

We can quantify the performance of blankets in 3 ways depending on how we wish to model them. These are

$$Q = kA(T_1 - T_0)$$

$$Q = \sigma A \varepsilon_b (T_1^4 - T_0^4)$$

$$Q = \sigma A \varepsilon_e T_1^4$$

where Q is the heat flow
           T_1 is the temperature of the inner surface
           T_0 is the temperature of the outer surface
           ε_b is an equivalent emissivity
           ε_e is the effective emissivity
           k is the effective conductance ($Wm^{-2}K^{-1}$)

From the form of the equations we see that the first models blankets conductively and the other two radiatively.

The equivalent emissivity is given by

$$\frac{1}{\varepsilon_b} = \frac{2n}{\varepsilon} - n$$

where ε is the emissivity of the internal surfaces
and n is the number of blanket layers.

The effective emissivity is given by

$$\frac{1}{\varepsilon_e} = \frac{1}{\varepsilon_o} + \frac{2n}{\varepsilon} - n$$

where ε_o is the emissivity of the outer surface.

It should be noted that if for some reason the internal layers have different emissivities ε_1 and ε_2 for their two surfaces, then ε in the above equation is replaced by

$$\frac{1}{\varepsilon} = \frac{1}{\varepsilon_1} + \frac{1}{\varepsilon_2} - 1$$

as we have seen before.

By equating pairs of the above, the interrelations between the overall conductances or emissivities of the blankets can be established.

$$\frac{1}{\varepsilon_e} = \frac{1}{\varepsilon_o} + \frac{1}{\varepsilon_b}$$

$$k = \varepsilon_b \sigma (T_1^4 - T_o^4)/(T_1 - T_o)$$

We see from this that the effective conductance as defined here is a function of temperature. This is an artifact which unfortunately limits its usefulness in hand calculations.

♦ *Example 3.1. MLI calculation*

Consider a 20 layer blanket for which the emissivity of the inner layers is 0.6 and for which the outer surface emissivity is 0.8. Assuming a net radiated power of 10 W for a unit radiating area. Calculate values of the inner and outer temperatures, the effective and equivalent emissivities, and the effective conductance.

The outer temperature T_o follows from Stefans Law

$$10 = 0.8\sigma T_o^4 \quad \therefore T_o = 121.7 K$$

The equivalent and effective equivalent emissivities are given respectively by

$$\frac{1}{\varepsilon_b} = \frac{2 \times 20}{0.6} - 20 \quad \therefore \varepsilon_b = 0.0214$$

$$\frac{1}{\varepsilon_e} = \frac{1}{0.8} + \frac{1}{0.0214} \quad \therefore \varepsilon_e = 0.0209$$

enabling us to calculate the inner temperature T_i

$$10 = 0.0209\sigma T_i^4 \quad \therefore T_i = 302.7 K$$

The effective conductance is then given by

$$k = 0.0214\sigma \frac{(302.7^4 - 121.7^4)}{(302.7 - 121.7)} \quad \therefore k = 0.055 \ Wm^{-2}K^{-1}$$

This is shown pictorially

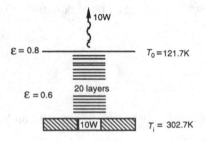

♦ *Example 3.2.   Further MLI calculations*

Determine the effect of decreasing the emissivity of the outer layer from 0.8 to 0.4.

As before,

$$10 = 0.4\sigma T_o^4 \quad \therefore T_o = 144.7\text{K} \qquad , \qquad \frac{1}{\varepsilon_b} = \frac{2 \times 20}{0.6} - 20 \quad \therefore \varepsilon_b = 0.0214$$

$$\frac{1}{\varepsilon_e} = \frac{1}{0.4} + \frac{1}{0.0214} \quad \therefore \varepsilon_e = 0.0203 \quad , \quad 10 = 0.0203\sigma T_i^4 \quad \therefore T_i = 304.9\text{K}$$

$$\text{and} \qquad k = 0.0214\sigma \frac{(304.9^4 - 144.7^4)}{(304.9 - 144.7)} \quad \therefore k = 0.062 \ \text{Wm}^{-2}\text{K}^{-1}$$

giving

In the above example it can be seen that the temperature of the outer layer of the blanket increases by 23 degrees as a result of the change in emissivity whereas the inner surface temperature increases by only 2 degrees. There is a 3% decrease in the effective emissivity whereas the effective conductance increases by about 14%. Clearly in the second case the blanket is a better insulator and this result simply illustrates the apparent dependence of the effective conductance on blanket temperature.

This example serves to show that the insulating effect of blankets is not greatly affected by changing the emissivity of just one of its layers, in this case the outer layer. We have discussed this in the context of an internal source of power, but in the case where some of this power derives from an external source, the above property can be used to good effect to limit the absorption of heat without significantly affecting the ability of the blanket as a whole to radiate heat. This is of importance when considering the absorption of IR radiation from the Earth since this will be controlled by the emissivity of the outer surface. Similarly the choice of absorptivity for the outer surface affects the absorption of power at visible wavelengths, either from direct sunlight or from reflected sunlight, but the absorptivity has no influence on the emission properties of the blanket at their typical in−orbit temperatures.

In the same way that an effective emissivity can be ascribed to a blanket so also can an effective absorptivity. It can be shown that the ratio of these two is the same as the emissivity and absorptivity of the outer surface of the blanket, that is, this ratio is unaltered by the way we choose to model the blanket.

The production of blanket material itself is a quality controlled, reproducible process. The fabrication of blankets however is something of an art and is much more difficult to control. This is simply because when blankets are fabricated they do have to be supported, adjacent layers do touch each other and the blankets do have edges which radiate like cavities. Thus it is not an easy task to reproduce the properties of an idealised blanket. Because of this it is usually advisable to procure blankets through firms which have the necessary specialist experience.

The blanket material comprises a plastic film such as Kapton (polyimide) or Mylar (polyester) which is metallised either on one side or both sides with, for example, aluminium. The thickness of the metallised film is usually several hundred Angstroms and the thickness of the plastic is about 0.3mm for internal layers and 1 to 2 mm for the outer layers. It is the metal film which gives the blanket its properties, the plastic serving simply to support the metal with a very smooth surface and to make the blanket flexible yet relatively rugged. Thus the weight of the blanket, which is an important consideration, is essentially that of the plastic itself. Sometimes, to prevent the metallised surfaces of individual layers from touching, a lightweight bridal veil spacer material such as Dacron (polyester) is used.

The choice of actual blanket material is based on a number of considerations such as effective emissivity, effective absorptivity, mass, cost, ruggedness, handling protection, the need for venting, outgassing, the need for an outer layer which is conducting, earthing, non–flammable, shaped or loose fitting, and susceptibility to particle, chemical or radiation damage as well as others. All of these are necessary and important considerations although they do not have a great impact on how a blanket should be modelled or how to take account of the fabricated blanket itself.

The loss of idealised performance of a fabricated blanket can arise from compression of the blanket layers either at seams or as result of configuration or handling, the presence of trapped gas, conduction down blanket support studs, and radiation from edges, holes or patches. These effects can be semi–quantified and measures such as overlapping, edge sealing etc. can be used in a methodical way to minimise their impact on performance. In general terms however the performance of a blanket is simply related to its planar area; large, flat blankets with few discontinuities have a better performance than small, shaped blankets with many discontinuities. Fig. 3.38 indicates what has typically been achieved for a number of different missions with blankets of different configurations.

As has been pointed out, blankets are difficult to fabricate and are costly but are a most important component for thermal control. Unfortunately their properties are not constant and can change as a result of handling and as a result of being exposed to the space environment.

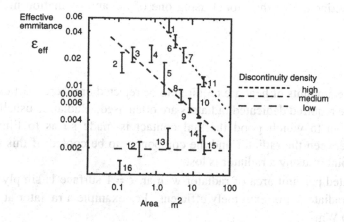

Fig. 3.38. MLI performance dependence on area
and discontinuity density.

1. MARINER 1969 SCI   2. PIONEER MAG   3. SATS   4. OGO

5. MARINER 1971 BUS   6. MARINER 1971 SCI 7. NIMBUS SERIES 8. PIONEER F&G

9. MARINER 1971 PROP 10. VICKING PROP   11. LUNAR MODULE 12. JPL CALOR

13. LeRC TANK CAPOR   14. LeRC TANK   15. MSFC TANKS   16. MSFC CALOR

There is not a great deal of published quantitative data but in general terms the effective emissivity of blankets can increase on the ground and the effective absorptance can increase in orbit.

The author has some experience of the former in that the blankets for the Wide Field Camera on ROSAT were tested twice in a thermal balance chamber over a period of 3 years. The sizes of the blankets were typically 1 m². During the 3 years the blankets were integrated and de–integrated from the hardware though not excessively so. Measurements indicated that the effective emissivity of the blankets increased from about 0.019 to 0.026 over that period.

Measurements carried out in orbit on various thermal control surfaces have shown that whereas the emissivity of many surfaces is apparently unchanged, the absorptivity of many surfaces can increase significantly, particularly for long–term missions. The mechanisms for these changes have not yet been established though proton radiation and UV exposure are suspected. In addition chemical effects resulting from exposure to atomic oxygen are known to erode plastics. Blankets therefore are very vulnerable components and the possibility of these effects has to be taken into account when considering the use of blankets in the thermal control of an experiment.

In conclusion it should be underlined that modelling blankets is to some extent simply a matter of judgement as to what is likely to be able to be achieved with reasonably conservative margins for error and, having thus selected a realistic specification, to enshrine this in the model using one of the approximation methods described earlier.

Radiators

Where large amounts of heat are required to be rejected or where particularly low temperatures are required dedicated radiators are often used. Radiators usually are simply large surfaces to which good thermal contact is made so as to limit the temperature drop between the radiator and the component to be cooled; if this is not the case then the point of using a radiator is lost.

The power radiated per unit area of radiator with an $\varepsilon = 1$ surface is simply σT^4 Hence in principle radiators are extremely efficient. For example a radiator at 290K can reject about 400 Wm^2.

If it is possible to design a system such that the radiator always views cold space then such heat rejections are possible. If however the radiating surfaces are able to see direct sunlight or, as is more usually the case, earth albedo and earth IR radiation, then radiators can be a source of heat input. To achieve an overall loss of heat therefore the surface finish of the radiator has to be carefully selected to give the most propitious ratio of α/ε; this may also require a knowledge of the orbital elements of the spacecraft, particularly the proportion of the orbital period spent near perigee where the earth radiation input is significant. In these cases it is usually necessary to carry out a transient thermal analysis (see later).

Clearly such heat inputs are of nuisance value for a cooling radiator. At times however it may be possible by careful design to ensure that earth heat inputs are avoided. For example, star trackers are designed to look at stars which is effectively the same as cold space, and they usually have long, stray–light baffles which are painted black and have relatively large apertures. The detector on the star tracker on the Wide Field Camera on ROSAT was a CCD and was cooled by thermally coupling it to its optical baffle.

The use of detectors that require to be cooled are now becoming quite commonplace. For required temperatures of –100C relatively simple passive radiators can be used whereas for lower temperatures it may be necessary to employ either more sophisticated radiators or to use some active device such as a Stirling cycle cooler (see later).

Clearly the parasitic heat loads into a cryogenic system at –100C can be considerable and often significantly greater than the power dissipated in the device which needs to be cooled. In these cases multistage radiators can be used. These comprise a system of nested radiators each with a view factor to space and such that each of the nested radiators is shielded by its neighbour. Fig. 3.39 shows an example of a 2–stage radiator.

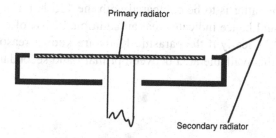

Fig. 3.39. A 2 - stage radiator.

At first glance, for a given radiating surface area to space, multistage radiators appear to offer something for nothing. This is not actually so because, although the heat rejected to space varies linearly with area, it varies as the fourth power of the temperature. This allows for some optimisation in the utilisation of the given radiating area. Consider the following example shown in Fig. 3.40.

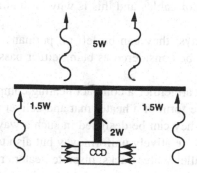

Fig. 3.40. A 1–stage radiator.

Suppose the internal dissipation and the parasitics through the signal and power leads to the CCD is 2 W and the parasitics into the radiator is 3 W. The radiator therefore has to radiate 5 W. Let the effective area of the radiator be 0.5 m². We can compare the performance of a single radiator of 0.5 m² with a 2–stage radiator with a primary of say 0.45 m² attached to the CCD and a secondary of 0.05 m² shielding the primary. Stefan's Law shows that

$$T \text{ (single radiator)} \quad\quad = 115 \text{ K}$$
$$T \text{ (primary radiator)} \quad = 94 \text{ K}$$
$$T \text{ (secondary radiator)} = 180 \text{ K}$$

Thus the 3W parasitic power is rejected by a relatively small area at the relatively high temperature of 180 K, whereas the 2 W primary power is radiated away at the lower temperature of 94 K. The latter is to be compared with the 115 K for a single radiator of the same total area and hence indicates that an additional 20 K of cooling has been achieved for the CCD. Clearly if the parasitic loads are known reasonably accurately then the division of the available area between primary and secondary can be optimised.

Heaters

Power dissipated in electronic boxes is a useful source of energy in terms of thermal design, especially if it is continuously dissipated at a constant level. However the positioning of such boxes is usually determined by considerations other than thermal and thus their heat value is not always of significant use. Indeed in themselves they sometimes represent an independent problem because the temperature at which electronics operates also has to be controlled within some band. In addition it may not be possible to readily transfer heat to locations where it may be required; the mass needed for significant heat transfer can be large. However, to transfer electrical power requires only a little mass in the form of cables and this is why heaters are of particular use.

Heaters can be used essentially in two ways; they can be left on permanently or they can be switched. In this sense they can be considered as being either passive or active elements.

Generally heaters are of two types. They are either a compact heating component and deliver power to a point or they are in the form of a heater mat and deliver power over an area. Mats have the advantage that they can be designed in such a way as to not only distribute the heat over a surface in a relatively uniform way but also to have inbuilt redundancy in the form of series – parallel heater tracks; discrete heaters require duplication to provide such redundancy. Obviously care has to be taken in physically mounting either device to ensure that the power is well coupled into the structure. The thermal modelling should reflect the difference in the way in which the power is input to the structure.

Heat pipes

A very efficient way of transporting heat from one place to another is by means of a heat pipe. Efficiency in this context is in terms of an effective conductance for a given mass.

An evacuated, sealed tube is partially filled with a working fluid such that at its nominal operating temperature the fluid and its vapour are in equilibrium. When heat is input (the evaporator) the heat is taken up as latent heat and the vapour pressure increases locally as a result. The pressure equalises by the transport of vapour to other parts of the pipe which are by definition cooler parts (the condenser). The vapour condenses and the latent heat is given up. The fluid is returned to its original location by capillary action through a wick or its equivalent.

Since the principle of operation of the pipe requires the working medium to exist both in the liquid and the gas phase, this will only be so over a limited range of temperature. Indeed this also places an upper limit on the maximum power that can be pumped. In order to extend the range over which heat pipes can be effective they can be filled with different operating fluids.

It is not the intention here to characterise heat pipes; the reader is encouraged to pursue this as necessary. It is useful however to give a brief account of some of the advantages and disadvantages of using heat pipes. Disadvantages include the relatively high cost, particularly the associated product assurance costs, and the limited operating range of a given working fluid, particularly if the operating temperatures are significantly different from the pipe storage temperature. The main disadvantage is that to achieve their full specification they require a zero gravity field. This can produce problems as far as testing the devices is concerned but sometimes a prudent mechanical design can circumvent this.

The advantages are that the pipe essentially represents a constant temperature component even when relatively large powers are being transported; the conductance of the pipe is independent of the power up to the pipe capacity (essentially determined by the pipe cross-section and wick capacity). The pipes can be bent. They can also be an integral part of the structure provided that the pipe 'hole' can be machined into the structure.

In order to make use of the properties of heat pipes, heat still has to be coupled into the pipe at the desired location and similarly extracted. If significant powers are to be transferred such considerations can dominate the usefulness of such devices especially if the power is to be transferred over relatively small distances. Standard thermal design procedures can be adopted for the application of heat pipes although the thermal properties of the working fluid do have to be taken into account but only in the regions where the heat is transferred to or away from the pipe.

3.5.3   Active elements

The elements described here are those for which the thermal properties can change or be changed in a controlled way in orbit. Some of these changes are autonomous, others are under external control.

Heaters

These are the same elements as described earlier but the power dissipation is variable. This can be done either by using a variable voltage or by interrupting a fixed voltage according to some variable duty cycle. Generally the latter is simpler to implement than the former. In order to monitor the power demand a temperature sensing device is required. When the temperature falls below some level power is supplied and is removed when the temperature rises above some upper level. The degree to which the levels differ depends on the degree of control required but it is usual to ensure that the two levels are not too close otherwise an excessive switching of the heater power results.

Variable conductance heat pipes (VCHP)

Ordinary heat pipes have a fixed conductance which is so high that the temperature difference across its ends is negligible and the variation of this temperature difference with heat load is insignificant provided the load is within the capacity of the pipe. The temperature at which the pipe operates is determined by the total heat load on the ambient system.

VCHPs however are designed to operate within a small, limited temperature range almost independent of the heat load. The smaller this temperature range the greater is the complexity of the heat pipe structure. Thus VCHPs can be used to control the temperature of the heat source over a wide range of loads. The variable conductance is achieved most commonly by the introduction of a non–condensing gas into the sealed pipe.

Louvre radiators

For a given temperature and surface emissivity, the power capability of a radiator is proportional to its aperture which, in the case of a plane radiator, is the actual area of the radiator. Hence in instances where the power delivered to the radiator varies, the temperature of the radiator can be controlled by varying the area. This is actually achieved by a system of louvres, rather like venetian blinds, which are placed over the outward facing surface of the radiator and can be rotated about their parallel, longitudinal axes. When the louvres are edge on to the radiator a larger fraction of the radiating surface is exposed compared with that when the louvres are face on. The control of the louvres is usually by means of bimetallic strips set to operate at a predetermined temperature.

Temperature control through husbanding heat losses is clearly more economic in power than by means of heaters.

Variable emissivity devices

Although not yet space qualified or commercially available, variable emissivity surfaces have been developed over the past few years. The low/high emissivity state is switched by the application of a small voltage. At present the emissivity values are about 0.3 and 0.8. but developments are continuing to extend this range. In order to achieve intermediate values such devices have to be arranged in a mosaic although this obviously makes the control more complicated.

Peltier coolers and Stirling cycle coolers

Before leaving this topic it is worth mentioning these two active coolers. They permit quite large temperature differentials to be established (50C and 200C respectively) but will only pump relatively small amounts of power (≤ 1 W and ~ 10 W respectively). In brief the Peltier cooler drives a current through two thermocouple junctions thus cooling one at the expense of heating the other, and Stirling cycle coolers pump a working gas through a compressor and displacer. They can be used to achieve relatively low temperatures but at the price of dissipating a disproportionately higher power at the high – temperature end of the system (e.g. efficiencies of ~ 15 % and ~ 2 % at 20C respectively). The temperature differentials are established only over a very short distance so the thermal problem translates into one of transporting a relatively large power but at the correspondingly higher temperature.

Peltier coolers are quite commonly used; Stirling coolers are relatively new devices which have now been space qualified but are expensive. In common with most active devices some form of redundancy is required in their application.

3.6   Thermal control strategy

We saw in the last section that the properties of thermal elements fall into three types; those that remain fixed, those that can vary and be controlled autonomously, and those that can be changed by some external control. The use of such elements therefore leads to the corresponding level of overall thermal control, though obviously a practical system will almost certainly contain some combination of types.

The advantage of passive elements is that they are reliable. In contrast the disadvantage of autonomous devices is that if they fail there is little that can be done to retrieve the situation. As regards control systems as a whole, active systems do bring a number of significant advantages. They can cope with a wider range of thermal environments and can adjust to changes in the performance of thermal elements; they can be used when very specific temperatures or temperature stability are required; they obviate the need of the design engineer to precisely predict the in–orbit performance and hence to implement a passive system to meet the requirement. As mentioned, their main disadvantage is that they are more susceptible to failure, but this can often be accommodated by building redundancy into the system.

The most usual form of active thermal control uses heaters to provide the power input and thermistors to provide a temperature monitor signal for regulation. Since all systems have a thermal capacity, which we can think of as 'thermal mass', then they also have 'thermal inertia'. This means that their control is not instantaneous. Hence it is important that any temperature control sensors are placed as close to the source of heat as is practicable. This however can cause some design problems if the component whose temperature is to be controlled is remote from the heaters. In this case the local temperature sensor will need to control to a higher temperature than that actually required at the remote component. It may be difficult to predict what the temperature differential needs to be.

Even though there is no inbuilt autonomy in the heaters themselves, they are often used in an autonomous way by linking them with a thermal switch. The thermal switch is a device which senses the temperature and operates a switch controlling the power to the heater. In its simplest form this comprises a bimetallic strip. The switch will have an associated nominal operating temperature range and a nominal operating temperature precision. In selecting these devices such considerations require to be carefully addressed.

Hence such a heating system can have either inbuilt control or can be controlled from some central processor. This processor can be a dedicated, prewired hardware unit or can be a dedicated or shared processor under software control. Both systems have their own advantages and disadvantages.

The autonomous system has the advantage that it requires minimal wiring and can be used to set local temperatures almost irrespective of the temperatures of neighboring regions. Its main disadvantages are that it offers no flexibility, has to be predetermined and, in the case of a failure, it is not possible to adjust other parts of the system to compensate. In addition, for the reasons given above, the thermal switches have to be placed close to the heaters and therefore have to be set to operate at some temperature higher than that of the component which it is designed to control.

The main advantage of the use of a central processor under software control is that it offers far greater flexibility, especially if new software can be uplinked from the ground in order to handle changes in the thermal environment or changes resulting from problems elsewhere in the experiment or spacecraft. However, such systems require additional looming runs since all sensor signals are returned to the processor unit and, because of the increased complexity, are more susceptible to problems. Of particular significance is the Single Event Upset (SEU) of the processor memory resulting from radiation damage which can place the system in a temporary unstable state if no provision is made for such an eventuality. This is discussed further in section 4.7.11.

Clearly it is up to the thermal design engineer to balance advantages and disadvantages of any particular control system in the context of the requirements and resources associated with their own particular experiment.

3.7   Thermal mathematical models (TMMs)

TMMs are self-explanatory; they are simply the mathematical representation of the thermal reality. The reality of the representation itself depends on the sophistication of the model in relation to the complexity and nature of the material object of the modelling.

The overall objective of TMMs is also simply stated. It is to initially support the development of the design but the finally evolved TMM is an important end product in itself because, if developed correctly, it enables the thermal characteristics of the experiment to be reliably predicted. In this context it is perhaps important not to lose sight of the fact that the internal structure of the model need not necessarily adhere to understood physical principles (though usually it does) provided that the predictions of the model adequately describe the overall thermal behaviour of the system. If a problem arises when the experiment is in orbit the experimenter needs to know the likely outcome of any correcting action taken. A TMM is a very convenient tool for achieving this quickly, cheaply and without risk.

TMMs can range from the very simple (as we have already seen in a number of examples in previous sections) to the very detailed which require computer solutions. This range is not necessarily a reflection of the lack of sophistication on the part of the simpler TMMs, after all it is not surprising that the quality of the output bears some relationship to the quality of the input; it is rather that different TMMs serve different purposes.

In general TMMs fall into three categories; the 'back-of-envelope' (b.o.e.), conceptual and detailed. The b.o.e phase represents the ball park feasibility stage in which the broad thermal philosophy can be addressed and from which the first estimates of the required thermal resources are made. The second stage is that in which the thermal design concept is fashioned and agreed and where the estimates of the resources made in the preceding stage are seriously addressed. It is in this stage that the specific thermal control strategy is determined and the reasonable contingencies needed to safeguard such a strategy are marshalled, often as the result of sensitivity analyses of the model to identify critical areas in the design. During this stage the thermal resources requirements are often changed but have to be frozen by the end of the phase. Also it is during this stage that the thermal/mechanical requirements are established in order that they be included in the detailed mechanical design. The final phase is the detailed modelling in which, ideally, all the previous decisions regarding philosophy, requirements and margins are confirmed.

In the next section we shall see how these TMMs play a part in the larger strategy in which the overall thermal design is evolved and verified

Most TMMs are now computer based regardless of their degree of sophistication. It should be emphasised that hand calculations should not be spurned; they can be used most effectively to quantify sound common sense in the face of a morass of computer output from a detailed model. The computer mathematical model comprises a set of 'nodes' which are thermal units representing the physical hardware. The node thus has material properties such as bulk thermal conductivity, thermal capacity and surface emissivity. However the most important property of nodes is that they are described by a single temperature, that is to say they behave as if, in reality, they were isothermal. In this respect it is a thermal 'point' but not a spatial 'point'. This apparent conflict has to be resolved by the thermal engineer since heat is transported only by courtesy of temperature gradients, that is, points of differing temperature. It is in this area that thermal modelling skills come to the fore; the choice of nodes and the charting of the 'thermal nodal map' can significantly affect the output of the model unless carefully assembled. As a simple example it is obviously unwise to represent as a single node a part of the hardware in which significant temperature gradients are expected.

The 'range' of TMMs can therefore be approximately expressed in terms of the number of nodes comprising the model. The b.o.e. is perhaps made up of a few 10s, the conceptual between 50 and 100 say, and the detailed a few 100s. The division is somewhat subjective but it should reflect the time required to construct the model, the time to run the model, and the ease with which the output can be digested.

Once the nodal map has been established (which should also include external or boundary nodes to represent the imposed thermal environment) the rest is relatively straightforward in principle though it can be quite difficult to perform. For each node it is necessary to specify:

a node identification number,

conduction links to other nodes to which it is conductively   attached, (characterised by dimensions and thermal conductivity),

radiative links to other nodes which it can directly 'see', (characterised by dimensions, surface emissivity and view factors),

any internal power dissipation,

an initial temperature for the iteration, and

a fixed temperature if it is a boundary node.

As an illustration let us take the earlier example of the two, thick discs separated by a thin–walled cylinder developed in section 3.4 (see Fig. 3.29). Although this was treated in the genuine b.o.e. style, it could equally well form a simple computer model comprising 3 nodes, one of which is the boundary node at 3 K (space) and one of which has an internal power dissipation. The model has 1 conduction link and 3 radiative links (see Fig. 3.30). These parameters are input to the computer mathematical model which essentially sets up a heat balance equation for each node in terms of the temperature of the node, as was shown in that section, and then iteratively solves the equations using the specified initial temperatures as starting values.

Table 3.4 shows the output for this particular model and from this particular program. What is listed is the equilibrium temperature of each node, the conductive and radiative powers and the precision of the thermal balance also for each node. The reader should compare these results with those shown in Table 3.3.

Larger models have more nodes, the number of links increasing quite rapidly with increasing numbers of nodes. The conductive links scale as a factor of about 2 to 3 with node numbers and the radiative links approximately as the square of node numbers. Indirect links, which can be characterised from the direct links, scale even more strongly and in principle can be infinite in number. In practice some lower limit is placed on the actual value of the link in order to keep the numbers within reasonable bounds since they affect the run time of the model.

Table 3.4. *Output from a 3−node model of configuration shown in Fig. 3.30.*

Node	Temp C	K	Power Conducted in	Power Radiated in	Power Imbalance	Isothermal Node
1	15.8	288.94	-5.6E+00	-4.4E+00	-3.3E-16	0
2	-2.9	270.24	5.6E+00	-5.6E+00	-1.1E-16	0
3	-270.1	3.00	0.0E+00	1.0E+01	4.4E-16	1

Conductive heat transfer (W)

1   2   0.561E+01

Radiative heat transfer (W)

1   2   0.441E+00   1   3   0.395E+01   2   3
0.605E+01

Although detailed models can and do play an essential role, it can be seen that as they develop they can become so large as to be unmanageable as a development tool. With the copious outputs which attend such models it often becomes difficult to 'see the wood for the trees'. This is why the smaller models are equally essential to a healthy thermal design programme.

There is a skill in constructing TMMs and like many skills it is best mastered through practise. This however is not an excuse for not attempting to define or exemplify good practice; we will do this in section 3.9. But first we should review the role of TMMs in the overall thermal design strategy.

3.8   Transient analysis

Thus far we have only considered the static thermal balance situation in which all the heat inputs to our system are constant and thus all the temperatures are steady. However for most spacecraft this is not the case since the thermal environment will change either as a result of internal or external changes to the heat input. We have considered active heaters as a means of controlling the temperature at some steady value, but here the control is usually organised such that only small increments of power are injected in any interval and the duty cycle kept short so that the system is in quasi–equilibrium.

There are other circumstances where this will not be the case. For example satellites in orbit will experience changes in heat input from the Earth as they go from day to night if they are in a circular orbit, or if they undergo a perigee passage in an elliptical orbit. Another important change is in the early orbits of a mission when the system is adjusting to the change from the Earth environment to the space environment. In the latter the temperature change will be in a single direction. In the former the temperature will cycle about some mean value in a periodic way. If the thermal time constants are long then the resultant temperature changes will be a result of both these effects. It is important that the thermal design engineer is able to predict such changes. This is achieved by performing a so called 'transient analysis'.

Towards the beginning of section 3.4 we set up a simple 2–node model comprising a surface radiating to space. If we generalise this as a node of mass m, specific heat capacity c for which the heat load is Q then, ignoring the heat received from cold space at temperature 3K, the heat equation becomes

$$mc\frac{dT}{dt} = -\varepsilon\sigma AT^4 + Q$$

Here Q is taken to comprise both the internal heat dissipation and external heat inputs. The equilibrium temperature is obtained by setting dT/dt to zero, thus

$$T_E = \left(\frac{Q}{\varepsilon\sigma A}\right)^{\frac{1}{4}}$$

Hence we can express the differential equation in terms of the equilibrium temperature

$$mc\frac{dT}{dt} = -\varepsilon\sigma A(T_E^4 - T^4)$$

What this equation is telling us is, if some load were applied to the node at some arbitrary temperature T then the equation describes the transient behaviour of the node as it changes its temperature in coming to its equilibrium temperature. We should note that we are considering just a single node to represent our experiment which is, by definition, isothermal. Hence this implies that there are no internal temperature gradients within the experiment or, alternatively, it is possible for heat to travel infinitely quickly across the experiment. This clearly is an over simplification which we should return to later.

This equation can be solved although the solution is somewhat involved and does not easily give us any particular insight into what is occurring. By making some approximations however the situation becomes rather clearer. Here we are going to make the same approximations for $(T_E^4 - T^4)$ as we did at the beginning of section 3.4.

If we make $T \sim T_E$

then the equation becomes linear in T

$$mc\frac{dT}{dt} = -4\varepsilon\sigma AT_E^3(T_E - T)$$

or

$$\tau\frac{dT}{dt} = (T_E - T)$$

where $\tau = \dfrac{mc}{4\varepsilon\sigma AT_E^3}$

which leads to the solution

$$T = T_E + (T_0 - T_E)e^{\frac{-t}{\tau}}$$

This shows that at $t = 0$, T is at its initial temperature T_0, that at $t = \infty$, T is at its equilibrium temperature, and that T approaches its equilibrium value exponentially with a time constant τ. That is to say, for every interval of time equal to the time constant, the difference in the spontaneous temperature of the body and its equilibrium value is decreased to a factor of $1/e$ (about 33%) or conversely by a factor of about 66%. We should note that this time constant is a function of T_E, that is to say it is indeed a constant in this case.

If we make the alternative assumption

$$T >> T_{\mathrm{E}}$$

then the equation reduces to

$$mc\frac{\mathrm{d}T}{\mathrm{d}t} = \varepsilon\sigma AT^3(T_{\mathrm{E}} - T)$$

which is similar to the earlier equation (ignoring the factor 4) but with T and T_{E} interchanged.

We can therefore see our way to a 'solution' to this equation again in terms of a time constant where the time constant is now

$$\tau = \frac{mc}{\varepsilon\sigma AT^3}$$

Here we can see that the time constant is a function of T which is itself changing and so the time constant in this sense is not constant. However if we multiply the top and bottom of the expression by T we then get

$$\tau = \frac{mcT}{\varepsilon\sigma AT^4}$$

The numerator is simply the instantaneous stored energy of the body (J), and the denominator is the instantaneous power, P, radiated by the body (J / s). Hence the time constant can be seen as the rate at which the stored energy of the body is being radiated away. This is a very general, but a very useful, definition of a time constant.

In order to interpret this, consider a body in thermal equilibrium at some temperature T_1 under conditions where it is dissipating say $2Q$ and then, at some instant, the dissipation is reduced to Q. In the first time interval, t, the body loses (by radiation) Pt joules of energy where P is sized by the initial temperature of the body, and gains Qt joules from its internal dissipation. Since the former exceeds the latter there is a net loss of energy ΔE which results in a decrease in temperature ΔT given by $\Delta E / ms$. Thus, at the end of this time, t, the temperature of the body is T_2 where

$$T_2 = T_1 - \Delta T$$

The body now radiates energy at a reduced rate because its temperature is lower. Although its stored energy is reduced, this scales as T whereas the radiation scales as T^4. Hence the net effect of this process is that the time constant is increased. The cycle can be repeated until such time that, in the given time interval, t, the number of joules radiated exactly matches the number of joules input from its internal source. The body will have now reached its new equilibrium value. Such an iteration can readily be programmed in a computer, but it is also a very simple hand calculation to do by selecting some sensible time interval.

A third approach is simply to ignore T_E altogether in which case the differential equation has a straightforward solution which is given by

$$\frac{1}{T^3} = \frac{3\varepsilon\sigma A}{mc}t + \frac{1}{T_0^3}$$

In the limit, this is analogous to setting T_E to zero which corresponds to the situation where there is no heat input to the body but that it has some stored energy. Clearly we could get the same result by performing the above iteration with Q set to zero.

Thus far we have only considered a single node. To some extent this can be of some use to us because we could consider our experiment (or complete satellite) as comprising a single node of some total mass of some representative, averaged thermal heat capacity which radiates to space through some surface area. This would give us a lower limit for the time constant. This is simply because in a real system heat cannot be transferred infinitely quickly (as it can in a 'node') and so the system takes longer to give up its excess heat.

Hence in a system represented by a number of nodes the situation is much more complicated; the transient behaviour of the system will not be simple and the time constant will comprise a mixture of different time constants which will not result in some constant value. What is clear however is that the time taken to reach equilibrium will be determined by the longest, component time constant. In the theoretical situation equilibrium is never reached in a finite time and in a real situation this time could be extremely long. In practice therefore some criteria for reaching 'equilibrium' have to be formulated.

Computer programs exist which will perform transient analyses. The only additional data that has to be input over and above that for the steady–state models are the thermal heat capacity and the mass for each node and power profiles for any of the variable heat sources whether they be internal or external. These will require time intervals to be specified since changes will only be implemented at the end of the time intervals. In addition if any of the thermal boundaries are likely to change then temperature profiles for these also need to be input.

There are a number of caveats and checks that may be useful in setting up, running and analysing transient models;

- transient analyses require a lot of processor time so do not make the time intervals shorter than is necessary

- avoid the use of nodes which have a very small mass since, for a small exchange of energy, there will be a very large temperature change predicted for that node. In reality large temperature differences are not encountered and are self–correcting due to other heat links. In terms of the model, such an artificial situation is likely to lead to instabilities in the model. This can be avoided by combining nodes to give an acceptable aggregate mass.

•   for in–orbit, periodic variations let the model run for a sufficiently long (mission) time such that the transient changes during each orbit are reproduced for successive orbits. This is not to say that the absolute temperatures are exactly reproduced because some very long time constants could be operative. What is often important are the relative temperatures between nodes and their time profiles, and typical temperature swings for each node.

•   the above implies that for such an analysis it is efficient to set the initial temperatures to the values that can be obtained by running the steady–state model

•   check that the transient temperatures from the above straddle the steady–state temperatures, with orbitally averaged power inputs.

3.9   Thermal design strategy

The overall strategy for thermal design is one of the development of an evolving TMM which is increasingly representative. We have already seen this in terms of the complexity of the TMMs themselves but this reflects only the quantitative aspects of the model, not its qualitative aspects. There is another vital component to the overall development and that is the verification and test of the TMM and how it is updated as a result.

The three most important stages in the verification of a model are the thermal balance test of the experiment (or subsystem), the thermal balance test of the spacecraft (or system) and the early orbits test. If the thermal design development has been carried out properly the three verifications should in principle become decreasingly traumatic. As the thermal balance test features in its own right it warrants a slightly more detailed account.

A thermal balance test is a test in which the experiment is permitted to come to a thermal steady state in a thermal vacuum chamber. It forms part of a qualification programme and is therefore performed on one of the development models of the experiment (the structural thermal model) and is not necessarily repeated on the flight model. The temperature of the experiment is determined by the internal dissipation of power and by the external thermal environment. In the chamber the latter is usually simplified by making it an enclosing thermal shroud maintained at a single temperature although in more complicated tests external heaters can be strategically placed to simulate additional heat inputs. For the thermal balance test to be performed satisfactorily sufficient time is required for the temperatures to equilibrate and a good knowledge is required of all essential temperatures and powers (see section 3.9). This usually calls for additional temperature sensors and power supplies. That the thermal balance test is a simplified test does not undermine its importance, indeed it is advantageous to test the TMM under simplified conditions; it is the TMM which is being tested not the environment. However the environment should be such as to ensure that representative routes and representative mechanisms for heat transport are established. Thus the vacuum achieved should ensure that heat is not transported convectively or conductively through residual air, and the temperature differences should be such as to establish the main radiative links. For the latter it is not necessary for the shroud to be at the temperature of cold space (3 K). Liquid nitrogen temperatures (77 K) will normally be sufficient if the experiment is typically at room temperature since $(300^4 - 88^4)$ only differs from $(300^4 - 3^4)$ by about 1% (see section 3.4)).

Clearly it is the first thermal balance test where any serious problems will first became manifest, usually as a result of either deficiencies in the TMM or of inadequacies in the build standard.

The second thermal balance test is important because the local thermal environment of the experiment is modified by the presence of the rest of the spacecraft, even though the spacecraft environment itself is still kept at a simplified level.

The post launch test is obviously important because it is here that the experiment is first exposed to its in–orbit environment and it is usually the first time that any transient thermal data is obtained. It may be thought that this is too late a stage to be able to effect any changes. This is not true. It is at this stage that the final modifications can be made to the TMM if required. Such an updated model should then allow for an optimised thermal control or, if a problem is subsequently encountered, for an effective and hopefully risk free solution to be implemented.

It can be seen that the three major 'correlations' between the detailed TMM and the experimental data represent important and necessary stages in the overall development. Good correlations can often be quite difficult to accomplish and it is therefore important that the practical aspects of the verification are addressed in advance to optimise the correlation process. The flight model of the experiment as launched will have temperature sensors for monitoring the overall thermal performance in space. These monitors form an important part of the thermal subsystem and as such should be placed to maximise the information that they give about the thermal system as a whole. Since the numbers of such sensors is limited their locations have to be selected with care. This is even more the case since these details should be decided at an early stage of the thermal design in order to provide bench mark data to carry through the various stages of verification, especially between the structural thermal model and the flight model. For thermal balance testing additional temperature monitors can be used to provide a fuller and more detailed set of thermal data. To this end an understanding of the model can greatly assist in deciding the locations of additional sensors; of particular importance are the significant thermal interfaces and/or those elements across which significant gradients might occur. For the thermal balance tests the model will have to be reconfigured to reflect the particular thermal environment of the chamber and, before starting the test, it is important to have available the results of this model.

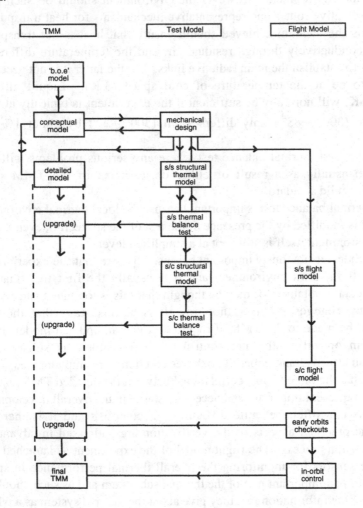

Fig. 3.41. A strategy flow chart.

If a TMM is to be realistic it will predict the thermal response of the system in different internal thermal states and different external states. Even though a thermal balance chamber represents a simplified environment it is still possible to set up different thermal scenarios. This can be done externally by the use of additional heaters and can be done internally by different levels of power and different distributions of power. Each set of results will require a correlation with its corresponding TMM and makes not only for more work but makes the task more difficult because the numbers of free parameters in the fitting is reduced; indeed a full correlation to a desirable level of accuracy may not be achieved. However, testing over a range of scenarios is often helpful in producing a better TMM than would otherwise have been achieved and the tests themselves bring with them a deal of useful experimental data in their own right.

There are other stages in the development of TMMs but these are arguably not as significant as those already discussed. The flow chart in Fig. 3.41 shows a representation of a viable thermal design strategy and illustrates how the different TMMs, the different hardware models and the different verification tests all combine to produce the final TMM.

3.10   Thermal design implementation

Thus far we have described some of the thermal design tools and have discussed the overall structure and strategy of a thermal design programme in which these tools are used. What we need to address now is how the tools are used. As any craft, that of thermal design is a rather subjective skill and does not lend itself very easily to a written description. It is a craft which is very much acquired by practice and personal experience. However, in common with many crafts, there are a number of hints and a number of do's and don'ts which can often prove very useful. What follows is such a list. It is personal, it is by no means exhaustive and the items are in no particular order. They have however been divided into groups which correspond to some of the different stages in the thermal design strategy discussed in the last section.

3.10.1   B.O.E. models

- calculate the mean temperature of the experiment given the total internal power dissipation and the total radiating area to space. This should indicate if your experiment requires overall cooling or overall heating

- if overall cooling you may need to propose a radiating surface separate from the experiment that it serves

- if overall heating is required quantify this bearing in mind that you will probably need to minimise radiation losses by selection of a low emissivity outer surface. MLI may need to be proposed

- consider what are the likely materials from which the experiment will be constructed. If possible decide if the model is likely to be predominantly conductive or predominantly radiative

- set up a small approximate model with estimated conductive and radiative links but without indirect radiative links

- determine the main heat flow paths

- determine where the main thermal gradients occur

3.10.2   Conceptual models

- carefully consider the nodal designations

  - do not have too many nodes; the model needs to have a short run–time and to be understandable

  - ensure you have adequate numbers of nodes characterising regions of significant thermal gradients

  - give some thought as to the likely nodal map of the detailed model; it is useful to have nodal boundaries which coincide with those of other models where possible and greatly simplifies the compatibility analysis between any two models

- consider at an early stage if any subsystem of the experiment can be thermally autonomous; if so, this makes for a tidier, more secure thermal design because it minimises interdependence and simplifies subsystem testing. If subsystems can be autonomous build in the appropriate thermal isolation between system and subsystem

- if heaters are required consider the practicalities of how and where they will be located

- if heater mats are to be used and are of a size comparable with the nodes on which they are to be located, reduce the conductive links to the heater nodes appropriately since the heaters will short out part of the links. It may be necessary to reappraise the local nodal network

- perform a common sense check on the output from the model. Do not proceed with the model development until you are satisfied with your physical understanding of the model; propagated errors in a model become increasingly difficult to identify

- construct a temperature map and a heat flow map as a format for presenting, summarising and understanding the model

- for any heat flow map, check that the net heat flow to any node which is *not* a boundary is zero to within the power resolution of the map

- divide large models into sub–models wherever possible. Represent sub–models by isothermal boundaries in the larger model. This makes for a more manageable design in the early stages

- minimise the changes between successive models; ideally limit this to one change so that the effects of that single change are manifest

- keep a log of all model runs and note the changes from previous models

- have a baseline model against which development models are tested. Upgrade the baseline model when it is acceptable. Keep a record of input and output data for all the significant (i.e. baselined) models

- remember to model the external heat input for steady–state models. This can be done as an orbitally averaged heat input to the appropriate nodes

- generate a transient model, particularly if the orbit is highly eccentric

- to assist in the sizing of heaters (or coolers) set the appropriate node at the required temperature and make it an isothermal or boundary node; the model should then indicate the constraining power required to hold that node at that constant temperature

- set up boundary nodes to represent external surfaces. It is useful to have a number of these to avoid their 'shorting' any parallel nodes and are useful in exploring the sensitivity of the system to asymmetries in the external thermal environment

- perform sensitivity analyses in order to identify critical areas of the design

- use the model to identify the locations of inflight thermal sensors that representatively monitor the thermal health of the experiment

3.10.3   Detailed model

- attempt to make the nodal map of the detailed model compatible with that of the conceptual model

- ensure that the critical areas and the sensitive elements as identified by the conceptual model receive particular attention in the detailed model

- keep detailed notes as to all calculations of all links

- if there are any autonomous thermal subsystems, model these separately and perform consistency checks against the corresponding conceptual model

- at the earliest stage carry out compatibility tests between the full detailed model and the full conceptual model. Update both as necessary since it is useful to carry forward a representative conceptual model even after the detailed model has been constructed

- use temperature maps and heat flow maps as an aid to understanding the characteristics of the model

- if the detailed model indicates that changes in the design are required, use the conceptual model to develop the changes and then transcribe into the detailed model

3.10.4   Thermal balance tests

- use the detailed model to predict the thermal performance of the experiment in the particular thermal balance test facility

- if boundary temperatures have to be established as part of the hardware conductive interface, carefully consider how these temperatures are to be established in terms of the power flows required to maintain these temperatures

- ensure that boundary temperatures are accurately measured and monitored, particularly if the power flows to the boundaries are high.

- if power flows can be measured (e.g. by the use of calibrated thermal impedances), then design the test hardware to include this

- carefully consider the locations of additional thermistors to ensure a good knowledge of the thermal characteristics in regions of large temperature gradients, large heat flows and important mechanical/thermal joints and interfaces

- use the test data to constrain the boundaries of a sub part of the full model to check the model in that vicinity and to determine the likely major heat flow paths

- ensure that all the thermistors are adequately calibrated in the temperature region over which the particular thermistors are to be used

- with the system in thermal equilibrium in air at room temperature and with no power dissipations ensure that all thermistors are recording the same room temperature

- ensure that the positions and types of all thermistors are properly documented in a log and on drawings prior to the test; photographs are particularly helpful

- model any mechanical support for the experiment and any electrical loom to the chamber walls particularly if they represent significant heat flow paths

- carefully select in advance the thermal stability criteria to be used as the basis for terminating the test.

- ensure that these criteria are met or that there are good grounds for changing them; it is better to have steady – state temperatures from one test than non – steady – state temperatures from a number of tests

- remember that the thermal balance test is to test the model; do not be over concerned if the test is unrepresentative of the flight conditions provided that it will maximise the return of the test in terms of understanding the experiment through the model.

- notwithstanding the above make one of the thermal balance test runs as fully representative of flight to obtain an experimental verification of the baseline thermal characteristics of the experiment independently of the model

- if at all possible ensure that there is sufficient time in the thermal balance test to get results from more than one power set–up

- keep real–time time plots of the key temperatures; these can be used to judge if the predicted power inputs are adequate and for extrapolation purposes to predict when the test termination criteria are likely to be met

- remember that a successful outcome from a thermal balance test is essential; if the implications of the results are inconclusive do not baulk at the prospect of a re–test

3.10.5 Spacecraft thermal balance

- if circumstances permit, perform the same tests as carried out in the subsystem thermal balance tests

3.11   Concluding remarks

In this chapter we have attempted to give an account of how you go about making a thermal design from the standpoint of the principles involved. The approach adopted is rather unconventional in as much as its aim has been to give a physical insight into what is going on and what lies behind the conventional set of engineering rules that could equally well have been applied. What the chapter is not is either a set of instructions (although it offers some ideas for a strategy and a list of hints) or a compilation of source material (although some data are given to provide some quantitative framework for the discussion). The chapter is therefore rather devoid of references which might be somewhat unsatisfactory as far as the reader is concerned. There is no shortage of heat engineering books or of basic science data books of the thermal properties of materials. However two references which are extremely comprehensive are singled out as being particularly useful: Morgan, W.L. and Gordon, G.D., [1989], *Communications Satellites Handbook;* and ESA PSS–03–108, [1989], *Spacecraft Thermal Control Design Data.*

As additional reading the reader may find the following of interest:

Siegel, R. and Howell, J.R., [1992], *Thermal Radiation Heat Transfer.*

Agrawal, B N., [1986], *Design of Geosynchronous Spacecraft.*

Berlin, P., [1988], *The Geostationary Applications Satellite.*

Leuenberger, H. and Person, R.A., [1956], Compilation of Radiation Shape Factors for Cylindrical Assemblies.

Stimpson, L.D. and Jaworski, W., [1972], Effects of Overlaps, Stitches and Patches on Multilayer Insulation.

As has been said a number of times in this chapter, the best way to learn how to make a thermal design is to do it. Armed with this chapter and the two references singled out above, the reader may now feel confident enough to do so.

4

Electronics

4.1 Initial design

4.1.1 In the beginning

Designing electronic subsystems for space vehicles can be considered in two overlapping phases. The circuitry has to carry out the required signal processing functions but also has to be capable of overcoming the particular problems associated with the subsystem existing and operating in the environment associated with the spacecraft.

In the early stages of a design, estimates have to be made of mass, volume and power consumption to determine what is feasible, within the constraints imposed by the spacecraft. Estimates are also required for cost, time and manpower to ensure that the flight hardware can be realistically produced by the required delivery date. Thus it is important to consider as soon as possible what problems associated with the space environment are seriously going to effect these estimates compared to a ground – based design.

A long life mission will have a significant impact on costs due to the requirement for increased reliability of components and manufacturing techniques, and for the introduction of component or system redundancy.

Apart from the requirements of telemetry transmitter power and type of antenna, the orbit can have a very significant effect on cost if it is associated with a high radiation environment. This may require the use of highly specialized, radiation tolerant components which may be difficult to procure. A high radiation environment can also have a major impact on mass where the wall thickness of the structure is no longer defined by structural and electrostatic screening requirements but by its ability to absorb radiation. The orbit will also define the requirements for on–board data storage facilities if contact between the spacecraft and ground station is intermittent.

The launch vehicle will specify the level of vibration that is put into the structure and the design of the electronics boxes and harnesses will have to meet this specification including amplification factors which the design must ensure do not become too severe.

Operating in the vacuum of space, without the advantage of convection currents, may require complex conduction paths for the dissipation of heat from the component to the spacecraft's main structure or to radiator panels. The subsystem may also have to endure substantial temperature cycling due to the sequence of sunlit and eclipse periods. Having to function in a vacuum introduces corona problems for high voltage systems and imposes restrictions on the use of many common materials which have a significant vapour pressure. These can outgas and either upset the balance of the spacecraft or redeposit onto sensitive optical surfaces. It also presents problems of cold welding with electromechanical devices.

The geometry of the spacecraft will probably define whether the subsystem can be built within one unit or has to be constructed in several smaller interconnected units. Multiple units will inevitably increase mass and volume and introduce additional harnesses and connectors. This problem is frequently found with spin stabilized spacecraft built to a cylindrical geometry. Here the peripheral volume will allow significant space for sensing devices such as antennae, cameras, photon and particle detectors but as one progresses inwards the available space for reasonable shaped packages becomes very restricted.

Estimating the requirements for harnesses needs serious thought in the early design phases. Consideration has to be given to the type of connectors to be used, their location, the size of the cables and their minimum bend radii, the space necessary for the reliable insertion and removal of the connectors and the access volume required for tools used to mount and demount the connectors and their locking devices.

As spacecraft often contain magnetometers either as attitude sensing devices or as scientific experiments, minimising the spacecraft's magnetic properties is a common requirement. This places a severe restriction on the use of normal ferromagnetic materials in structures, nuts and bolts, connectors pins and shells etc. requiring alternative materials which are usually more expensive and difficult to procure.

The general level of ElectroMagnetic Radiation (EMR) in a spacecraft is usually quite high due to the large number of subsystems packed into a small volume and the presence of the telemetry transmitters and possibly radars. In this situation harnesses have to be considered as receiving and transmitting antennae and precautions have to be taken to minimize the amount of EMR that is received or radiated by these cables. These precautions include the use of twisted pairs or groups of wires, the screening of individual or groups of wires and the use of additional screens over the entire harness. Consideration also has to be given to the provision of line drivers and receivers to preserve the integrity of the signals when loaded with the impedance of these harnesses.

Harness routes have to be specified so that adequate channels and holes can be incorporated in the mechanical design. Due to the requirements of minimizing the weight of the spacecraft, aluminium honeycomb and other esoteric materials are often used as part of the structure. As this requires the use of special inserts for any fixtures such as harness clamps, the number and location of these clamps have to be decided as soon as possible.

Another problem that can seriously effect the overall structure of the subsystem arises from the number of independent packages which ultimately have to be connected to the same telemetry and power systems.

The mechanical structure of a subsystem will be electromagnetically coupled to its contents and hence the existence of a potential difference between them may well introduce noise. The optimum situation occurs when the subsystem's structure is at the same potential as the common reference line of the most sensitive part of the system which is usually that of the detector. It would appear that the simplest solution to this problem is to electrically connect this line to the subsystem's structure at the detector. As the subsystem's structure will generally be electrically connected to the spacecraft's main structure, both for thermal and mechanical integrity, this implies the detectors reference should be directly connected to the spacecraft's structure. However, in a multi – subsystem spacecraft, this would lead to the introduction of multiple ground loops which in the presence of the EMR environment could cause serious noise problems.

One solution to this problem is to electrically isolate the structure of the detector from the main spacecraft structure so that electrical connections between the detector's reference and its immediate structure can be made without also introducing connections to the spacecraft's structure. Although this technique may solve electrical noise problems it can causes severe mechanical problems as it requires the provision of electrically isolating mounting structures which may not provide adequate strength or accuracy in location. Items such as electrically conducting thermal blankets, which need to be wrapped around the outer part of the spacecraft and be electrically connected to the spacecraft structure, must not then come into contact with the isolated subsystem.

An alternative solution is the introduction of electrically isolating components in the signal and power lines such as transformers and opto couplers, but allowances will then have to be made for these additional complexities.

As many of the above estimates will have to be made long before detailed designs have been finalized, it is very sensible to study previous spacecraft and their subsystems and profit from the experience gained during their manufacture and flight operations.

4.1.2   Typical subsystem

The block diagram of Fig. 4.1 shows a typical subsystem composed of a Sensor, its Front End Electronics (FEE), Main Electronics (ME) and Power Unit, and its relationship to the other subsystems on the spacecraft.

Basically all information arriving at a spacecraft is either in the form of EMR or particles. Hence although there are many types of sensors there is only a limited number of parameters which need to be measured. This includes, amplitude (energy), count rate (frequency), time of arrival and the angular distribution of incoming particles or photons. Thus by discussing a limited range of detection systems one can cover most of the techniques and problems associated with most types of sensors and their electronic interfaces. The range of sensors we shall be considering can all be used for attitude sensing and so this provides the platform for the next section of this chapter.

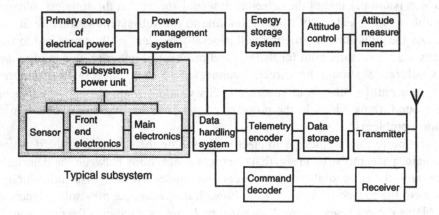

Fig. 4.1. Subsystem with interfaces to spacecraft.

We shall then look at the analogue and digital processing of signals produced by such sensors and see how this processed data is collated with data produced by other spacecraft subsystems, ready to be transmitted back to Earth. As this processing needs electrical power, we shall investigate how this is generated, controlled and distributed around the spacecraft together with the problems associated with ensuring that ElectroMagnetic Compatibility (EMC) requirements are met. EMC is an important feature of spacecraft subsystem design and as harnesses and their connectors can be major contributors to this problem, these components will be considered in detail.

Finally we shall consider how the electrical subsystems of the spacecraft are designed and fabricated to ensure that they function correctly over the expected life of the spacecraft in the unusual environment in which they have to operate.

4.2   Attitude sensing and control

4.2.1   Sun sensors

The Sun is an important attitude reference as many spacecraft have sensors which measure solar related parameters, have Sun dependent thermal constraints and depend on the Sun for electrical power.

Attitude sensing with respect to the Sun is fairly simple as it is bright enough to permit the use of relatively simple sensors and processing electronics, and is much brighter than any other astronomical source. It also has a relatively small angular diameter of 0.53° from the Earth and hence can be considered for some applications to be a point source.

There are two main types of Sun sensors, analogue sensors which produce an output whose amplitude is related to the angle between the Sun and a spacecraft reference, and digital sensors which produce a constant amplitude signal whenever the Sun is within the angle of view of the sensor. The analogue sensors can rely on the electrical output of the solar sensitive surface being linearly related to the flux of the incident radiation and hence to the cosine of the angle between the vector representing the surface and the source of radiation.

To increase the sensitivity of the detector's output with angle, masks can be added to vary the area of the sensor illuminated. This technique can be refined to make a two dimensional detector using the configuration shown in Fig. 4.2, where the outputs from all four cells are only identical when the solar vector is normal to the cells.

Sensors are typically photodiodes or photo emissive cells, their electrical outputs being sufficiently large to make the analogue signal processing and subsequent analogue to digital conversion simple. However they both degrade under the influence of radiation, the photodiode, being a reverse biased diode, suffers from an increase in leakage current and the photo emissive cell becomes less sensitive. Therefore as these sensors rely on the surface sensitivities remaining constant for accurate results, it is sensible to include some form of on-board optical calibration system which is protected from radiation damage when not in use.

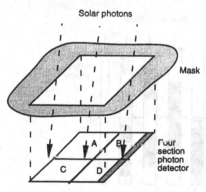

Fig. 4.2. Analogue two dimensional solar sensor.

Digital detectors rely on collimation (mechanical or optical) which in the simplest case is the two slit device shown in Fig. 4.3. These are often used on spinning spacecraft to produce a Sun pulse when the Sun is at a particular angle relative to a spacecraft vector. To obtain attitude information over a much wider angle of view using digital techniques, the single lower slit can be replaced by a coded pattern of masks above several detector strips. The attitude is then determined by which set of detectors has been illuminated. This is illustrated in Fig. 4.4.

The use of the Gray code is preferable to a Binary code as this minimizes coding errors occurring at the edges of the slits caused by irregularities in the masks. In the example shown mask irregularities could produce all possible codes from 0000 to 1111 in the Binary system but only sequential two codes, 0100 and 1100, in the Gray system.

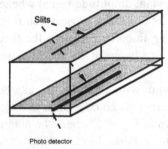

Fig. 4.3. Digital solar sensor.

As the cosine law applies to all the strips, a continuous strip is included in the system and the output of this is used as a reference. This strip is made half the width of the others so that the electrical output of a detector strip is defined as a 'one' when its output is more than the output of the reference, which is equivalent to a standard strip being half illuminated. This also allows for a reasonable variation in the sensitivity of the bars with age. The limiting accuracy of this system is the finite angular width of the Sun.

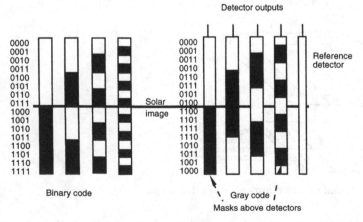

Fig. 4.4. Digital solar sensor with wide field of view.

Fine Sun sensing to an accuracy of a few arc seconds can be achieved by optically focussing the Sun's image onto a pinhole above a photo detector and registering the edge of the Sun's disc as it moves across the pinhole.

4.2.2   *Earth sensors*

A knowledge of the orientation of the spacecraft to the Earth is necessary for several reasons. These include the alignment of the antennae to ground–based receiving stations, and the ability to point on–board instrumentation to Earth bound features or Earth related features of the ionosphere and magnetosphere. As the Earth presents an extended source particularly to low Earth orbiting spacecraft, the sensor is usually designed to detect the horizon. For a spin stabilised spacecraft the natural motion of the spacecraft can be used to scan the sensor across the horizon, but for a three axes stabilised spacecraft some form of mechanical or electronic scanning will be required.

Horizon detectors for near Earth orbiting, spin stabilized satellites can be infrared sensitive as the Earth is a fairly uniform blackbody emitter at a temperature of about 290 K. Such sensors will function in both the sunlit and eclipsed parts of the spacecraft's orbit and can produce attitude data with errors of a few arc minutes. However for highly eccentric orbits, due to the relatively low temperature of the emitter, the signal to noise ratio at the output of the detector could become so low that cooled sensors would be required. This problem can be avoided by making use of the Earth's albedo which produces a large optical output (Cruise *et al.*, [1991]). The definition of the Earth's albedo is the ratio of the total solar radiation reflected from the Earth to the radiation incident on the Earth; has a typical value of 0.4 and produces a radiation density of about 240 Wm^{-2} at a Low Earth Orbit (LEO).

A basic photodiode behind a mechanical collimator or lens system can be used to sense the horizon. Although the signal from the sensor will vary significantly as it passes over the Earth's disc due to variations in the Earth's albedo, these variations can be reduced by making the angle of view large in the direction tangential to the horizon. This also increases the sensor's output and hence reduces problems associated with spurious solar reflections which will occur in the collimator. Accuracy in the required direction is obtained by collimating the angle of view in that direction and this technique can sense the horizon with an error of about half a degree.

Fig. 4.5 shows the angle of view of the detector relative to the horizon together with the detector's output vd for one spin of the spacecraft. The offset shown is due to the leakage current of the photodiode.

Referring to Fig. 4.6, if the gain of the first stage amplifier is set so that its output saturates at a small fraction of the nominal minimum response of the detector to the Earth's albedo, a well defined horizon can be obtained as shown by $v1$. This allows for degradation in the sensitivity of the photodetector, variation in the albedo across the Earth's surface as well as the effects of the elliptic orbit.

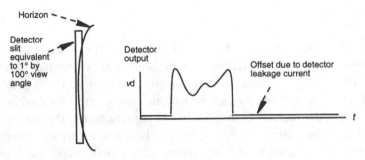

Fig. 4.5. Horizon detector with output waveform.

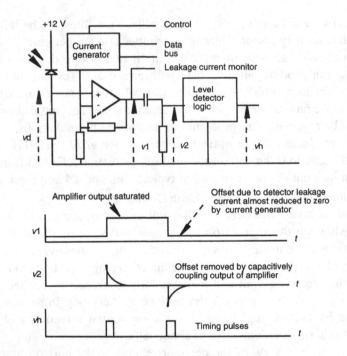

Fig. 4.6. Horizon detector circuit and waveforms.

Setting the amplifier's gain this high could cause the system to fail if the leakage current of the sensor became sufficiently large to saturate the amplifier permanently. This type of problem can normally be overcome by capacitively coupling the sensor to the amplifier. However as the sensor will see the Sun as well as the Earth, the very large Sun signal would cause baseline shift problems due to this capacitive coupling (see Fig. 4.25). Reducing the time constant of the coupling network to such a value that baseline shifts were not significant would reduce the signal to noise ratio significantly and also reintroduce the variation in the amplifier's output due to the Earth's variable reflectivity. Thus it is better to directly couple the sensor to the first stage amplifier as shown, and provide commandable current backing off facilities to allow for the increase in the detector's leakage current with mission life. The detector's leakage current can be measured as $v1$ immediately after the spacecraft goes into eclipse before its temperature has fallen significantly.

The amplifier's output is capacitively coupled to the level detector to remove the small offset voltage caused by the difference between the leakage current and the backing off voltage. The capacitive coupling is now acceptable as both the Sun and Earth signals are the same amplitude namely the saturated output of the amplifier. The time constant of the waveforms indicated by $v2$ should be made sufficiently small to allow the signal to recover to its baseline in time for the next horizon signal. The risetime of the signal should be made as long as possible to reduce noise but not so long that timing accuracy is impaired. The horizon pulse vh can be used to read the spacecraft's clock and transfer this data into the subsystem's telemetry format. The occurrence of a pulse due to the Sun can easily be determined from the periodic nature of the expected signals.

The disadvantages of optical horizon detectors are their inability to operate during eclipse and their sensitivity to spurious internal solar reflections. The relative intensity of the Earth's albedo to the Sun, per unit solid angle at optical wavelengths is about 1 to 30000 whereas the relative intensity of the Earth as an infrared emitter around 12 microns to that of the Sun at similar wavelengths is about 1 to 400 (Trudeau et al., [1970]). However as far as the detector is concerned, both these ratios are improved by a factor of about 100 with a collimator designed to have an angle of view of about $1°$ perpendicular to the horizon and about $100°$ parallel to it. As the infrared detector senses a well defined horizon rather than that influenced by the atmosphere its angular accuracy can be a factor of ten better than that of an albedo detector.

The materials used for infrared sensors are thermistors, thermocouples and pyro–electric crystals. The first two devices are very stable but have rather slow response times, the third is significantly faster but is less stable. More complex systems using photo–multipliers can be used but should be avoided due to power, weight and complexity if a more simple system is adequate.

4.2.3   Star sensors

These are the most accurate type of attitude sensing devices, as stars are essentially point sources of light, but due to their limited photon flux, the sensing system is complex and hence consumes considerable power and weight. If this form of attitude sensing is used in a pointed non–spinning spacecraft, a one or two dimensional position sensitive detector (PSD) is required to determine the position of the star. However before this can be carried out, the stellar image must be converted to photo electrons using a photo cathode material and some process is required to raise the level of this signal above the background noise. This either requires a very low noise amplifier or an integrating sensor which will allow sufficient charge to accumulate to produce the required signal. Low noise amplification can be achieved using channeltrons or Micro–Channel Plates (MCP) (Smith, [1986]).

The MCP is basically a collection of fine bore lead glass tubes each about 12 μm diameter and about 1mm long. The internal surfaces of the tubes are treated so that they are electron emitters. With about 1 kilovolt across the array of tubes, a single electron striking the upper surface of any one tube will be accelerated down the tube and each collision with the walls will produce more electrons resulting in a few thousand electrons being emitted from the far end of the tube. Usually two arrays are used in tandem producing a relatively noise free amplification of several million and a charge sufficiently large to be processed by standard electronic techniques.

As the number of electrons emitted at each collision increases, the voltage gradient down the second array becomes distorted, resulting in the final collisions not producing electron multiplication. Thus there is an upper limit to the size of the final charge cloud produced. The variation in the size of the charge cloud up to this limit is caused by the angle at which electrons are emitted at each collision and hence the amount of energy they can acquire before their next collision. There may also be a variation in the electron emitting sensitivity of the walls along their length. The array can usually be designed to produce a pulse height distribution for the final cloud which has a full width at half maximum which is about the same as the mean cloud size. Thus the charge amplifier only has to deal with signals which vary in amplitude by a factor of about two times larger or smaller than the nominal. If the nominal MCP charge output is N electrons, the level detector associated with the amplifier can be set around $N/4$ which will allow for a significant degradation in the gain of the MCP.

There will be some spurious signals caused by thermally generated electrons, but unless these occur close to the top of the array they will not produce an output pulse large enough to be detected. To ensure that the number of thermally generated electrons is minimised the temperature of the array must not be allowed to exceed about 50 or 60°C and this implies an upper limit to the power which can be dissipated by the array. If we assume that the parameters associated with each array are an accelerating voltage V of 1 kV and a power limit of 100 mW, V^2/R defines the minimum array resistance R to be $10^6/0.1 = 10$ MΩ. As this may be composed of several million individual tubes, the resistance of each tube will be greater than 10^{13} Ω. Treating each tube as a distributed resistance capacitance network and making the assumption that the capacitance associated with each tube is about 0.005 pF, gives a recovery time for each tube of about $10^{13} \times 5 \times 10^{-15}$ s equals 50 ms. Thus if the array is only going to be stimulated over a very small area, the maximum count rate will be severely limited.

The position of the stellar image is finally determined by allowing the output of the MCP to fall on a PSD such as a Crossed Anode Array, a Charge Coupled Detector (CCD) or a Wedge and Strip System.

4.2.4   Crossed anode array system

A two dimensional PSD can be produced by allowing the output charge cloud from microchannel plate to be collected by a crossed array of isolated anode wires as shown in Fig. 4.7. If the diameter of the MCP charge cloud is allowed to spread out so that it just exceeds the spacing between two anodes then at least two orthogonal

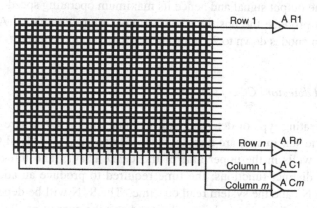

Fig. 4.7.   Crossed anode PSD.

will be stimulated. Thus with $n + m$ amplifiers and level detectors the coordinates of the star image can be determined. The resolution of this system is limited by the size and number of the anode strips and the volume and power limitations imposed by the analogue electronics. As the capacitive loading associated with each anode is relatively slight, the speed of response of this system can be very high allowing random count rates up to 10^6 pulses per second to be measured.

To reduce the number of amplifiers, the anodes can be interconnected in a pattern to produce coarse and fine position data as shown in the one dimensional array of Fig. 4.8. The electron charge cloud must cover two anodes, and in this case, will produce an output on C4 and F1. Thus with 32 anodes only 8 output lines are required but

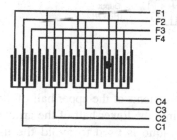

Fig. 4.8.   Coarse and fine position coding.

resolution is halved due to the increased electron cloud size required. This system is equivalent to a 16 anode array but only requires 8 sets of electronics and so is much more power efficient. As the number of anodes connected to each fine and coarse line is increased, the power efficiency also increases but the interconnection pattern in the detector becomes more complex, resulting in an increased capacitive load for each anode. This effect and the division of the charge between the anodes reduces the signal to noise ratio of the output signal and hence its maximum operating speed. Also the address decoding to produce the position data becomes more complex. Arrays with the spacing between anodes down to 25 micrometres have been produced.

4.2.5   *Charge coupled detector (CCD) system*

This is an integrating type of detector, as illustrated in Fig. 4.9, where photo electrons produced by the star image, are directly accumulated in MOS (Metal Oxide Semiconductor) charge wells in the upper half of the CCD array. The integration period depends on two distinct functions; the time required to produce an adequate signal to noise ratio (S / N) and the system read out time. The S / N will be dependent on the magnitudes of the particular stars being observed and the energy of the initial photons. For high energy photons a single event may produce adequate charge. However the system readout time is usually the dominant factor in defining the integration period.

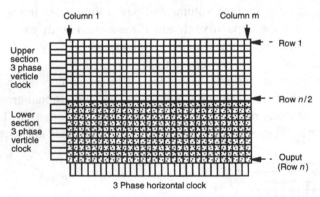

Fig. 4.9. CCD array.

At the end of an integration period, the entire contents of the upper half of the array are transferred, using a phased clocking system, into the lower half of the array which is shielded from external radiation. This technique is used to avoid the use of a mechanical shutter to block the stellar photons during the readout period.

If complete rows can be shifted down at a rate of V rows per second, with an array of n rows, the time taken to shift the upper half of the array to the lower half is given by $n/(2V)$ seconds. The bottom row is so constructed that it can be shifted out serially. As the horizontal clock only has to shift m cells at a time, as opposed to the vertical clock which has to shift $m \times n/2$ cells at a time, its shifting rate of H pixels per second can be considerably faster. Thus the time taken to read out the bottom row is m/H seconds. Once the data in the upper half of the array has been transferred to the lower section, the upper section of the vertical 3 phase clock is disabled.

After data in the bottom row has been shifted out serially it is replaced in parallel using the lower 3 phase clock to shift the entire bottom half of the array one row down. Thus the time taken to read one row is $m/H+1/V$, and the time taken to shift out the entire image from the lower half of the array is given by $(1/V+m/H)n/2$ seconds. For an array of 512 by 512 pixels operating at a vertical shift rate of 1 M rows per second and a horizontal shift rate of 3 M pixels per second, the time taken to shift the image from the upper half of the array to the lower half is 256 microseconds whereas the time taken to shift out the entire image from the lower half of the array is 44 milliseconds. Thus very little blurring will occur during the time taken to shift the image into the shielded lower half of the array compared to the time taken to read out the whole image. The mechanism of charge transfer and the associated phases of a 3 phase system is shown in Fig. 4.10.

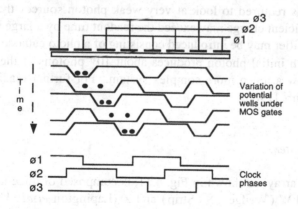

Fig 4.10.   Charge transfer in a 3 phase system.

The output stage is basically a Field Effect Transistor (FET) which monitors the voltage on the last cell of the output row of the array. To ensure that a measurement is only made of the charge that is transferred to the last cell each cycle, a process called 'correlated double sampling' is often employed. This involves shorting the last cell before each new charge is clocked in and storing this voltage as an 'empty cell voltage'. Then the difference between the 'empty cell voltage' and that which occurs when the new charge is transferred across is measured with a differential amplifier. Thus the output of this amplifier is only due to the new charge even if the last cell is not fully discharged each cycle. As this process is carried out for every data sample, it extends the processing time beyond that just calculated.

The advantage of a CCD system is the simplicity of the single readout stage and the fact that no high voltages are required for electron multiplication. The number of gates that need to be pulsed simultaneously to shift the contents of the upper half of a 512 by 512 array to its lower half is about 250000 presenting a very significant capacitive load to the clocking system. Hence the main disadvantage of this system is the peak power dissipated during the vertical clocking period and the associated problem of maintaining the precision required for the phasing of the clocks so that the individual charges are efficiently transferred across the array. Other disadvantages are the time resolution, limited by the integration and data processing period and the degradation in performance of CCD arrays with radiation.

If one is only interested in data from a limited area in the array, only the rows containing that data need be shifted out, and only the relevant pixels in those rows need to be processed, reducing the overall processing time considerably.

If lower resolution at a higher speed is required, several rows can be shifted down and hence accumulated in the bottom row before it is shifted out. Similarly several horizontal shifts of the bottom row can be provided without a reset of the output cell allowing charge to accumulate from several horizontal cells.

If this type of system is required to look at very weak photon sources the time required to accumulate sufficient charge may exceed the readout time by a large factor. In this case an image intensifier may be introduced consisting of a photo cathode, MCP and scintillator so that each initial photon produces about 10^6 photons at the CCD array. However this is now a much more complex system. For further reading on CCDs, see Beynon and Lamb, [1980].

4.2.6   Wedge and strip system

In this system the array, as shown in Fig. 4.11, is composed of three isolated metallic anodes designated W (Wedge), S (Strip) and Z (Lapington *et al.,* [1986]). The array is placed at a sufficient distance below the microchannel plate that the diameter of the charge cloud will cover several elements of both the wedge, strip and Z anode patterns. The distribution of the charge in this manner does not decrease the signal amplitudes as all the charge is still collected by the three anodes. This allows the positional resolution, defined by the centroid of the charge cloud, to be much smaller than the diameter of the charge cloud.

The fractions of the charge acquired by the Wedge and Strip anodes indicate the two dimensional position of the charge cloud. If w is the charge collected by the Wedge, s the charge collected by the Strip, and z the charge collected by the intermediate anode, the two positions are given by $2w/Sum$ and $2s/Sum$ where

$Sum = w + s + z$. The factor of 2 is required to ensure the output range is nominally 0 to 1 for both wedge and strip systems, as shown in Fig. 4.12, where the total charge deposited on the array is 2 units.

The ratioing technique is essential to allow for the variation in the *Sum* signal caused by the original particle or photon striking different parts of the front end of the MCP, the variation of the gain of the MCP with position and the change in these parameters during the lifetime of the spacecraft.

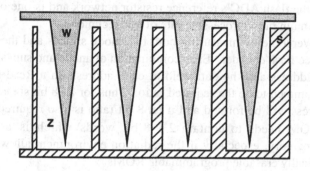

Fig. 4.11. Wedge and strip anode array.

A typical array size would be 50 mm square with a pattern pitch of 0.2 mm and the charge cloud arranged to be about 1mm diameter to produce a position resolution of about 35 μm.

The accuracy to which the anode patterns can be produced and the S/N of the wedge and strip amplifiers limit the accuracy to which the ratios can be determined and hence the positional resolving power of the system.   The S/N is critically dependent on both the total and inter–electrode capacities of the anode array as these will reduce the signal output and increase the cross talk between the anodes.

	Min s	Max s
Max w	A	C
Min w	B	D

	s	w	z	2s / Sum	2w / Sum
A	0	1	1	0	1
B	0	0	2	0	0
C	1	1	0	1	1
D	1	0	1	1	0

Fig. 4.12. Wedge and Strip position indication.

There are several ways of forming the required ratios $2w / Sum$ and $2s / Sum$. The accuracy of simple analogue division is limited by the offset voltages associated with the dividing circuit and the latter's high–frequency response may also be inadequate. Conversion of the anode signals to digital data using Analogue to Digital

Converters (ADC) followed by binary division can be used but this consumes a significant amount of power to produce the required accuracy at a reasonable speed as digital division is essentially sequential. If flash ADCs are used to encode the anode signals, with the reference of the ADC driven by the sum signal, the encoded output will automatically provide the required ratio. However if the ADC is to attain its specified accuracy, the reference signal has to be held constant for a time significantly greater than that required for the signal inputs. This time is required to charge the stray capacities associated with the flash ADC's reference resistor network and is one of the factors that limits the maximum pulse processing rate of this technique.

The problem can be overcome by measuring both the anode signals and the sum signal with a fixed reference flash encoder. Each combination of anode and sum signal is combined to form an address and the data held at each address, in a Read Only Memory (ROM), is programmed to be the required ratio. Thus for an 8 bit signal and sum system, a 16 bit address will be formed and if an 8 bit ratio is also required, the look–up table in the ROM needs to contain 2^{16} 8 bit words which is not an unreasonable amount of memory, especially if the radiation environment allows the use of ultraviolet or electrically erasable programmable ROMs.

In these systems, the position data are used to address random access memory in a read increment write system. Each address will contain the number of counts accumulated in a specific period at a specific position. Although this type of detector cannot operate as fast as the crossed anode array, random counting rates up to 100000 counts a second can be achieved.

As the number of level detectors required in an N bit flash encoder is 2^N, high speed, high resolution devices consume significant amounts of power, but devices operating at a few MHz are available at power levels of about 100 mW. Flash encoders which operate on a two pass system; encoding the most significant bits on the first pass and the remaining least significant bits on the second pass; reduce the number of level detectors to the square root of the number of resolution levels required ($2^{N/2}$). This produces a very significant saving in power, but at the expense of reducing the encoding rate by at least a factor of two.

4.2.7   *Magnetometers*

These devices are widely used in attitude measuring systems due to their stable operation and the low power consumption of the simplest system. They do not deteriorate with radiation dose and operate over a wide range of temperatures. The precision of attitude data produced by magnetometers is limited by inaccuracies in the knowledge of the Earth's magnetic field, and their range of operation is limited to an altitude of a few thousand kilometers above the Earth's surface. This is because the Earth's magnetic field strength is proportional to $1/r^3$ where r is the distance between the spacecraft and the centre of the Earth and as *r* reaches the above altitude the residual field due to magnetic materials within the spacecraft dominates that due to the Earth. This effect can, to some extent, be mitigated by mounting the magnetometer at the end of a fixed boom. The use of a deployable boom would not be sensible for an attitude sensing magnetometer due to the additional complexity of the deployment mechanism.

Search coil magnetometer

This is the basic system which functions according to Faraday's law of Magnetic Induction which states that an emf *V* is induced in a conducting coil placed in a time varying magnetic flux ϕ. Thus if a coil of *N* turns wound on a material with a relative magnetic permeability μ_r and cross–sectional area *A* is placed in a variable magnetic flux density *B* which has a component B_p along the axis of the coil

$$V = -N\frac{d\phi}{dt} = -N\frac{d(A\mu_r B_p)}{dt}$$

If the coil is placed in a spacecraft, whose spin rate is ω radians per second, with its axis perpendicular to the spacecraft's spin axis and the angle between the Earth's magnetic field *B* and B_p is θ, as shown in Fig. 4.13

$$B_p = B\cos\theta\sin(\omega t) \quad \text{and}$$

$$V = -AN\mu_r Bw\cos\theta\cos(\omega t)$$

Thus dependent on the accuracy to which the magnitude and direction of *B* and the spin rate of the spacecraft are known, the angle θ can be determined from the amplitude of *V*. The angle between the axis of the coil and the component of *B* in the plane perpendicular to the spin axis of the spacecraft can be deduced from the phase angle of *V* relative to its peak value.

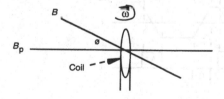

Fig. 4.13.   Search coil magnetometer in a spinning spacecraft.

Apart from the errors already mentioned, inaccuracies also occur due to non–linearities in the permeability of magnetic materials at low levels of magnetic induction.

As this type of magnetometer requires a changing magnetic flux relative to the coil, it can only be used on a spinning spacecraft. For a three axis stabilised spacecraft, magnetic attitude measurement can be obtained using a fluxgate magnetometer.

Fluxgate magnetometer

This device consists of a magnetic core which is driven into saturation by a triangular current waveform applied to the primary coil as shown in Fig. 4.14.

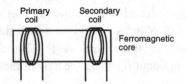

Fig. 4.14.   Fluxgate magnetometer.

If the ambient magnetic field is zero the secondary coil produces a symmetrical waveform. However with an ambient field the magnetic intensity in the core becomes unsymmetrical and this changes the phasing of the secondary coil's output as illustrated in Fig. 4.15. Thus by measuring the phase difference between the A and B pulses relative to the period of either set of pulses enables the magnitude of the external field to be calculated relative to the saturating field. By the use of three orthogonal coils, a three axis attitude determination of the spacecraft can be made.

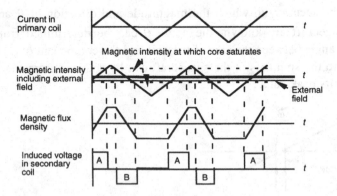

Fig. 4.15.   Fluxgate magnetometer waveforms.

The accuracy of this device is also limited by the spacecraft's residual magnetic field, the precision to which the Earth's magnetic field is known and also the ability of the core driving circuitry to produce the required symmetrical current waveform with no offset errors. The latter is quite a significant requirement as the amplitude of the magnetic field required to saturate the magnetic core is orders of magnitude higher than that due to the Earth's field. The non–linearity error at low levels of magnetisation associated with the search coil magnetometer does not occur here as the core material is always driven into saturation. However the magnitude of the current required to saturate the core may preclude the use of this type of magnetometer in low power spacecraft and the presence of such magnetic material may not be acceptable for magnetically clean spacecraft.

4.2.8   *Thrusters and momentum wheels*

The attitude of a spacecraft in orbit needs to be defined for reasons discussed in the last section. However without active control the attitude will drift due to a variety of reasons which may include the effect of residual atmosphere, solar wind, magnetic fields, gravity gradients and energetic photons. It may also need to be changed for operational requirements. Thus attitude stabilization and control are necessary (Wertz, [1978]). There are two common methods of attitude control both being dependent on gyroscopic action. In one case the major part of the spacecraft is spun up to between 5 and 100 rpm and in the second case the spacecraft contains one or three momentum wheels.

With one momentum wheel the attitude can be controlled by altering either the spin rate or the position of the axis of the wheel and with three wheels mounted orthogonally, the attitude of the spacecraft can be controlled by changing the spin rate of the appropriate wheel. Once the momentum wheels have reached their maximum angular velocity, momentum can be dumped, without altering the attitude of the spacecraft, by simultaneously activating a pair of suitably positioned gas thrusters. These may also be used for manœuvring the spinning spacecraft. Major attitude changes will normally be carried out by ground command with fine control determined by the on–board attitude measuring subsystems.

Gas jets or thrusters using compressed gases provide limited attitude control, but large manœuvres usually require the higher thrust associated with the chemical decomposition of the fuel using a catalytic converter. Ion thrusters using electrostatic particle accelerators with their much higher exhaust velocities can also be employed with significant advantage over chemical thrusters but they require very high electrical power for their operation. However all these systems use limited resources which have a significant weight impact on the spacecraft.

Momentum wheels present the standard problems of bearings in space (lubrication and mechanical wear). The electronic control of thrusters is limited to supplying pulses to operate valves and relays.

A technique for attitude control that can be used for near Earth orbiting spacecraft is to use the reaction between the Earth's magnetic field and an internally generated magnetic field (Schmidt, [1978]).

4.2.9   *Magnetic attitude control*

The technique of magnetic torquing has the advantage of no moving parts and no consumption of limited resources. If we have a coil of wire radius r, with N turns, giving an area A through which a current I is flowing, the magnetic moment **m** is given by

$$\mathbf{m} = NIA\mathbf{s} \quad (\text{Am}^2 \text{ or Wm})$$

where **s** is a unit vector in the direction normal to the plane of the coil. The torque **T** applied to this coil when placed in a magnetic field **B** is given by

$$\mathbf{T} = \mathbf{m} \times \mathbf{B} = NIA\mathbf{s} \times \mathbf{B}$$

This would cause the coil to rotate with an angular acceleration ω given by $\mathbf{T}/\mathbf{I}_S$ where $\mathbf{I}_S$ is the moment of inertia of the coil about the $\mathbf{s} \times \mathbf{B}$ axis. However if the coil were positioned in a spinning spacecraft as shown in Fig. 4.16, as **s** would tend to line up with **B**, the spacecraft would precess around **B**. Given the angular momentum of the spacecraft about its spin axis is $\mathbf{L} = L\mathbf{s}$ then

$$\mathbf{T} = \frac{d\mathbf{L}}{dt} = L\frac{d\mathbf{s}}{dt}$$

where the angular precession of the spacecraft caused by the magnetic torquer is given by

$$\frac{d\mathbf{s}}{dt} = \frac{\mathbf{T}}{L} = \frac{NIA}{L}\mathbf{s} \times \mathbf{B}$$

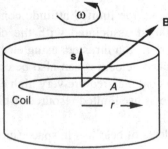

Cylindrical Spacecraft

Fig. 4.16.   Magnetorquer aligned to spin axis of spacecraft.

♦ *Example 4.1. Magnetorquer calculations*
Typical coil parameters for a small LEO spacecraft might be $N = 160$ turns, $I = 1$ amp, $A = 0.5$ m^2, and with the Earth's field $\mathbf{B}$ at 30×10^{-6} tesla, a maximum torque $\mathbf{T}$ of $160 \times 1 \times 0.5 \times 3 \times 10^{-6} = 0.0024$ Nm would be produced. If I_m, the moment of inertia of the spacecraft about its spin axis, is 15 kgm^2 and its spin rate is 12 rpm its angular momentum $\mathbf{L}$ is given by

$$\mathbf{L} = I_m \omega$$

$$\omega = \frac{2\pi \times 12}{60} = 1.26 \text{ radians s}^{-1}$$

$$\mathbf{L} = 15 \times 1.26 = 18.85 \text{ kgm}^2 \text{ s}^{-1}$$

Hence the maximum precession rate, given by

$$\mathbf{T/L} = \frac{0.0024}{18.85} = 127.3 \text{ microradians s}^{-1}$$

$$= 127.3 \times 10^{-6} \times \frac{180}{\pi} = 0.0073 \text{ degrees s}^{-1}$$

Allowing for the variation in the Earth's magnetic field strength and direction around the orbit; the restriction on usage of the torquer to full sunlight as the current has to be large to produce a significant torque against the Earth's magnetic field; and operation for a fraction of a revolution to maintain high torque due to the vector cross product term; would produce an angular displacement of a few degrees per orbit.

Ferromagnetic material which could be used to increase the dipole moment of the coil is excluded both from weight and magnetic cleanliness considerations.

Ideally a constant current generator should be used to stabilize the torque as the power may be taken directly from an unstabilised main power bus whose voltage may vary significantly. This also guards against the variation of the coil's resistance with temperature.

To enable the direction of the torque to be changed as required, a current reversal system, as shown in Fig. 4.17, may be included. Such a magnetorquing system results in a change in distribution of the angular momentum relative to the spacecraft's axes.

For a coil wound in the plane containing the spin axis, if the current is pulsed at twice the spin rate of the spacecraft with the polarity of the current changed at each pulse, this results in maintaining the torque in a given direction and so can be used to change the total angular momentum of the spacecraft.

As the torque produced by the coil is proportional to NIA, consideration has to given as to how this can be maximized. A is limited by the diameter of the spacecraft. Increasing N or I will impact both the weight and power of the system. Hence it is necessary to determine the relationship between $\mathbf{T}$ and these two parameters before proceeding with the design of the coil. Let R be the coil resistance, a the cross-sectional area of the wire, σ its resistivity, ρ its density and M the mass of the coil whose radius is r and area A, P the power dissipated in the coil, I the current through the coil and V the voltage across the coil.

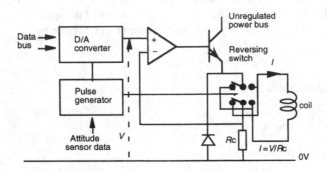

Fig. 4.17. Magnetorquer drive system.

$$R = \frac{2\pi r N \sigma}{a} \quad \therefore \quad N = \frac{aR}{2\pi r \sigma}; \quad M = 2\pi r N a \rho \quad \therefore \quad N = \frac{M}{2\pi r a \rho}$$

$$P = I^2 R, \quad I = \sqrt{\frac{P}{R}} \; ; \quad A = \pi r^2;$$

$$T = NIAB = N\sqrt{\frac{P}{R}}\, AB = \sqrt{\frac{PN^2}{R}}\, AB$$

$$= \sqrt{\left(\frac{P}{R}\frac{aR}{2\pi r \sigma}\frac{M}{2\pi r a \rho}\right)} \, \pi r^2 B = \sqrt{\frac{PM}{\sigma \rho}}\, \frac{\pi r^2 B}{2\pi r} = \frac{Br}{2}\sqrt{\frac{PM}{\sigma \rho}}$$

Thus for a given coil diameter, one can trade power against mass for the required torque, depending on which parameter is in shorter supply. Having decided on the values for M and P, the values of N and a can then be determined for a given V.

As the product $\rho\sigma$ for Copper is 1.8×10^{-4} kgΩm^{-2} and for Aluminium is 0.8×10^{-4} kgΩm^{-2}, for a given mass power budget

$$\frac{T_{Al}}{T_{Cu}} = \sqrt{\frac{(\sigma\rho)_{Cu}}{(\sigma\rho)_{Al}}} = \sqrt{\frac{1.8 \times 10^{-4}}{0.8 \times 10^{-4}}} = 1.5$$

which shows that an aluminium coil will give a 50% greater torque than the corresponding copper one.

As operating the magnetorquer would magnetize any magnetic materials on the spacecraft, a demagnetizing programme would be needed. This consists of rapidly changing the polarity of the current in the coil whilst gradually reducing its amplitude to zero. The frequency of this change will be limited by the switching speed of the reversing relay and the time constant associated with the inductance of the coil. A typical demagnetization programme would take a few minutes.

4.3   Analogue design

4.3.1   *Introduction*

The electrical outputs of spacecraft sensors are of two main types; charge pulses from individual particle or photon detectors and continuously variable currents or voltages from integrating detectors. One of the main design constraints on the sensor and processing electronics is power as more power implies larger batteries, larger solar arrays and hence higher overall costs. This constraint also limits the maximum speed at which the system can operate as higher speeds tend to demand more power.

4.3.2   *Charge sensitive amplifiers (CSAs)*

Many sensors produce packets of charge Q where this represents the energy of the incoming photon or particle. The simplest way of measuring this charge is to deposit it on a known capacitor Cs and measure the voltage V this produces, where $V = Q/Cs$. However the capacity of the detector and the wiring to its preamplifier is probably not well defined in the early stages of the design and may be temperature dependent. Thus a design which can measure Q independent of Cs is highly desirable. A diagram of such a preamplifier is shown in Fig. 4.18.

The open loop gain of the amplifier is A; Cs represents all the stray input capacities due to wiring, sensor and amplifier connections; Cfb is a stable, well defined capacitor and Vi and Vo represent the input and output voltages of the amplifier. When the charge Q arrives it will be distributed between Cs and Cfb in the following way

$$Q = Cs Vi + Cfb (Vi - Vo)$$

$$= Vi (Cs + Cfb) - Vo Cfb$$

$$= - Vo (Cs + Cfb) / A - Vo Cfb \quad as \ Vo = -A Vi$$

$$= - Vo (Cs / A + Cfb (1 + 1 / A))$$

$$= - Vo (Cs / A + Cfb) \quad as \ A \gg 1$$

$$\therefore Vo = - Q / (Cfb + Cs / A) = - Q / Cfb \quad if \ Cs / A \ll Cfb$$

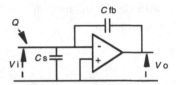

Fig. 4.18. Charge Sensitive Amplifier.

If we design the system so that $A \times Cfb \gg Cs$, the output voltage will be independent of Cs. With high speed amplifiers A will be limited to a few thousand and if Cs is large, the requirement for $A \times Cfb \gg Cs$ can only be met by making Cfb fairly large. As this will reduce Vo for a given Q and hence require extra subsequent amplification, the sensor design should try to minimize Cs. Minimizing Cs is also vital for S/N requirements as will be shown later.

Another property of this type of preamplifier is its ability to reduce cross talk between adjacent inputs on a multi–channel system as illustrated in Fig. 4.19. A charge Q deposited on Anode 1 will produce a signal $Vi1$ at the input of the upper amplifier. This will induce a charge Qc onto the input of the lower amplifier due to the inter–electrode capacitance Cc, where $Qc = Cc (Vi1 - Vi2)$. Thus the output from the lower amplifier due to Qc deposited on the upper electrode is given by

$Vo2 = -Qc / Cfb = - Cc (Vi1 - Vi2)/ Cfb$

as $Vi1 = - Vo1 / A$ and $Vi2 = -Vo2 / A$

$\therefore Vo2 = Cc (Vo1 - Vo2)/ ACfb$

$Vo1 Cc / ACfb = Vo2 (1 + Cc / CfbA) = Vo2$ as $Cc / CfbA \ll 1$

$\therefore Vo2 / Vo1 = Cc / ACfb$

Thus the cross talk $Vo2 / Vo1$ can be reduced to the required level by increasing the product $ACfb$.

This facility can also be considered in terms of the effective input capacitance of the charge sensitive amplifier Cin.

$Vo = - Q / Cfb = - AVi1$

$Cin = Q / Vi1$

$= ACfb$

Thus the input voltage Vi caused by Q is reduced by this large capacity Cin, and it is Vi which causes the cross talk.

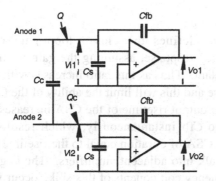

Fig. 4.19.   Cross talk due to inter-electrode capacity Cc.

CSA noise considerations

The basic closed loop gain G of the amplifier shown in Fig. 4.18 is given by

$$G = -\frac{Zfb + Zs}{Zs} = -\frac{Cs}{Cfb} - 1$$

where Zfb is the impedance of the feedback capacitor and Zs is the impedance of the stray input capacitance. If the amplifier input noise is v_n, the output noise v_{on} is given by

$$v_{on} = -v_n\left(\frac{Cs}{Cfb} + 1\right)$$

and the S/N for an input charge dump Q is given by

$$\frac{v_0}{v_{on}} = \frac{-\frac{Q}{Cfb}}{-v_n\left(\frac{Cs}{Cfb} + 1\right)} = \frac{Q}{v_n}\frac{1}{(Cs + Cfb)}$$

As the input capacity Cs will usually be significantly larger than Cfb, minimizing Cs is a prime requirement for low noise charge sensitive preamplifiers. This requirement implies that the preamplifier is placed as close to the sensor as possible to minimize the stray wiring capacity.

As the main source of noise is the channel of an N junction FET which is usually used to form the first stage of the preamplifier, a device with the highest transconductance (ratio of change in channel current to change in gate voltage) should be chosen so that this source of noise, when referred to its input (gate) is minimised.

Risetime of CSA output

As the ratio of $(Cs + Cfb)/Cfb$ defines the closed loop gain of the amplifier, this will also affect its frequency response, assuming the gain bandwidth product of the CSA is approximately constant. Thus as this ratio increases with Cs so the output risetime of the CSA will increase and this will limit the ability of the CSA to resolve closely spaced events. Also as the output risetime of the CSA increases, it no longer transfers the input charge Q onto Cfb instantaneously, which results in a voltage spike occurring at the input of the CSA of duration equal to the risetime of the CSA. This spike can then induce cross talk into adjacent amplifiers. The degree of cross talk will depend on whether the frequency components of this spike occur within the bandwidth of the subsequent amplification. This may well occur if the system has been designed to accommodate a high event rate.

Practical CSA circuit

The circuit shown in Fig. 4.18 is idealized and lacks many of the components required to make the system operational. A more realistic circuit is shown in Fig. 4.20 incorporating a sensor with an anode at high voltage. For those not familiar with looking at detailed circuit diagrams, I have adopted the convention that wires that cross directly do not inter – connect.

The major problem associated with such an amplifier is the possibility of high voltage transients, generated within the sensor, damaging the first stage of the amplifier. For this reason the charge sensitive amplifier should be designed with its first stage external to the main amplifier as shown in Fig. 4.20. If damaged this stage can then be replaced without having to unsolder a complete integrated circuit (IC).

The channel current of the first stage FET should be chosen so that the gate to source voltage Vgs is slightly negative. This makes the gate to source leakage current, and the noise associated with it, small and ensures that the the FET operates near its maximum transconductance. Ideally Vgs would be made zero but any drift in the circuitry could then cause Vgs to become slightly positive (forward biased instead of reverse) which would have disastrous results on the gate input current. To allow for changes in Vgs due to temperature, aging and radiation an initial value of about -1 V is reasonable.

Rfb is necessary to provide a path for the gate input current; to allow a discharge path for the charge built up on Cfb, and to define the gate voltage. However as it introduces a noise current into the circuit given by $\sqrt{4kTB/Rfb}$ where k is Boltzmann's constant $= 1.38 \times 10^{-23}$ JK^{-1}, B is the circuit bandwidth and T its absolute temperature, its value should be kept as high as possible within the limitations specified above.

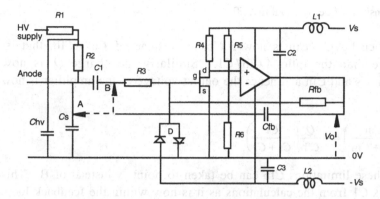

Fig. 4.20. Practical Charge Sensitive Amplifier.

$C1$ is the high voltage coupling capacitor from the sensor's anode. Its value will be defined by the efficient coupling of the sensor's output to the amplifier. However if any high voltage breakdown does occur at the sensor the energy stored in $C1$ will be dissipated in the protection network consisting of $R3$ and the two diodes D. If $C1$ is too large the energy dissipated will destroy this network. $R3$ must be capable of dissipating a large peak power (several kilowatts for a few microseconds) and hence the more robust carbon composition type is preferable to the thin film variety. Tests on carbon composition resistors show a slight reduction in resistance during short high power dissipation tests whereas the film type gradually vaporize and eventually fail open circuit. Although R3 needs to be relatively large to reduce the power dissipation in the protection circuitry it cannot be made too large as it both limits the rise time of the input signal to the FET and increases the input voltage noise to the amplifier by $\sqrt{4 \times R3 \times kTB}$.

The effect of $C1$ can be calculated from Fig. 4.21. If the feedback capacitor is taken to point B, the effective capacitance at A is $Cin + Cs$ where Cin equals $C1$ in series with $ACfb$.

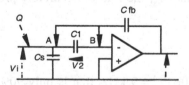

Fig. 4.21. CSA with input coupling capacitor $C1$.

$$Cin = \frac{1}{\frac{1}{C1} + \frac{1}{ACfb}} = \frac{AC1Cfb}{ACfb + C1} = \frac{C1}{1 + \frac{C1}{ACfb}} < C1$$

Thus $C\text{in} < C1$ instead of being equal to $AC\text{fb}$

The cross talk given by $Cc/C\text{in}$ is now $> Cc/C1$ instead of $Cc/AC\text{fb}$, that is, increased by more than the ratio $A\,C\text{fb}/C1$. Similarly the charge Q is now distributed between Cs and $C\text{in}$ and hence the output voltage of the amplifier is now reduced to

$$-\frac{Q}{C\text{fb}}\left(\frac{C\text{in}}{Cs+C\text{in}}\right)=-\frac{Q}{C\text{fb}}\left(\frac{C1}{Cs+C1}\right)$$

To overcome these limitations $C\text{fb}$ can be taken to point A instead of B. This effectively removes $C1$ from the calculations as it is now within the feedback loop. However this does require $C\text{fb}$ to be a high voltage capacitor as well as $C1$. This is disadvantageous as the number of components subject to high voltages should always be kept to a minimum and high voltage capacitors are usually much larger than their low voltage equivalents. It also allows high voltage transients to be directly coupled into the output of the amplifier, although as $C\text{fb}$ will be a small capacity, this will not be such a serious problem and it can be overcome with additional protection diodes at the amplifier's output terminals. An alternative solution is to maintain the B connection and make $C1 \gg Cs$ but this exacerbates the high voltage breakdown problem and increases the size of $C1$ significantly. Hence the A solution is usually preferred.

4.3.3   *Pulse shaping circuits*

The design of the pulse shaping network following the preamplifier is determined mainly by the repetition rate of the charge dumps, the accuracy to which their amplitude or timing needs to be measured and noise considerations. Assuming that the event rate is random but the average rate is F and the circuit can accurately process a pair of pulses if they are spaced T_s or more apart, then, as a fraction of the pulses will occur with a separation of less than T_s, the rate of accurately processed events F_a is given by

$$F_a = Fe^{-FT_s}$$

Thus if T_s is 10 µs and $F = 10$ kHz, $F_a = 10e^{-0.1} = 9$ kHz; that is only 90% of the events will be accurately processed. If the system requirement is to measure the amplitude of the accurately processed charge dumps to within a specified error, say 1%, then Ts is defined by the time taken for the amplitude of the shaped pulse to fall to 1% of its peak value. Pulses occurring within this timescale will be sitting on the tail of the previous pulse and so have an erroneous amplitude. A simple shaping network which produces a pulse whose amplitude falls fairly rapidly is shown in Fig. 4.22.

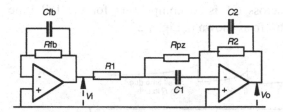

Fig. 4.22.   Unipolar pulse shaping with pole − zero cancellation.

If the output of the preamplifier Vi is a step function, the time constant $C1R1 = C2R2 = T$, and ignoring the pole−zero cancellation component Rpz, the unipolar pulse amplifier produces a waveform defined by:

$$\frac{Vo}{Vi} = -\frac{Z2}{Z1}$$

Applying Laplace transforms to these parameters results in

$$Z2 = \frac{1}{1/R2 + pC2} = \frac{R2}{R2C2(p + 1/R2C2)} = \frac{R2}{T(p + 1/T)}$$

$$Z1 = R1 + \frac{1}{pC1} = \frac{1 + pR1C1}{pC1} = \frac{R1C1(p + 1/R1C1)}{pC1} = \frac{R1}{p}\left(p + \frac{1}{T}\right)$$

$$\therefore \ \frac{Vo}{Vi} = -\frac{p}{T}\frac{R2}{R1}\frac{1}{(p + 1/T)^2}$$

As the Laplace transform for the input step function Vi is Vi / p, Vo is the inverse function of

$$-\frac{Vi}{T}\frac{R2}{R1}\frac{1}{(p + 1/T)^2} = \frac{Q}{Cfb}\frac{R2}{R1}\frac{1}{(p + 1/T)^2} \quad \text{as } Vi = -Q/Cfb$$

$$\therefore \ Vo = \frac{Q}{Cfb}\frac{R2}{R1}\frac{t}{T}e^{-\frac{t}{T}}$$

This waveform peaks at $t = T$ with the time to peak being independent of the peak amplitude which is $0.37 \times Vi \times R1 / R2$. The waveform recovers to 1% of its peak value in $7.6T$. The choice of making $T1 = R1C1 = T2 = R2C2 = T$ is governed by the requirement to optimize the S/N. Assuming that the noise associated with the input signal is broadband noise, if $T1$ is made less than $T2$, the signal output amplitude is reduced more than the noise is reduced, and if $T1$ is made larger than $T2$, the noise is increased more than the signal amplitude is increased. The optimum occurs when $T1 = T2$. The resistor Rpz across $C1$ is to compensate for the fall time of the preamplifier defined by $t = RfbCfb$ as shown in Fig. 4.23.

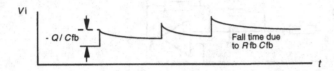

Fig. 4.23. Charge amplifier output voltage.

This fall time extends the recovery time of the unipolar waveform considerably as illustrated in Fig. 4.24. The output waveform from the unipolar shaping amplifier is shown to the left of Fig. 24 for $t = \infty$, and to the right for realistic values of t.

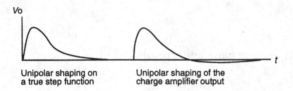

Fig. 4.24. Result of fall time of CSA on subsequent amplification.

As the output voltage of the CSA falls exponentially, this waveform, which is applied to the shaping amplifier, can be expressed in Laplace format as

$$Vi = -\frac{Q}{pCfb}\frac{Rfb}{Rfb + \dfrac{1}{pCfb}} = -\frac{Q}{pCfb}\left(\frac{p\tau}{1+p\tau}\right) \quad \text{where } \tau = CfbRfb$$

If we now include Rpz in the calculations and make $RpzC1 = \tau$

$$\frac{Vo}{Vi} = -\frac{\dfrac{1}{1/R2 + pC2}}{R1 + \dfrac{1}{1/Rpz + pC1}} = -\frac{\dfrac{R2}{1+pR2C2}}{R1 + \dfrac{Rpz}{1+pC1Rpz}} = -\frac{\dfrac{R2}{1+pT}}{R1 + \dfrac{Rpz}{1+p\tau}}$$

$$= -\frac{R2(1+p\tau)}{(1+pT)}\frac{1}{\big(R1(1+pC1Rpz)+Rpz\big)} = -\frac{R2(1+p\tau)}{(1+pT)}\frac{1}{\big(R1+pC1R1Rpz+Rpz\big)}$$

As $Rpz \gg R1$ by design, we can neglect $R1$ in the denominator.

$$\therefore \; \frac{Vo}{Vi} = -\frac{R2(1+p\tau)}{Rpz(1+pT)^2}$$

Substituting the result deduced for Vi from the CSA in the above equation results in

$$Vo = \frac{Q}{pCfb}\left(\frac{p\tau}{1+p\tau}\right)\left(\frac{R2(1+p\tau)}{Rpz(1+pT)}\right)$$

Thus the zero term cancels with the pole term yielding

$$Vo = \frac{QRpzC1}{Cfb}\frac{R2}{RpzR1C1T(p+1/T)^2}$$

$$= \frac{Q}{Cfb}\frac{R2}{R1}\frac{1}{T(p+1/T)^2}$$

This expression is identical to the one previously calculated for a simple unipolar amplifier without pole – zero compensation when driven by a true step function. Thus by making $RpzC1$ equal to $RfbCfb$, optimum recovery time is ensured. However this system still has the disadvantage associated with all capacitively coupled unipolar networks at high event rates, namely the depression of the baseline, as illustrated in Fig. 4.25. As the entire system is, by design, capacitively coupled, its average voltage output must be zero. The ideal output shown in A can be considered to be the sum of B which has a mean level of zero plus the offset shown in C. As the C output is blocked by capacitive coupling, the actual output is B and it is clear from the figure that the baseline shift is a function of count rate.

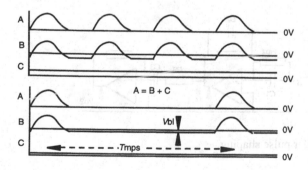

Fig. 4.25. Baseline shifts.

An approximate calculation will indicate when this problem becomes significant. The unipolar waveform is given by

$$Vo = Vi\frac{t}{T}e^{-\frac{t}{T}} \quad \text{for a step input of amplitude } Vi.$$

Thus the area under a single pulse is given by

$$Vi\int_0^\infty \frac{t}{T}e^{-\frac{t}{T}}dt = Vi\left[-te^{-\frac{t}{T}} + \int_0^\infty e^{-\frac{t}{T}}dt\right]_0^\infty = -Vi\left[te^{-\frac{t}{T}} + Te^{-\frac{t}{T}}\right]_0^\infty = ViT$$

This must approximately equal the depressed area of the baseline $Vbl \times Tmps$, where Vbl is the baseline depression and $Tmps$ is the mean pulse separation.

$$\therefore \; Vbl = \frac{ViT}{Tmps} = \frac{v_p}{.37}\frac{T}{Tmps}$$

where v_p is the peak amplitude of the unipolar pulse.

♦ *Example 4.2. Baseline depression*
If the shaping time constant T is 1µs and the average pulse rate is 10k pulses per second, ($Tmps$ = 100 µs), the baseline will be depressed by $v_p / (0.37 \times 100) = 0.027v_p$. This is a 2.7% error in the measurement of v_p even when the ratio of $Tmps$ to T is as high as 100 .

To overcome this problem the unipolar pulse can be converted to a bipolar pulse by a second CR coupling network, with $C3R3 = T$, as shown in Fig. 4.26.

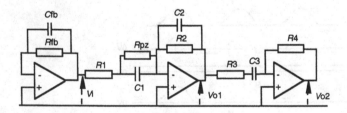

Fig. 4.26.   Bipolar pulse shaping.

The output waveform $Vo2$ from this network due to the same input voltage Vi can be deduced by multiplying the Laplace transform of $Vo1$, by the transfer function of the final stage which is given by

$$-\frac{R4}{R3+1/pC3} = -\frac{pC3R4}{1+pC3R3} = -\frac{pC3R4}{R3C3(p+1/T)} = -\frac{R4}{R3}\frac{p}{(p+1/T)}$$

Thus Vo2 is given by the inverse function of

$$-\frac{Q}{Cfb}\frac{R2}{R1}\frac{1}{T(p+1/T)^2}\frac{R4}{R3}\frac{p}{(p+1/T)} = -\frac{Q}{Cfb}\frac{R2}{R1}\frac{R4}{R3}\frac{1}{(p+1/T)^3}$$

$$\therefore Vo2 = -\frac{Q}{Cfb}\frac{R2R4}{R1R3}\frac{t}{T}\left(1-\frac{t}{2T}\right)e^{-\frac{t}{T}}$$

This waveform peaks at $(2-\sqrt{2})T(=0.59T)$ from the start of the pulse, with an amplitude of $0.23 \times Vi \times R2 \times R4/(R1 \times R3)$ (where $Vi = -Q/Cfb$), crosses zero at $2T$ and recovers to 1% of its peak amplitude in $9.7T$. Typical waveforms for comparison purposes are shown in Fig. 4.27.

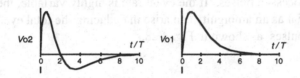

Fig. 4.27. Bipolar and unipolar waveforms.

Although bipolar waveforms take longer to recover to zero and have a lower S/N ratio than an equivalent unipolar system, these disadvantages are usually outweighed by the fact that the total area under the bipolar waveform is zero and so baseline shifts do not occur.

4.3.4   Pulse processing system design

If the amplitude of a charge dump needs to be measured with an error of less than 1%, then to ensure that the amplitude error due to one pulse sitting on the tail of the previous one, for a bipolar system, is less than this value, the minimum allowable separation between pulses T_s has to be $9.7T$. If we specify that 90% of the input pulses need to be processed to this accuracy then the maximum acceptable random count rate can be deduced from

$$F_a = Fe^{-FT_s} \text{ where } F_a/F = 0.9 = e^{-FT_s} \text{ and } T_s = 9.7T$$

$$\therefore e^{9.7TF} = 1/0.9$$

$$9.7TF = \ln(1.11) = 0.11$$

$$T = 0.01 / F = 0.01 T\text{mps}$$

Thus the circuit's time constant T has to be 100 times smaller than Tmps, the mean separation of the input pulses which implies a much higher bandwidth with the attendant increase in noise.

Although one cannot prevent pulses from arriving with separations of less than T_S, one can prevent them from being measured by using a pulse pile–up limiter circuit. This is basically a retriggerable monostable of pulse length $9.7T$ (assuming the previously specified conditions still apply) which is used to inhibit the processing of any pulses which arrive within a time $9.7T$ from a previous pulse. If a second pulse does arrive within this time not only will it be ignored but a further $9.7T$ (dead time of system) will be allowed for the amplitude of this pulse to fall to 1% of its peak level. Hence all pulses processed will be processed accurately. As this system ignores a fraction of the input pulses it is sensible to monitor the total event rate and then apply a correction factor to the processed pulses. If the event rate is highly variable, the total event rate monitor is essential as an ambiguity can arise in deducing the total event rate from the rate of processed pulses as shown in Fig. 4.28.

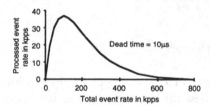

Fig. 4.28.   Effect of pulse pile–up limiter on throughput of processed pulses.

To prevent low level noise pulses and high level pulses caused by unwanted events such as cosmic rays from being processed, the analogue pulses will normally be interrogated by a window detector. This is formed from two level detectors, the Lower Level Detector (LLD) set to trigger only if a pulse reaches a certain minimum amplitude and the Upper Level Detector (ULD) set to trigger when a maximum amplitude has been reached. The outputs from these two detectors are logically coupled to inhibit the measurement unless the pulse amplitude is between the two limits. Fig. 4.29 shows the complete system.

From Fig. 4.30 it can be seen that to generate the ADC strobe pulse, triggered by the delay pulse, precisely at the peak of the bipolar pulse, the start time of the analogue pulse must be determined as accurately as possible. The simplest way of ascertaining this is to incorporate a lower low level detector (LLLD) set just above the noise level. As this may occasionally be triggered by noise, only pulses whose amplitudes fall within the window's range, LLD to ULD, are processed.

From the equation of the bipolar waveform, one can deduce that the peak of the waveform is only constant to 1% for $0.15T = 0.0015\ T$mps. Therefore the timing of the ADC strobe must be positioned to this accuracy unless a peak and hold circuit is employed.

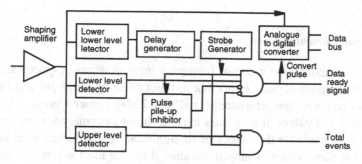

Fig. 4.29. Block diagram of peak amplitude measurement.

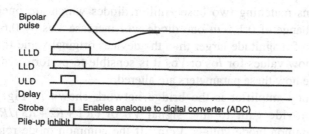

Fig. 4.30. Peak amplitude measurement and pulse pile–up limitation.

If the rise time of the original charge dump is not fast compared with T and is variable, the position and amplitude of the peak of the shaped pulse is not well defined. This will then determine the minimum value for T and hence define the maximum counting rate for the system.

If the mean event rate and required accuracy are such that several microseconds may elapse between valid events, a sample and hold circuit can be used to maintain the peak amplitude for a sufficient period for a successive approximation amplitude measuring system to operate. However if a much faster response is required then a flash encoder can be used. To reduce power consumption, two pass flash encoders can be used, but these require the analogue pulse to be constant for a substantially longer time than the single pass devices and so reduce the maximum rate at which events can be processed.

In all shaping amplifiers the shaping network should be determined in terms of the optimum S/N as well as recovery time and baseline stability. The networks discussed above are amongst the simplest but the basic concepts apply to all types of pulse amplifiers. More complex designs involving inductances as well as capacitors and resistors, or equivalent circuits, with several poles and zeros in their transfer functions can produce waveforms with recovery times about twice as fast as those described, but significant improvements beyond these are difficult to achieve.

4.3.5   Low frequency measurements

The problems encountered with low frequency measurements in spacecraft are the removal of common mode signals, errors due to input offset currents Ios and input offset voltages Vos, errors in the reference signal and EMR induced noise. Some amplifiers have significant values of input bias current Ib but extremely low values of Ios. Age and radiation can alter the values of Ib significantly, producing very large changes in Ios. Thus these types of amplifiers should not be used where the initial very low value of Ios is an important feature of the operational requirements of the circuit.

Low values of Vos are produced by careful laser trimming of the input devices. For bipolar ICs this means matching two base emitter diodes which are operating around 0.65 V. Thus a change of 0.1% in one diode will increase Vos by 0.65 mV which may be two orders of magnitude larger than the devices original value for Vos. Hence if circuits rely on low values for Ios or Vos it is sensible to perform radiation tests on the ICs used to see how these parameters are altered.

A typical low frequency amplifier is the bridge network shown in Fig. 4.31. Amplifiers A1 and A2 provide gain for the signal Vb of $(R1+R2+R3)/R2$ but only unity gain for the common mode voltage Vcm. If the common mode rejection amplifier A3 has unity gain, ($R4=R5=R6=R7$), the accuracy and stability requirements of $R4$, $R5$, $R6$ and $R7$ are minimized as any fraction of the common mode voltage introduced by errors in the equality of these four resistors is only amplified by one.

If there is a significant lead length between the circuit and its ADC, the reference resistor $R6$ should not be returned to the main current return line, designated power ground, but to its own reference line as shown. This ensures any potential difference along the power ground line is excluded from the output. $R8$ can also be included to minimize any current in the signal ground line due to the common mode signal.

The filtering of power supply lines to amplifiers is an important feature of the analogue design as the power supply rejection capability of the amplifier may be very good at low frequencies (100 dB), but deteriorates rapidly at high frequencies. Minimizing the effects of EMR on harnesses connected to the amplifiers will be discussed in sections 4.5.11 and 4.6.

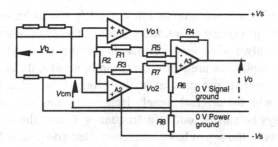

Fig. 4.31. Instrumentation amplifier.

If the measurements are required to be more precise than the offsets will allow, the original signal will have to be chopped and some form of phase sensitive detection system employed. The chopping has to be carried out at an amplitude where errors introduced by the chopper are adequately small. A typical system is shown in Fig. 4.32 where both the common mode voltage and the first stage amplifier's offset voltage are treated as common mode signals for the second stage amplifier.

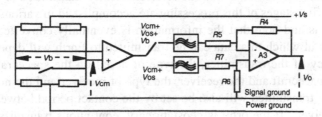

Fig. 4.32. Phase sensitive amplifier.

In most cases it is advantageous to provide power line voltage stabilization at the subsystem to minimize the effects of low frequency noise on the supply lines particularly at the preamplifier stage. Thus the subsystem power supply may provide ± 15 V for the main analogue system, but this can stabilized at card level down to ± 12 V for the more sensitive amplifier networks and may be further stabilized to ± 6 V for the preamplifier in the detector area.

4.3.6   *Sampling*

When analogue signals are converted to digital, an adequate sampling rate must be provided. The purpose of sampling is subsequently to be able to reconstruct the original signal without loss of information. Nyquist states that to extract all the information from a waveform, the minimum sampling rate f_s must be twice the frequency of the highest frequency component of the signal f_h. This sampled signal will contain the original signal plus sidebands up to $\pm f_h$ around f_s and its harmonics. However due to the impossibility of making an infinitely sharp cut – off filter above a specified frequency there will always be components above the nominal upper frequency limit f_h, and if their amplitudes are significant, sampling at only twice f_h would produce spurious (alias) signals not present in the original data due to the overlapping of these sidebands with the original signal. Hence sampling in excess of the minimum rate should always be carried out at a frequency where the nominal maximum frequency components of the signal have been attenuated adequately by the bandwidth limiting filter.

4.4   Data handling

4.4.1   Introduction

Scientific spacecraft are designed to collect information from on–board sensors and process this information so that it can be efficiently transmitted to Earth. We have seen how the first stages of this processing are accomplished in various types of electronic subsystems and note that the information is eventually converted to a digital format. The rate at which this data can be transmitted to Earth will depend on the power and efficiency of the transmitter, the type of aerial on the spacecraft, the distance between the spacecraft and the receiver, the type of receiver and the noise in the link. The total data transmitted will also be set by the contact period between the spacecraft and the receiver. If the orbit is geostationary, continuous transmission is possible but for most other orbits only short contact periods are possible, which usually requires that the spacecraft has an on–board mass data storage system such as a tape recorder. In either case maximising the efficiency of the communication system within the spacecraft is most important and this is accomplished by applying additional digital processing within the subsystem and having an efficient on–board data handling system.

4.4.2   User subsystem digital preprocessing

Data from a subsystem has to be produced so that it can be accommodated by both the telemetry's maximum data processing rate and the time it has access to the telemetry system. However the rate at which data are produced by the sensor will in general be completely different from that required by the telemetry system.

If the input data rate temporarily exceeds that acceptable by the telemetry, the process of queuing can to be introduced but if the data rate continuously exceeds this value, then data compression has to be introduced. If the data has to be associated with several other parameters, then the process of histogramming may be invoked as well.

4.4.3   Queuing

Suppose a sensor produces N_1 events a second and each event is associated with some parameter such as wavelength, energy etc whose value is specified in M bits. This situation is compatible with the spacecraft's data transmission rate R so long as $MN_1 \leq R$. However if MN_1 exceeds R for a period t, but this is followed by a period T of low activity ($MN_2 < R$), this can still be accommodated with a queuing system. The requirement for the system is that the queue length, in bits, must be at least $M(N_1 t + N_2 T)$ and the average data rate $(MN_1 t + MN_2 T)/(T + t)$ does not exceed R.

Although this could be accomplished using a microprocessor and memory array, as this is such a common situation, a special type of Shift Register (SR) called a FIFO or First In First Out register has been designed for this purpose. This functions by allowing the data to be clocked into the SR at the rate at which it is generated but, by using an addressing system internal to the FIFO, allows the data to be effectively stored at the output end of the FIFO and to be shifted out using an independent clock. In a standard SR, the following sequence occurs:

Input data	24 bit SR
10110101	XXXXXXXX,XXXXXXXX,XXXXXXXX
and after 8 clock pulses	1 0 1 1 0 1 0 1,XXXXXXXX,XXXXXXXX
In a FIFO, after 8 clock pulses	XXXXXXXX,XXXXXXXX,1 0 1 1 0 1 0 1

The FIFO can be made with byte wide inputs and outputs so that it can deal with complete 8 bit words per clock cycle.

Data	Byte wide 8 bit FIFO	After one clock pulse
1	XXXXXXXX	XXXXXXX1
0	XXXXXXXX	XXXXXXX0
1	XXXXXXXX	XXXXXXX1
1	XXXXXXXX	XXXXXXX1
0	XXXXXXXX	XXXXXXX0
1	XXXXXXXX	XXXXXXX1
0	XXXXXXXX	XXXXXXX0
1	XXXXXXXX	XXXXXXX1

So long as the FIFO is not completely filled, it can accept data at the rate at which it is generated by the sensor, but can present it to the telemetry system at the rate specified by the telemetry clock. The first two waveforms in Fig. 4.33 show the data arriving in the FIFO at a higher data rate than it is being read out. The second pair shows how the data rates equalize in the long term.

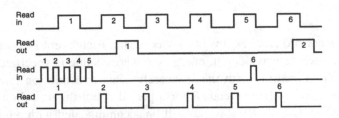

Fig. 4.33.   Queuing using a FIFO.

4.4.4   *Data compression*

In a simple random pulse counting system where the data rate can vary from say a thousand to a million events a second, if the required counting error is to be less than 3%, how many events N must be counted? As the event rate is random, from Poisson statistics, the fractional rms error is specified by $\sqrt{N}/N$. Therefore $1/\sqrt{N}$ must be less than 0.03 requiring N to be at least $(1/0.03)^2 \approx 1000$. This requires the minimum Count Acceptance Period (CAP) to be one second at the lowest count rate and also specifies that the minimum resolution required by the counter is 33.

As the count rate increases up to 10^6 events per second, the counter would record up to $10^6/33 = 3 \times 10^4$ resolution elements producing a counting error of one element in 3×10^4 which is one thousand times smaller than the required 3%.

To improve the efficiency of the system, simple logarithmic data compression can be used. The CAP is defined by the accuracy required at the lowest count rate, which in this case is one second and a 20 bit counter is provided to accommodate the 10^6 counts as $2^{20} \approx 10^6$

The system operates by transferring the number accumulated in the 20 bit counter during the CAP to a 20 bit SR. The number of zeros ahead of the first significant bit set to a '1', is determined by shifting the number to the left and counting the number of shift clock pulses required to produce a '1' at the LH edge of the SR. However as we need to preserve the next 4 digits plus this '1' as the mantissa ($33 \approx 2^5$), the maximum number of clock pulses need never exceed $20-5 = 15$. This number of clock pulses can be expressed by a 4 bit number. Thus the total number of bits needed to specify the count with an error of 3% or less is the 5 most significant bits plus the 4 bit exponent, compressing the 20 bit number to $4+5 = 9$ bits.

Let us look at a few examples of this type of compression starting with decimal counts in the range 1024 to 1087 in the 20 bit SR which will have 9 leading zeros.

$$00000000010000000000 = 1024 \text{ coded as } 1001\ 10000, \text{ and}$$
$$00000000010000111111 = 1087 \text{ also coded as } 1001\ 10000$$

Thus the compressed number 100110000 will indicate counts in the range 1024 to 1087 or 1055 ± 32 which has a maximum error of 3%.

Now let us consider counts between 2016 and 2047

      00000000011111000000 = 1984 coded as 1001 11111, and
      00000000011111111111 = 2047 also coded as 1001 11111

Thus 100111111 will indicate counts in the range 1984 to 2047 or 2016 ± 32 which has a maximum error of 1.5%. These examples cover the octave range from 1024 to 2047. From 2048 to 4095 the only difference in the compressed number will be that the first 4 bits will reduce by one from 1001 to 1000, and this cycle will repeat every octave. Thus the accuracy will cycle from 3 to 1.5% throughout each octave.

In general the number of bits required for the mantissa Nm is $\log_2(1/\text{error})$ and if the number of bits required by the counter is N, the number of bits required for the exponent Ne is $\log_2(N - Nm)$. Both logarithms need to be rounded up to the next whole number. Thus the total number of bits in the compressed number is Nm + Ne.

Low numbers will be coded at a higher accuracy that statistics require, as numbers up to 31 will be transmitted precisely whereas the statistical errors are $\sqrt{N}$. However maintaining constant accuracy for compression of large numbers and not providing greater accuracy than required for very low numbers requires more complex hardware than for the scheme outlined above, which only needs simple counters and SRs.

This type of compression technique only operates where we have a sequence of '1's or '0's at the beginning or ending of data words. If this is not the case a technique can be used in which each successive data word is compared with a library of expected patterns and the code specifying the identical pattern is transmitted. Thus for patterns with a high probability of usage the code would be short and for infrequent patterns the code would be longer resulting in data compression (Shannon, [1948], Fano, [1961], Huffman, [1952]). Morse code is an example of such a scheme. However we need an *a priori* set of patterns.

If such patterns are not known an adaptive scheme can be used which takes blocks of data and determines the relative frequency of repetitive patterns and assigns code on the above basis. This is a much more complex scheme and takes time to build up and transmit the library during which time the data is not compressed. However with careful use of such a system significant compression can be achieved. Variable length codes such as these suffer from problems if synchronisation is lost or data is corrupted. In such cases, the whole block of data may be lost.

4.4.5   *Histogramming*

Consider a particle energy analyser (Coates, [1985]) on a spinning spacecraft.
The information that is required about the particles is their energy and their direction
relative to a fixed vector on the spacecraft.

How is this done? If the total range of energies that can be accommodated by the
analyser is divide into 16 levels and for every spin of the spacecraft the energy
accepted by the analyser is changed, the total energy range will be accepted over 16
consecutive spins.

If one spin period is divided into 16 timed sectors, an angular resolution of
$360 / 16 = 22.5$ degrees is obtained. Thus data can be sequentially accumulated in an
array of 16 by 16 or 256 cells with each cell or memory address containing counts
associated with a particular energy and angle as shown in Fig. 4.34. If each cell can
accommodate up to 255 counts $= 8$ bits, the total array will take $256 \times 8 = 2048$ bits of
memory or 256 8 bit words. This is typical of a preprocessing scheme although more
than two dimensions are often required and the word length may be different.

The above technique incorporates queuing as the memory array can accept bursts of
data at a higher rate than the telemetry can accept. The queuing is accomplished by
duplicating the array rather than using FIFOs. While one array is being read to the
telemetry system, data is being accumulated in the second array and then the second
array will be read while the first accumulates data and so on.

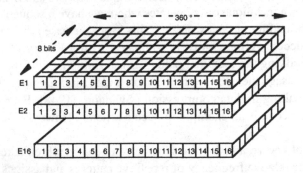

Fig. 4.34. Histogramming.

As well as the user subsystem's internal preprocessing it is clear that an overall
spacecraft processing scheme must also be applied to enable the data from all the
spacecraft's subsystems to be handled in an efficient manner. This is normally
accomplished by a process called time division multiplexing in which each
subsystem's data is regularly presented to the telemetry system for a limited period of
time. This process and the distribution of data and commands throughout the
spacecraft is accomplished by the On – Board Data Handling system or OBDH system.

4.4.6   On-board data handling system

This system is divided into two main channels; the telemetry channel which controls the transfer of data from the various subsystems to the telemetry encoder and the telecommand channel which distributes commands from the telecommand decoder to the users' subsystems. Although there are many types of OBDH systems, as they all basically serve the same functions we shall consider the operation of a type which is common to many European spacecraft. A block diagram of such a system is shown in Fig. 4.35.

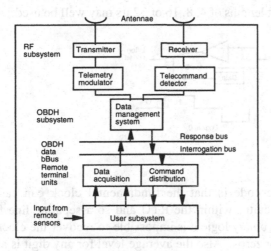

Fig. 4.35.   Spacecraft data paths.

Signals produced by all the subsystems in the spacecraft are introduced into the OBDH system by means of the OBDH bus. To enable data and commands to be handled simultaneously, the bus is a fully duplex digital network but it does not directly interface with the user subsystem. The link between the user and the OBDH bus is the Remote Terminal Unit (RTU) which is part of the OBDH system rather than part of the user's subsystem. Although each RTU's interface with a user subsystem is unique, its interface with the OBDH bus is identical. This removes the necessity of each user having to design and build his own interface to the vital OBDH bus. Each RTU has a unique address and so this technique identifies every subsystem.

The OBDH bus consists physically of two twisted pairs of wires; the Interrogation bus and the Response bus. When data is required from a particular subsystem, the Interrogation bus is used to transmit the address of the RTU associated with that subsystem. It also supplies the commands which enables that RTU to extract data from its user subsystem and transmit these data back along the Response bus. An RTU can only be activated when it has received its correct address from the interrogation bus.

The structure of the data on the interrogation bus is a continuous stream of 32 bit instructions. When no RTUs are required to be interrogated, a dummy 32 bit instruction, to a non–existent address, is generated. The 32 bits are divided into synchronisation bits, RTU address bits, data bits and an error check bit. The format of these data is 'Litton Code' which is a ternary code consisting of three distinct levels H, O and L where level O is midway between H and L. The method of generating Litton Code and the waveforms for logical '0' and '1' are shown in Fig. 4.36. There are always 4 transitions per bit interval, with a logic '1' being HOLO and a '0' being LOHO. The minimum length for a data bit is typically 2 µs with the basic clock frequency at 1 MHz although bit lengths of 4, 8, 16 or 32 µs may well be used.

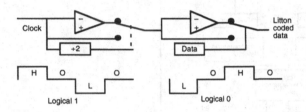

Fig. 4.36.   Litton code.

The reason for this strange code is that the fundamental clock can easily be recovered from the data by circuitry within the RTU and so a separate line for the clock is not required. In normal binary logic it is impossible to recover the clock from a continuous sequence of ones or zeros. Also the average level for any digit is always at the 'O' level and so baseline shift problems do not occur. The clock is regenerated by the simple process of data rectification as shown in Fig. 4.37, which explains why the interrogation bus is always kept active. Converting the Litton coded data back to binary logic is the reverse of its generation using the divided clock for phase comparison. See Fig. 4.38.

To ensure accurate interpretation of the code the first 4 bits of the 32 bit word are synchronisation bits which form an invalid Litton code (HOHOHOHOLOLOLOLO) and so are easily recognisable from the data.

The data transmitted by the RTU back to the OBDH is also Litton coded and is normally in 8 or 16 bit words, with 4 bits at the start to specify the destination address and a final check bit. All zeros for this address implies transmission directly to the data management system. Non zeros allows communication between RTUs. Thus a subsystem requiring attitude information to be embedded within its data can receive it from the attitude measurement subsystem. An RTU can receive, as well as transmit data via its response bus input, but this complexity has to be built into the associated RTUs.

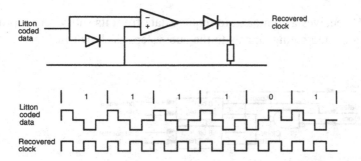

Fig. 4.37.   Clock recovery by rectification of Litton coded data.

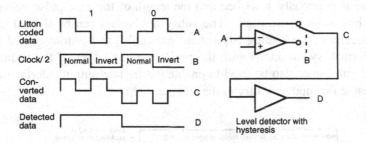

Fig. 4.38.   Recovery of binary data from Litton code.

A typical sequence of events at the OBDH bus RTU interface is shown in Fig. 4.39. The 64 bit delay is to allow time for the transmission of data from the subsystem to the RTU.

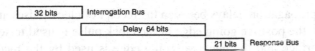

Fig. 4.39.   Timing sequence on OBDH bus.

4.4.7   Subsystem to RTU interface

Of the 32 bits arriving at an RTU, 4 are synchronization bits, the next 8 bits are the RTU address, and the final bit is an error check bit leaving 19 bits for the actual command. Depending on the pattern of these bits, the command may be a request for data from the subsystem or a command to change some operating mode within the subsystem.

When an RTU receives a command to request data from its subsystem it provides the acquisition clock and gating signals to the subsystem as shown in Figs. 4.40 and 4.41.

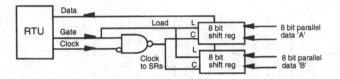

Fig. 4.40.   RTU to subsystem interface.

The clock, which is usually of equal mark to space ratio, is derived from the interrogation bus as previously described and the length of the gate pulse covers the number of data bits to be transmitted. The subsystem's data can be shifted into its output SRs on receipt of the leading edge of the gate pulse and is transmitted to the RTU in serial format, synchronous with the positive going edges of the acquisition clock. The gate pulse may also be used to enable the tristate output interface of the subsystem or enable the input circuitry to the telemetry.

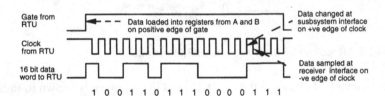

Fig. 4.41.   Data transfer waveforms.

To accommodate the propagation delays between the receipt of clock pulses and the transmission of data back, the positive going edge of the clock pulse is used to set the data onto the output interface and the negative going edge is used by the receiving interface to sample the data. Thus delays of up to half the period of the clock may be tolerated. Although the acquisition clock is generated continuously within the RTU, the power wasted driving the clock harness continuously could be significant and so the clock is usually gated to the subsystem, where it may well be gated again to improve noise immunity. The 21 data bits sent back along the response bus would typically consist of 4 address bits, 16 data bits and an error check bit. The 4 address bits would normally be 4 zeros indicating direct transmission to the DMS, but allowing for inter subsystem communication if that were designed into the system

The digital signals may be single ended or biphase, the latter allowing for significant common mode interference without degradation of the signals, and minimizing EMR. Terminating the lines with resistors equal to the characteristic impedance of the harnesses reduces problems due to reflections but consumes power. However as the timing of the signals' edges is usually the most significant feature, introducing a capacitor in series with the terminating resistor can reduce this power depending on the duty cycle of signal. A Receive/Transmit signal may be provided to the subsystem by the RTU to indicate whether the subsystem is required to receive commands or transmit data. This allows sequential transmission of commands and data within one gating pulse.

The receiving end of a biphase signal will be a level detector with in−built hysteresis designed to accommodate the expected common mode signal. This may be as high as ±5 V on a 5 V digital system. Adequate filtering at the receiver must ensure that noise spikes, possibly as wide as 20% of the expected signal, are not registered as valid signals. This filtering is one of the contributory causes to the propagation delays mentioned above. The subsystem to RTU interface must be made as electrically robust as possible with both voltage and current limits included. The highest reliability components should be used and the possibility of single point failures should be avoided where possible.

4.4.8   Data management system (DMS)

This subsystem is the heart of the OBDH. It generates the spacecraft clock, used for data and command synchronisation and as a clock, in the normal sense, for indicating precisely when data was received. It organises the collection and formatting of the data which is presented to the telemetry modulator and processes the commands received from the command detector so that they will be sent to the correct subsystem at the correct time.

The data from the RTUs are usually presented to the OBDH response bus as 8 or 16 bit words, with the first bit being the Most Significant Bit (MSB). The DMS arranges the sequence of data collection using Time Division Multiplexing. This is illustrated in Fig. 4.42 which shows words from the various subsystems A, B, C, D etc being grouped together to form a frame and groups of frames being linked together to produce the telemetry format. As the length of both the frame and format are fixed this is called a Fixed Frame Format Time Division Multiplex system.

The first two words in a frame S1 and S2, contain synchronising bit patterns so that the start of a frame can always be identified. The next word contains the frame counter which is incremented by one for each new frame so that the position of any frame within the format can be identified. Let us assume that the complexity of the spacecraft dictates that a 64 word frame is adequate. This would allow, in the simplest case, about 60 different subsystems access to the telemetry, provided only one word per subsystem per frame was an adequate data rate for all the subsystems.

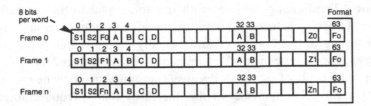

Fig. 4.42.   Fixed frame format telemetry system.

To accommodate subsystems which require more or less than this basic rate, the processes of super and sub commutation can be introduced. As an example Fig. 4.42 shows subsystems A and B being super commutated at two words per frame. Sub commutation is provided for subsystems which require low sample rates and particularly for reading housekeeping data. This is carried out by multiplexing a different subsystem or sensor output to the same word position in each consecutive frame. Thus in Fig. 4.42, with $n = 127$, word 61 could be multiplexed to 128 different data inputs so that each input is sampled once per format rather than once per frame. To complete the identification process, the last word in each frame is the format count which is incremented at the start of each new format.

From a knowledge of the position of the subsystems' data in the frame, the frame and format counts, the identification and time of receipt of that data is unique so long as the format counter does not recycle. With a data rate of 8 kbits / s, each basic subsystem would be sampled once per frame $(8 \times 64 / 8k = 64 \, ms)$ and one format would take $128 \times 64 \, ms = 8.2 \, s$ to transmit. Thus the format counter would recycle every $8.2 \times 256 = 35$ minutes. This would not be adequate for a LEO spacecraft recording data on tape during eclipse periods, if telemetry contact between the spacecraft and its receiving station did not occur every orbit. Reserving 12 bits for the format counter rather than the 8 would allow it to recycle every 9.3 hours which would be more than adequate. However it is now common practice to require no ambiguity, even in archived data, for the whole duration of the mission requiring recycle times of decades .

If the DMS clock which is normally a very precise oscillator, is provided with say a 32 or 40 bit counter then an absolute time code can also be included within the format.

4.4.9   Packet telemetry

In 1984 the European Space Agency's (ESA) Consultative Committee for Space Data Systems recommended as an alternative to fixed frame format telemetry, a Packet telemetry system which would not have the rigid structure of the old system. This was suggested to overcome the limitations of requiring the receiving station, set up for a particular frame format combination, to be manually reset to receive data from another spacecraft with a different combination.

Thus the telemetry packet would contain, apart from the subsystem data, a precise description of how the data was formatted. This information would be contained in a Header frame H which would be followed by the subsystem's data of variable length and finally an optional error detection code E as shown in Fig. 4.43.

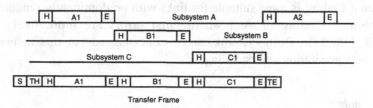

Fig. 4.43. Packet telemetry.

Table 4.1. *Contents of packet*

Function	No data bits	Description of function
Version No	4	Version of packet
Secondary header flag	1	Shows if secondary header is used
Subsystem identifier	11	Includes flag to indicate use of packet error control
Packet structure	2	Indicates type of commutation
Source sequence count	14	Associates this packet with other packets from the same subsystem
Packet length	16	
Optional secondary header		Variable - could contain spacecraft clock
An, Bn	X	Subsystem data of variable length
E	16	Optional packet error check

The packets from the various subsystems are embedded within a transfer frame by the DMS for transmission to the telemetry encoder. For standardization of telemetry error detection codes, it was suggested that the frame length be 8920 bits. The embedding of packets within a frame may be such that several packets from one subsystem occur within one frame or one packet may be time multiplexed into several frames. Thus the transfer frame has to contain information about how the packets are distributed within it. Details of this structure for ESA spacecraft can be found in ESA

PSS – 04 – 106, [1988] but essentially it starts with a block of synchronization bits S, followed by the Transfer Frame Header TH which contains the transfer frame version number, spacecraft identification, information on size of subsystem data packets, number of packets and information on error detection codes. This is then followed by the subsystems' packets and finally terminates with the error code TE as shown in Fig. 4.43. The information to be provided in the packet by the DMS is shown in Table 4.1.

Convolutional Coding is most suitable for links with predominantly random noise and Reed Solomon Coding is used where burst errors are more likely (ESA PSS – 04 – 103, [1989]). Both systems can be concatenated but this incurs more complexity in both generation and decoding.

4.4.10 Commands

Command transmission

Commands will normally originate from a ground–station where they start life as a standard binary digitally encoded signal. A typical command packet will consist of a header, describing the command and its destination, the actual command and optional error check bits. Precise details of the ESA packet structure can be found in ESA PSS – 04 – 107, [1992]. A simple mode change command will probably be smaller than the telecommand frame length specified for the spacecraft and so several packets from the same subsystem or different subsystems can be multiplexed into the same command frame. If the command packet is a complex software patch, it may well be kilobytes long and so have to be segmented between several command frames. Examples of these two processes are shown in Fig. 4.44. In the second case the segment headers S1 and S2 which describe how the command is segmented, have to be included within the length of the telecommand frame. Each telecommand frame is now embedded between a frame header (FH) and a frame error control (FEC) word using cyclic redundancy coding, to form a frame block.

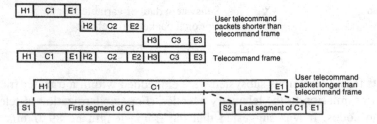

Fig. 4.44. Examples of telecommand frames.

To improve reliability still further, each frame block is broken down into codeblocks, as shown in Fig. 4.45, where a codeblock consists typically of 7 bytes of data followed by one byte of error correcting code. This is typically generated using the cyclic block code system of which a subset known as Bose, Chaudhuri and Hocquenghem (BCH) coding is commonly used (Hocquenghem, [1959]; Bose and Chaudhuri, [1960]). Such coding systems will be discussed in the next section.

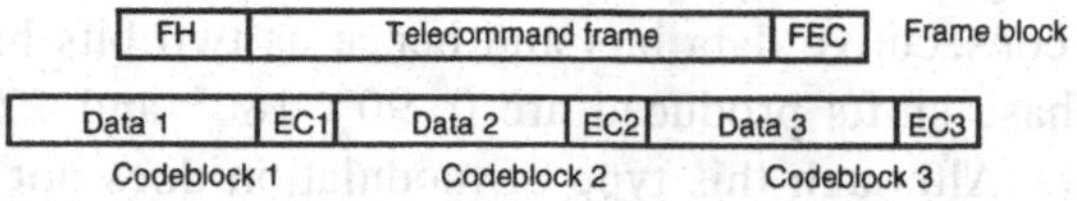

| FH | Telecommand frame | FEC | Frame block |

| Data 1 | EC1 | Data 2 | EC2 | Data 3 | EC3 |
| Codeblock 1 | | Codeblock 2 | | Codeblock 3 | |

Fig. 4.45.   Formation of codeblocks.

To enable synchronisation to be established the frame blocks are preceded by a known regular pattern of bits called the acquisition sequence. Between each frameblock, when several are being transmitted sequentially, an idle sequence of bits is generated to maintain synchronisation. These frame blocks plus their synchronising signals are now ready to be converted into a form suitable to up link to the spacecraft.

As the sharp edges of digital signals imply an unnecessarily wide bandwidth, the first process is to convert the signal into a more 'sinewave' like form. The common choice of conversion for spacecraft communication systems is Phase Shift Keying (PSK) where, in the simplest form, a logic '1' is defined as a phase change resulting in a positive slope and a logic '0' as a phase change resulting in a negative slope. The frequency of the sinusoidal subcarrier on which the modulation is performed is a harmonic of the command clock and is typically 8 or 16 kHz for the up link telecommands. An example of such modulation is shown in Fig. 4.46 together with the generation process. Finally the PSK signal is used to phase modulate the Radio Frequency (RF) link. The RF frequencies used with most LEO ESA spacecraft for both telemetry and telecommand are in the range 2 to 2.3 GHz.

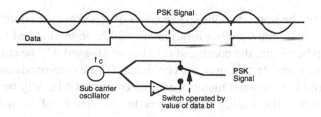

Fig. 4.46.   Phase shift keying.

One of the properties of PSK encoding, similar to that of Litton coding, is that by the process of rectification, filtering and division the original command clock subcarrier waveform can be recovered. This can then be used for phase comparison and by means of a simple product detector, to reconstitute the original data. As such a scheme will inherently have a 180° phase ambiguity, this defines the requirement for the acquisition and idle synchronising sequences. Other advantages of this type of modulation are the efficient use of bandwidth and the formation of a symmetrical signal with no DC component.

A variation of PSK called quadrature PSK (Spilker, [1977]) is in common use as it enables twice the data rate for a given RF bandwidth. The phase shift in this type of modulation is defined by two consecutive data bits and hence as two bits have four possible combinations, the phase shifts produced are 0, 90°, 180° and 270° with respect to the reference carrier. Although this type of modulation does not produce such a good S/N at the detector, for a given RF noise, as basic PSK, the higher through−put more than compensates for this disadvantage.

Command reception

At the start of a command signal from a ground station, the RF signal with no data or subcarrier modulation is frequency swept around the nominal centre frequency with a sufficient swing to ensure that the spacecraft's frequency lock system will be efficiently activated, allowing for all possible drifts and doppler shifts. Once the spacecraft's frequency is locked and the lock status has been transmitted back to the ground station, the up link frequency is brought back to nominal. A similar process occurs at the start of the telemetry down link.

Once the RF link has been established, the modulated signal will be transmitted and this will be demodulated reproducing the PSK signal. This will then be decoded using the techniques described above, to restore the original code blocks within each frame block. If an error is detected within any code block the subsequent data within the frame will be discarded and a request initiated for the retransmission of the corrupt frame. If no errors are seen, the original command packets can be reconstructed and if the header data is correct, the command with its address will be put onto the OBDH bus.

As commands can only be sent when the spacecraft is over a ground station, if they need to be executed at a later time, timing information will have to attached to these commands. A flag to indicate that the command is to be time tagged will be contained in the header and also the absolute timing of the command. These commands and their timing data will be stored in command memory and the command will only be output onto the OBDH bus when the timing tag equates to the spacecraft's internally generated time code.

There are two types of commands that are sent to subsystems; High Power On – Off commands and the Memory Load command.

High power on – off commands

This type of command provides sufficient power to latch and unlatch relays without the use of the subsystem's own power. Their primary use is to switch power on to a subsystem or to operate systems prior to the normal power up sequence. A typical pulse would be 15 ms long, 15 V amplitude and capable of supplying 100 mA. As a latching relay is the typical recipient of such a command, two commands will normally be available along two pairs of wires; the latch and unlatch commands. Although these commands could be provided by the RTU, as voltages in excess of those required by normal digital logic are required, they would normally be generated in the power distribution unit on request from the DMS.

Memory load (ML) commands

For a simple mode change in a subsystem, the information provided by the telecommand packet will include the address of the subsystem's RTU and the command data. The latter will typically be 19 bits, as previously described, which can be considered as 3 local address bits and 16 data bits. As one combination of the 3 local address bits has been designated as the data request code, the remaining combinations can address 7 locations within the user's subsystem with 16 bit command words. However the 16 bit command word could also be treated as a 4 bit address (16 locations) with 12 bits of data thus providing $7 \times 16 = 112$ locations which should be adequate for most subsystem's requirements.

The RTU will usually provide the user subsystem with the command in serial form together with the ML clock, gate and address on four twisted pairs of wires. Within the subsystem the gate enables the clock and commands to all its SRs whereas the decoded ML address only enables the required SR. The ML address could be decoded within the RTU and several command lines supplied to the subsystem but this would be inefficient in terms of the additional harness required for these command lines unless only one or two commands were required. To prevent spurious command generation as the command is being received, the trailing edge of the gate can be used to initiate the received command.

Fig. 4.47 shows the 16 bit command word, clock, gate and 3 bit memory load address supplied to the user subsystem from an RTU. In this example one decoded address G1 enables the command to be loaded into a SR where it can be used for discrete commands. A second address G2 is used to set up a DAC to produce an analogue output which might be used to set the level of a programmable power supply. If the command packet is required to modify some software within the subsystem, many 16 bit words may be transmitted after each RTU address. To facilitate reception and processing by the subsystem, the data, clock and gate waveforms will normally be presented in a burst mode as shown in Fig. 4.48. The idle period simplifies the timing procedures for shifting the data from the input SRs to the subsystem's internal data handling system. For more details see ESA TTC − B − 01, [1979].

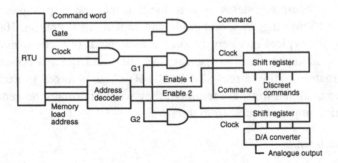

Fig. 4.47.   Command reception and processing.

Fig. 4.48.   Burst mode protocol.

4.4.11   *Error detection*

Introduction

As the transmitter, in common with all other spacecraft subsystems, is power limited, the quality of the signal received at the ground station via the noisy RF link will frequently be such that the demodulated data will contain erroneous data. This is particularly true when the spacecraft is close to the ground station's horizon. Although power limitation is not a problem with the ground based transmitter, the requirement for accuracy in the reception of commands is also very important. Thus it is desirable that all telemetered data are formulated in such a way that it is possible to detect when errors occur.

There are a variety of techniques used on spacecraft systems for error detection. These include linear code blocks, cyclic redundancy checks and convolutional error checking systems. The basic operation of an error checking system is to add redundant bits to the data stream before transmission over the noisy link so that a combination of the redundant bits and the original data forms a known pattern which is independent of the data. Thus analysing the data and check bits at the receiving end of the noisy link can enable corrupted data to be rejected if the known pattern is not reformed. Systems that not only error detect but can specify which bits have been corrupted are called Error Correcting systems. It is possible to design a system which could correct every error in the data stream, but the number of additional redundant bits would make the ratio of useful data to total data too low for efficient use of the communication link.

Block error detection systems

The simplest system allows a single error generated within a block of $k+1$ bits to be detected by adding a single parity or check bit onto the end of the k data bits. The value of the parity bit is given by the modulo -2 sum of the k data bits, which makes the modulo -2 sum of the $k+1$ bits in the codeword zero.

k data bits	parity bit	codeword	modulo -2 sum of codeword
0101101	0	01011010	0
0101010	1	01010101	0

If a single error occurs, the modulo -2 sum of the codeword will no longer be zero and the error will have been detected. However if two errors occur no error will be indicated. In general this system will show errors if the number of errors is odd but not if the number is even. Although this does not seem very advantageous at first glance, one has to consider how significant is the probability of two errors occurring in one block of data.

In a block of data n bits long the probability $P_e(r)$ of r bits being in error is given by the binomial probability function:

$$P_e(r) = {}^nC_r p^r (1-p)^{n-r}$$

where p is the probability of a single bit being erroneous. In the above example, the probability of an undetected error occurring in these 8 bits requires two errors to occur, making $r=2$. The probability of this occurring is given by

$$P_e(2) = {}^8C_2 p^2 (1-p)^{8-2} = 28p^2 (1-p)^6$$

As p is small the probability of 4 errors occurring can be neglected with respect to the two error case. For most spacecraft communication networks, a bit error rate (BER) of more than 1 in 10^5 would be unacceptable, making p less than 10^{-5}.

Without parity the probability of an undetected error in a 7 bit word is given by the probability of one error occurring in 7 bits which is

$${}^7C_1 p (1-p)^6 = 7p(1-p)^6$$

The improvement in error detection is $7p/28p^2 = 1/4p$, which is 25,000 for $p = 10^{-5}$. This great improvement is very simple to provide as the value of the parity bit can be produced electronically using 'Exclusive Or' (EOR) gates. Error detection is performed in a similar manner on the n bit codeword which will produce a zero for correct data (or for an even number of errors).

Linear block codes

Instead of a single check bit per word or block of bits, we can transmit k data bits plus c check bits to form an n bit code block where $n = k + c$. The n bit block can be generated by multiplying the k data bits by a generator matrix G which is chosen to reproduce the data plus the check bits. These are called n, k Hamming Codes (Hamming, [1950]). The data efficiency of such a system is defined by k/n. For example, let the k data bits be 111 and the G matrix be defined by

$$G = \begin{bmatrix} 1 & 0 & 0 & 1 \\ 0 & 1 & 0 & 1 \\ 0 & 0 & 1 & 1 \end{bmatrix}$$

The n bit code block (T) to be transmitted is formed, using modulo–2 arithmetic, from $T = [D][G]$ where D is the original data.

$$T = \begin{bmatrix} 1 & 1 & 1 \end{bmatrix} \begin{bmatrix} 1 & 0 & 0 & 1 \\ 0 & 1 & 0 & 1 \\ 0 & 0 & 1 & 1 \end{bmatrix} = \begin{bmatrix} 1 & 1 & 1 & 1 \end{bmatrix}$$

which is a single bit parity system or a 4, 3 Hamming code. The only rules for generating the encoding matrix are that it must have k rows and n columns to enable the matrix multiplication to produce an n bit result and the first k columns must be able to produce the k data bits. Thus the encoding matrix G can be divided up into two sections, the Identifier Matrix I_k and the Parity matrix P, that is $G = [I_k P] n$, k. The I_k matrix is always a diagonal of ones in order to regenerate the original data and the P matrix is decided empirically on the usefulness of the results it produces. A typical encoding matrix G for an 8, 4 Hamming code ($n = 8$, $k = 4$) is shown below.

$$G = \begin{bmatrix} \overset{I_k}{} & & & & \overset{P}{} & & & \\ 1 & 0 & 0 & 0 & m_0 & n_0 & o_0 & p_0 \\ 0 & 1 & 0 & 0 & m_1 & n_1 & o_1 & p_1 \\ 0 & 0 & 1 & 0 & m_2 & n_2 & o_2 & p_2 \\ 0 & 0 & 0 & 1 & m_3 & n_3 & o_3 & p_3 \end{bmatrix}$$

This is called a block coding system as each data word has a unique set of check bits.

Let us now consider how to implement a 8, 4 Hamming Code in hardware. As the coding reproduces the original data we only have to produce the check bits. If the original data are *a b c d* the check bits are given by

$$[a \quad b \quad c \quad d] \begin{bmatrix} m_0 & n_0 & o_0 & p_0 \\ m_1 & n_1 & o_1 & p_1 \\ m_2 & n_2 & o_2 & p_2 \\ m_3 & n_3 & o_3 & p_3 \end{bmatrix}$$

$$= (a.m_0 \oplus b.m_1 \oplus c.m_2 \oplus d.m_3), (a.n_0 \oplus b.n_1 \oplus c.n_2 \oplus d.n_3),$$
$$(a.o_0 \oplus b.o_1 \oplus c.o_2 \oplus d.o_3), (a.p_0 \oplus b.p_1 \oplus c.p_2 \oplus d.p_3)$$

If we let the parity matrix be defined by,

$$P = \begin{bmatrix} 1 & 1 & 1 & 0 \\ 1 & 1 & 0 & 1 \\ 1 & 0 & 1 & 1 \\ 0 & 1 & 1 & 1 \end{bmatrix}$$

the check bits will become $(a \oplus b \oplus c)$, $(a \oplus b \oplus d)$, $(a \oplus c \oplus d)$, $(b \oplus c \oplus d)$ which can be formed with six EOR gates. If we let the original 4 data bits be 1010, the check bits become 0101 and the transmitted code word 10100101.

We need a method of determining whether the recovered *n* bits are erroneous and this requires a parity check matrix *H* which is defined by

$$H = [P_T : I_{n-k}]$$

P_T is the transposition of *P* obtained by interchanging its rows and columns, and I_{n-k} is an identity matrix for the check bits of the encoded data. For 4 check bits $n - k = 4$.

$$I_{n-k} = \begin{bmatrix} 1 & 0 & 0 & 0 \\ 0 & 1 & 0 & 0 \\ 0 & 0 & 1 & 0 \\ 0 & 0 & 0 & 1 \end{bmatrix}$$

Multiplying the received encoded data *R* by the transposed parity check matrix H_T gives the error syndrome S.

Using the specified parity matrix, the parity check matrix is given by

$$H = \begin{bmatrix} 1 & 1 & 1 & 0 & 1 & 0 & 0 & 0 \\ 1 & 1 & 0 & 1 & 0 & 1 & 0 & 0 \\ 1 & 0 & 1 & 1 & 0 & 0 & 1 & 0 \\ 0 & 1 & 1 & 1 & 0 & 0 & 0 & 1 \end{bmatrix}$$

and the transposed check matrix by

$$H_T = \begin{bmatrix} 1 & 1 & 1 & 0 \\ 1 & 1 & 0 & 1 \\ 1 & 0 & 1 & 1 \\ 0 & 1 & 1 & 1 \\ 1 & 0 & 0 & 0 \\ 0 & 1 & 0 & 0 \\ 0 & 0 & 1 & 0 \\ 0 & 0 & 0 & 1 \end{bmatrix}$$

From the laws of matrix multiplication $[T][H_T] = S = 0$. If an error is introduced into the transmitted data T so that the received data $R \neq T$ then $S = [R][H_T] \neq 0$ which indicates the presence of the error. S can easily be produced in hardware using simple gates. However if S does equal zero, it does not prove we have an error free transmission due to the possible combination of multiple errors. Calculating the syndrome;

$[R][H_T]$ gives us $S = [10100101][H_T] = [0000]$ as expected.

If $R = 11100101$, produced by an error in the second bit, $S = [1101]$ correctly indicating an error.

If $R = 11110101$ produced by two errors, $S = [1010]$.

If $R = 11111101$ produced by three errors, $S = [0010]$.

If $R = 11111111$ produced by four errors, $S = [0000]$ erroneously indicating an error free result.

In general this Hamming code can always detect 3 errors but may not detect 4. If the code not only detects the error but locates where it occurred the error can be corrected. From the single error example we can see that $S = 1011$ corresponds to the second row of the H_T matrix and the error does occur in the second bit of the received word R.

For single errors in an 8,4 Hamming code, the S matrix always indicates the position of the erroneous bit as shown above. Thus more than one error is necessary to produce an erroneous word as the first error is correctable, and 4 bit errors are required for the errors to be undetected. The probability of an undetected error, produced by 4 errors, occurring in 8 bits is $^8C_4\, p^4$ compared with $4p$ for an undetected error produced by a single error in a 4 bit word without an error correcting system. This gives an improvement of $4/70p^3$ which is approximately 5×10^{13} for a bit error rate of 1 in 10^5.

The maximum number of valid data words that can be formed with k data bits is 2^k and this is also the maximum number of valid codewords as they are just formed by adding check bits onto the valid data words. However as there are $n = k + c$ bits in the codeword the maximum possible number of codewords would be 2^n. This indicates that there are $2^n - 2^k$ invalid codewords. It is the ability of errors to generate invalid codewords which enables error detection to be carried out. Thus a significant property of the Linear Code Block is the 'Hamming Distance' (d) which is the number of bits between any two valid codewords (T), on a corresponding bit for bit basis, which are different. d_{min} is the smallest distance between any two valid codewords in all possible valid codewords, for a given order code, and this parameter defines how many errors a given code can detect or correct. As long as the maximum number of errors e is less than d_{min}, an invalid codeword will be formed. Thus $d_{min} - 1$ errors can always be detected. Hence the usefulness of the parity matrix can be assessed by determining which pattern produces the highest value for d_{min}.

If $d_{min} - 1$ is greater than or equal to twice the number of errors e that can occur then all those errors can be corrected. The argument for this statement is that if $d_{min} - 1 \geq 2e$, the erroneous codeword will always be closer to the one from which it originated than to any other one. For example the maximum number of valid codewords for the 8, 4 Hamming code is $2^4 = 16$, the maximum number of all codewords including the invalid ones is $2^8 = 256$ and the value for d_{min} is 4. This can be verified from Table 4.2 which shows the coded data T for all possible values of a four bit data word D using the specified encoding matrix G. Thus for this system, the number of errors per codeword that can be detected is $4 - 1 = 3$ and and the number of errors that can be corrected is $\leq 3/2 = 1$.

A simpler method of deducing d_{min} arises because it equals the minimum 'weight' of the Hamming code specified by the minimum number of '1's in any valid codeword from the total set (excluding the all '0's case). This can be deduced from Table 4.2 to be 4.

As can be seen from these examples the number of check bits that have to be added to the original data, so that more than one error can be detected, becomes a significant fraction of the transmitted block and hence reduces the efficiency of data transmission. To improve this factor, the ratio of n to c must be made much larger but this considerably increases the number of gates required to implement the coding and checking procedures. A simpler set of codes which have overcome this problem and which are fairly easy to implement in hardware are the cyclic block codes.

Table 4.2. *An 8, 4 Hamming Code*

D	G	T
0 0 0 0	1 0 0 0 1 1 1 0	0 0 0 0 0 0 0 0
0 0 0 1	0 1 0 0 1 1 0 1	0 0 0 1 0 1 1 1
0 0 1 0	0 0 1 0 1 0 1 1	0 0 1 0 1 0 1 1
0 0 1 1	0 0 0 1 0 1 1 1	0 0 1 1 1 1 0 0
0 1 0 0		0 1 0 0 1 1 0 1
0 1 0 1		0 1 0 1 1 0 1 0
0 1 1 0		0 1 1 0 0 1 1 0
0 1 1 1		0 1 1 1 0 0 0 1
1 0 0 0		1 0 0 0 1 1 1 0
1 0 0 1		1 0 0 1 1 0 0 1
1 0 1 0		1 0 1 0 0 1 0 1
1 0 1 1		1 0 1 1 0 0 1 0
1 1 0 0		1 1 0 0 0 0 1 1
1 1 0 1		1 1 0 1 0 1 0 0
1 1 1 0		1 1 1 0 1 0 0 0
1 1 1 1		1 1 1 1 1 1 1 1

Cyclic block codes

The name cyclic is given to these codes because if the combination of data and parity bits in any one codeword is given a logical rotation of one or more bits, another valid codeword is formed. For example if 1001011 is a valid codeword, then 0010111 is another valid codeword which can be derived from the same code generator but a different data word. The coding also uses modulo – 2 arithmetic whose logic is defined by the EOR function. Consider a 4 bit data word D ($k=4$) with 3 check bits ($n=7$). If we call the code generator G, the transmitted codeword T is given by $T = D \times G$. To make n equal 7, G must be $n - k + 1 = 4$ bits long. Let $D = 0101$ and $G = 1101$. Multiplying D by G gives the following result.

$$
\begin{array}{r}
0\ 1\ 0\ 1 \\
\underline{1\ 1\ 0\ 1} \\
0\ 1\ 0\ 1 \\
0\ 0\ 0\ 0 \\
0\ 1\ 0\ 1 \\
\underline{0\ 1\ 0\ 1\ \text{-}\ \text{-}\ \text{-}} \\
T = \quad 0\ 1\ 1\ 1\ 0\ 0\ 1
\end{array}
$$

The data is decoded by dividing T by G using modulo – 2 division. If any errors arise in the transmission of the coded word T then the received codeword R when divided by G will leave a remainder. Although this technique is adequate, it is not systematic as T does not contain D explicitly. A systematic cyclic block code can be achieved in the following manner. D is left shifted by $n-k$ bits and this shifted data is divided by G to produce a remainder R. R is then added onto the shifted data to form the codeword T. As an example with $D = 0101, n = 7$ and $k = 4$, the shifted data is 0101000 and R is formed by dividing this by G as shown below.

$$
\begin{array}{r}
1\ 1\ 0 \\
1\ 1\ 0\ 1\ )\ 0\ 1\ 0\ 1\ 0\ 0\ 0 \\
\underline{1\ 1\ 0\ 1} \\
1\ 1\ 1\ 0 \\
\underline{1\ 1\ 0\ 1} \\
1\ 1\ 0
\end{array}
$$

Thus the remainder $R = 110$ and the codeword $T = 0101000 + 110 = 0101110$ which contains the original data explicitly. T must be divisible by 1101 (G) as we have effectively removed the remainder from the original shifted data (addition and subtraction are the same in modulo – 2 arithmetic). D can now be extracted by three right shifts and division of T by G will give no remainder if no transmission errors have occurred, subject to the normal provision of multiple errors invalidating the error detecting system. However the only way a valid result can be indicated when it does include errors requires the errors to have been added in multiples of G. By the choice of G, the probability of this can be made very small. As before G is determined empirically.

Single errors will produce a received codeword R which can be corrected by changing R a bit at at time until a valid codeword is produced. This requires a maximum of 7 changes. If no valid codeword is produced within 7 changes, then more than one error has occurred and although these cannot be corrected the data is known to be invalid.

Cyclic Block Codes can be generated using simple gates and SRs. For example Fig. 4.49 shows a standard dividing block for $G = 1101$, with Fig. 4.50 showing how this is implemented electronically.

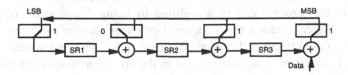

Fig. 4.49. Block diagram of basic 1101 divider.

The system encodes in the following manner. When the Code Control Line (*CCL*) is zero, the EOR gates act as buffers and so SR1, SR2 and SR3 function as sequential SRs, and their contents are shifted into the output SR.

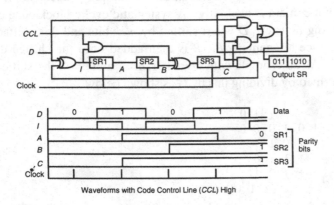

Fig. 4.50.   Cyclic block code generation.

When *CCL* is high, the output of SR3 is inhibited from entering the output SR, the data *D* is routed directly into the output SR and the EOR gates function normally. Thus with *CCL* held high for 4 clock pulses the data *D* is fed into the output SR and also into the 3 coding SRs modified by the 3 EOR gates. At the end of these clock pulses, the 3 SRs will contain the remainder bits associated with *D* shifted by 3 bits and divide by 1101.

CCL is then taken low, and 3 more clock pulses shift the data from the 3 coding SRs into the output SR, which will then contain the data *D* plus the three check bits. The waveforms in Fig. 4.50 show the sequence of events when *CCL* is high, the data *D* is 0101, the *MSB* being to the left and clocked in first.

Assuming only single bit errors, an extended version of the above circuitry can also provide error correction. If *CCL* is held high and the 7 bit encoded word *D* is shifted in with 7 clock pulses, the 3 coding SRs will contain all zeros unless the coded word is in error. Thus when an erroneous word is received (with only one error), if one bit is changed and the modified word is shifted in again at D and now produces zeros in the 3 coding SRs, that bit must have been the erroneous one. This process only needs to be repeated 7 times (maximum) to discover and hence rectify the error.

The circuitry for this process can be simplified by using the property of cyclic codes, namely that a logical rotation of a codeword produces another valid codeword. If the coding SRs are ORed together the output of this gate will show a 1 when an error is present and this 1 can be used to change a bit in the erroneous codeword until the coding SRs become zero again.

Fig. 4.51 shows the encoding network plus a data buffer which contains the received encoded data and the error correcting circuitry.

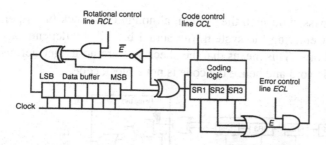

Fig. 4.51. Cyclic block code error correction network.

Assuming there is an error the first time the data in the data buffer is shifted into the coding network, the 3 input OR gate will make E a 1. With the *RCL* low and the *ECL* high for the first pulse of the next 7 clock pulses, the EOR gate inverts the *MSB* of the data as it is shifted into the encoding network again. If E now becomes a zero, that will indicate that the *MSB* was the erroneous bit. If E is still 1, a 1 bit rotation is carried out in the data buffer so that MSB - 1 will be inverted when data is shifted into the coding network. This process is repeated until E remains zero showing that the error is in the right hand bin of the data buffer. As $\bar{E}$ is now high the next single bit rotation will invert the erroneous bit in the data buffer. Seven rotations are necessary to restore the data in the data buffer to its correct sequence.

Codewords with $n \gg 7$ are more usual and subsets defined by Bose, Chaudhuri and Hocqenghem known as BCH codes, with N around 1000 are commonly used. The basic properties are similar to those already discussed but the data transmission efficiency is much higher as only about 10% of the codeword is used for the error detection bits. A specific subset of the BCH codes called the Reed Solomon code is commonly used in ESA telemetry systems. Details of this code are given in ESA PSS $-04-106$, [1988]. For further reading on this topic, consult Lin & Costello, [1983] and Peterson & Weldon, [1972].

Convolution encoders

One other type of error correcting network in common use in spacecraft systems is the Convolution Encoder Viterbi, [1971]; Forney, [1973]; Ristenbatt, [1973]; Spilker, [1977]. Check bits are formed sequentially as the data stream is shifted through the encoder and typically each data bit will be separated by a check bit. A typical coding system is shown in Fig. 4.52 with the clock controlling the switch between the data and check bits.

At the receiver, the data and check bits are separated, new check bits are generated from the received data bits and these are compared with the received check bits to verify the data and check bits. This is illustrated in Fig. 4.53. Errors will change E to a 1 and this will make F invert the erroneous bit to correct the data at I. Note that once 4 data bits have passed through the system, although one check bit is generated for every new data bit entering the system, the actual bit will be dependent on the pattern of the current 4 bits. This means that the check bits are no longer defined by a single independent codeword and hence the code is not a block code.

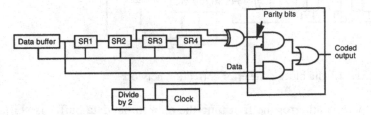

Fig. 4.52.   Convolution encoder.

In all decoding systems the decision to use the error correcting ability of the code or its error detecting ability will depend on the expected number of errors, their distribution and how vital it is to correctly interpret the data. There will always be the possibility that the errors are so distributed that the correcting network will introduce another error which exceeds the detection capability of the system and so incorrectly indicates that the data is valid.

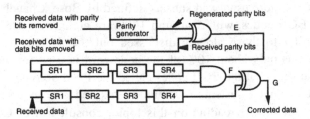

Fig. 4.53.   Convolution encoder error correction network.

4.5   Power systems

4.5.1   *Primary power sources*

The power to mass ratio for most scientific spacecraft is usually between 0.5 and 1.0 watt per kilogram. Examples from several spacecraft and their payloads are shown in Table 4.3.

This is a useful guide to determine whether an initial power and mass estimate is reasonable.

Table 4.3. *Power to mass ratios*

Spacecraft	Mass kg	Array Power W	Power to Mass ratio W/kg	Launch Date
HEOS A1	108	55	0.5	1968
TD 1A	476	330	0.7	1972
GEOS	575	124	0.2	1977
Ulysses	350	280	0.8	1990
ERS1	2400	2500	1.0	1991
SOHO	1350	750	0.6	1995
Cluster	550	250	0.5	1996

The power density for a typical signal processing printed circuit board (PCB) is usually about 0.75 mW cm^{-2}. Power consumption of subsystems will range from a few watts to a few hundred watts. This power will have to be provided throughout the spacecraft's life which may be up to ten years.

Primary sources of power are solar cells, batteries and radioisotopes. The majority of long – term Earth orbiting spacecraft use solar cells as the primary power source with rechargeable batteries providing power during eclipse periods and peak power requirements. Short– term missions can use batteries alone.

The flux of solar radiation through an area normal to the Sun Earth vector at an altitude of a few hundred km above the Earth's surface, where atmospheric attenuation is negligible, is 1370 Wm^{-2}, with the cosine law applying at other angles of incidence. Around Jupiter this has dropped to about 50 Wm^{-2}. Thus deep space probes which need to operate where the solar flux is very low, may need radioactive sources to generate electrical power from Seebeck effect devices (Radio–isotope Thermoelectric Generators RTGs), or carry auxiliary batteries which are only used for short periods of time during the vital parts of the mission. These auxiliary batteries can be primary cells such as those derived from lithium which have a high energy capacity and very low self discharge characteristics.

Radioactive power sources present problems both during pre–launch operations, and at the End Of Life (EOL) of the spacecraft. They have to undergo all the spacecraft environmental tests which presents risks to test personnel and as the duration of these tests is often several months, the source is already significantly depleted at launch. At the EOL of the spacecraft there is the possibility of radioactive material being returned to Earth in an unknown manner depending on whether the spacecraft completely disintegrates on re–entry on not.

As the electrical power generating density of an RTG, using a radio isotope with a half life of about twenty years, is around 5 Wkg^{-1}, it only becomes viable for orbits where the generating power density of a solar array has dropped to a similar value. This occurs for orbits extending to Jupiter and beyond. However for such orbits, relative to solar arrays, there are additional advantages to RTGs due to their insensitivity to attitude, external particle and photon radiation and eclipses.

4.5.2   Solar cells

Various materials derived from silicon, germanium, gallium and indium exhibit photovoltaic effects, but the techniques for doping silicon, developed by the semiconductor industry, have been advanced so much more than for the other materials, that, until recently, the performance of silicon cells per unit cost exceeded that of all other types. Thus Silicon (Si) has been used almost exclusively for solar cells on current spacecraft. However other materials such as gallium arsenide and indium phosphide are being developed due to several advantages they have over silicon.

Gallium Arsenide (GaAs) is less susceptible to damage caused by high energy photon and particle radiations particularly when cover glasses are used (Weinberg, [1990]). It produces a higher voltage per cell than silicon, with an open circuit output voltage of about 1 V as opposed to 0.6 V for silicon. Thus fewer cells are required to produce a given output voltage. At room temperature gallium arsenide is more efficient than silicon at converting solar flux to electrical power and this differential increases as the temperature rises. This can be an important feature as solar cells only convert about 15% of the solar power incident on them to electrical power. Most of the remaining 85% is converted to heat which can result in their operating temperature being significantly higher than that of the spacecraft depending on the degree of thermal coupling between the array and the spacecraft.

The historical disadvantage of gallium arsenide was cost. On a cell for cell basis, the cost is still several times that of silicon but due to the complexity of producing the arrays from the cells, the cost ratio, on an array for array basis, is nearer 1.5. As the power production per unit area for GaAs is about 30% higher than from Si cells, the cost per watt for GaAs is now now approaching that of Si (Khemthong *et al.*, [1990]). However there are still technical disadvantages associated with the use of GaAs. It is difficult to produce a backing material for GaAs which has a matching temperature coefficient of expansion although the use of flexible adhesives has somewhat reduced this problem. The fragile nature of GaAs necessitates the cell being made considerably thicker than its silicon equivalent and this has a considerable impact on the power output per unit mass. This problem may be solved by the use of germanium as a substrate for the GaAs as both materials have a similar coefficient of expansion but Ge is much lighter than GaAs (Gonzales, [1990]). The use of highly toxic materials like arsenic requires much more elaborate safety and waste disposal programmes for their manufacture and this deters new manufacturers. These problems have prevented the widespread use of GaAs in commercial spacecraft but have not deterred their use on several military systems.

The maximum conversion efficiency of the best silicon solar cells is about 18% in terms of the flux reaching the active area of the cell, but small losses occur due to surface reflections, shadowing caused by the front power pick off conductors and internal resistances. The surface reflections are aggravated by the requirement to have protective cover glasses cemented to the cell's external faces. These cover glasses are required to reduce the degradation of the cell's conversion efficiency due to ionizing particles and micrometeorides. However the reflections can be minimized by the use of an appropriate anti – reflective coating on the glass.

The cover glasses cause further problems as they are made of an electrically insulating material (glass or silica) and the continuous flux of charged particles and ionizing radiation on to the spacecraft results in them acquiring a significant charge and hence a potential difference from the main spacecraft structure. This potential difference can adversely effect the operation of on–board instrumentation and can become so large that an electrical discharge occurs, resulting in the generation of electrical noise and possible damage of sensitive components.

To overcome this problem the cover glasses can be coated in a transparent, electrically conductive and radiation resistant medium. One material which has these properties is indium tin oxide but the material and coating process is expensive and contributes further to the cells' losses.

Allowing for all these factors, the final power conversion efficiency for an undegraded silicon solar cell is about 14%. A typical silicon cell 2 cm by 4 cm (although cells up to 8 cm by 8 cm are available), in full sunlight, in a low Earth orbit, will produce about 160 mW of electrical power at its optimum operating voltage of about 0.53 V providing a current of about 300 mA. Its mass would be about 1 g including about 0.5 g for the mass of the cover glass.

Fig. 4.54 shows the construction of a typical silicon solar cell from a P type wafer about 200 microns thick with N type impurity diffused into it to a depth of about one micron, forming a shallow PN junction one micron below the surface. This dimension is important because solar photons striking the cell will generate hole–electron pairs at the point of absorption. If this point occurs more than a few diffusion lengths away from the depletion region of the junction then recombination will occur, but if it occurs close to this region then the electric field, due to the junction, will cause the electrons to drift to the N region and the holes to the P region where they can be collected to do external work. Hence the junction is fabricated at the path length where maximum absorption occurs and this is a function of the wavelength of the incident photons as well as the physical properties of the semiconductor.

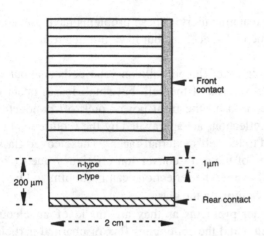

Fig. 4.54. Top and side views of n – on – p solar cell.

Referring to Fig. 4.55, the relationship between the voltage V applied across the PN junction and the current I flowing through it, under conditions of no illumination, is given by

$$I = I_0 \left(e^{\frac{eV}{kT}} - 1 \right) \quad \text{or} \quad V = \frac{kT}{e} \ln \left(1 + \frac{I}{I_0} \right)$$

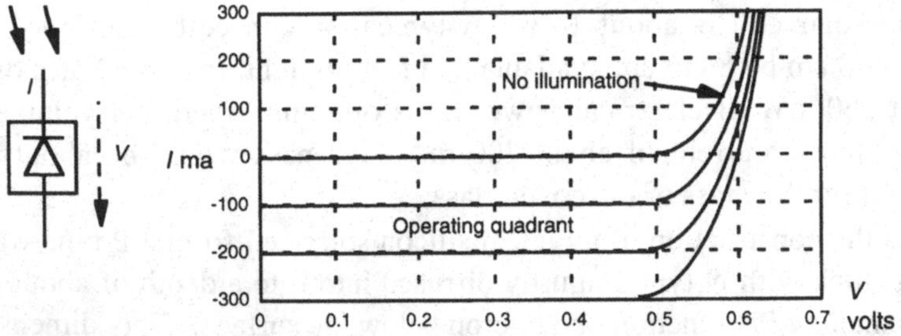

Fig. 4.55. IV curves of Si solar cell at various levels of illumination.

where I_0 represents the opposing drift and diffusion currents that flow in the diode under zero bias conditions; e is the electronic charge 1.6×10^{-19} C; k is Boltzmann's constant 1.38×10^{-23} J/K and T is the absolute temperature in K.

If we now illuminate the junction, the open circuit voltage Voc is given by

$$V\text{oc} = \frac{kT}{e} \ln \left(1 + \frac{I\text{pc}}{I_0} \right) \quad \text{where } I\text{pc is the photo current.}$$

Fig. 4.56 shows the equivalent circuit of the cell with the imperfections in the manufacturing techniques shown as Rs and Rp and the load as $R1$. Rs is the equivalent series resistance of the cell

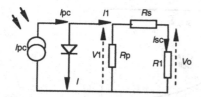

Fig. 4.56.   Equivalent circuit of solar cell with resistive load.

which is a function of its thickness, its resistivity, typically a few ohm cm, and the contact area of the pick–off conductors. The larger the front contact the less its resistance but this reduces the area exposed to the incoming solar flux. Thus a compromise is made between conversion efficiency and series resistance resulting in values for Rs of about 0.05 Ω for a typical cell. Rp is the cell's leakage resistance, caused principally by edge effects, and can be made to exceed 250 Ω for a similar cell.

In terms of the current $I1$ caused by this resistive network and the payload $R1$, the loaded voltage Vo becomes

$$Vo = \frac{R1}{Rs + R1} V1 = \left(\frac{R1}{Rs + R1} \right) \frac{kT}{e} \ln \left(1 + \frac{Ipc - I1}{I_0} \right)$$

producing the curves in the operating quadrant of Fig. 4.55. Rs is the most important parameter as the optimum value for the load $R1$ is about 1.8 Ω for a 2 by 4 cm^2 Si cell. The losses due to Rs and Rp can usually be limited to about 3% of the total power available.

For maximum conversion efficiency the spectral response curve of the solar cell should cover the spectral distribution of energy from the Sun. A comparison of the spectral sensitivities of silicon and gallium arsenide cells relative to the Sun is shown in Fig. 4.57 illustrating the peak output of the solar spectrum occurring at 0.5 micron (blue) with the silicon cell peaking further towards the red and the gallium arsenide cell with a more rectangular response peaking closer to the solar peak, but with a more limited wavelength range.

Solar cells can undergo large temperature changes from being in full sunlight to total eclipse. For a LEO spinning spacecraft with solar cells mounted on the cylindrical surface and hence thermally well coupled to the spacecraft structure, the thermal time constant of the structure prevents any large temperature fluctuations occurring even on the timescale of an orbit. Cells mounted on paddles have poor thermal contact with the structure and when the spacecraft passes from full sunlight to total eclipse, the cells' temperature may drop by 150°C. Thus both the cells and the adhesive used to fasten them to the spacecraft's panels have to withstand this rapid change in temperature.

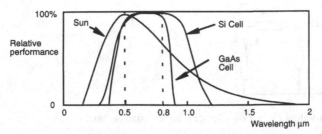

Fig. 4.57.   Comparison of solar cell outputs with that of the Sun.

The variation of the voltage–current characteristic curve of a cell, and hence its power production, with temperature is illustrated in Fig. 4.58.

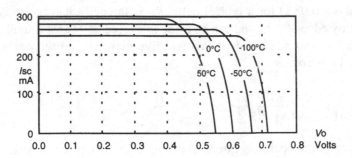

Fig. 4.58.   IV curves of Si cell at various temperatures.

These curves are for a 2 cm by 4 cm silicon cell when receiving the solar flux expected by a LEO spacecraft. Although this appears to be at variance with the equation on page 248 which shows voltage proportional to temperature, this is not so as Io also increases with temperature, but at a very much faster rate. Thus maximum power is available when the cells have just come out of eclipse and are cold, which is an important feature when the profile of the power production and consumption is being estimated for the overall power supply design.

Due to this improvement in efficiency at low temperatures the thermal design for the array should minimize the upper operating temperature. Typical designs produce temperatures in the range from + 50 to - 150°C.

High energy ionizing radiation impinging onto the cells forms dislocation centres in the crystal structure which become recombination centres for the hole–electron pairs created by the solar flux. If these occur near the junction then the conversion efficiency of the cells falls. Both current Isc and voltage Vo decrease for a given flux. This type of damage is usually irreversible on a spacecraft although it can be cured by annealing in a laboratory. A total flux of 10^{13} 1 Mev electrons per cm^2 (500 krad = 5 kgray) on a typical silicon cell will just produce a noticeable effect. If this flux is increased to 10^{15} (50 Mrad = 0.5 Mgray), the power available will be considerably reduced as shown in Fig. 4.59.

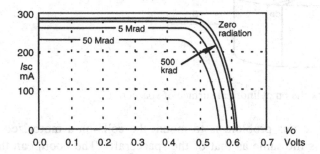

Fig. 4.59.   IV curves of Si cell after various doses of radiation.

4.5.3   *Solar arrays*

The location and shape of the solar cell array is entirely dependent on the type of spacecraft and its attitude control system. Spin stabilized spacecraft of a general cylindrical geometry with the spin axis normal to the Sun Earth vector will normally have the cylindrical surface covered with cells. This is the simplest system, both electrically and mechanically, as the final electrical connections are hard wired to the spacecraft's power subsystem and there are no solar paddle deployment systems required. However less than half the cells are illuminated at any one time and the cosine law is applicable to the power output of all those cells not normal to the spacecraft Sun vector.

If the solar flux incident on the cylindrical surface of a spacecraft of radius r and height h, as shown in Fig. 4.60, is S watts per unit area and we take a slice of the cylinder of area $dA = hr\,d\phi$, the flux incident on dA is given by $S\,dA\cos\phi$. Thus the total flux incident on the solar array is given by

$$\int_{-\pi/2}^{+\pi/2} S\cos\phi dA = \int_{-\pi/2}^{+\pi/2} S\cos\phi h d\phi = \left[Srh\sin\phi \right]_{-\pi/2}^{+\pi/2} = 2Srh$$

If the entire array was normal to the incident solar flux, the received flux would be $2S\pi rh$. Thus for a cylinder of cells the ratio of the actual power output to that which would be available if all the cells were evenly illuminated is given by $2Shr/2S\pi hr$ or $1/\pi$. This is further reduced if there are detectors which need to project through the cylindrical walls, reducing the total area available for the cells, and also if these detectors or other protuberances shadow the cells. The effective power loss of this type of solar array imposes a severe cost penalty relative to a three axes stabilized spacecraft.

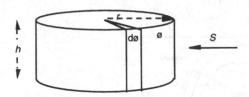

Fig. 4.60.   Solar cells on cylindrical surface of spacecraft.

One way to overcome this problem is to mount the cells on a motorized boom whose axis of rotation is the same as that of the spacecraft. The boom can then be rotated in the opposite sense to the spacecraft maintaining the cells in continuous full illumination. This was carried out on a series of orbiting solar observatories, but does require a very reliable motor drive system and slip rings to pick up the solar power to supply the rotating part of the spacecraft.

Fixed panels of solar cells may be used for non–spinning spacecraft but these severely restrict the pointing direction of the spacecraft and hence any on–board instrumentation. Steerable panels are used where the spacecraft has to be pointed in a particular direction. If the relative angular positions of the solar panels and spacecraft do not need to exceed $\pm 180°$ the final electrical contacts can be hard wired to the spacecraft's power control system. For a geosynchronous orbit where the spacecraft needs to rotate once a day relative to the solar array, the use of slip rings will be unavoidable but the associated contact wear problems are not severe due to the very low rotational speed.

Solar cell arrays are usually constructed from a combination of series and parallel blocks of cells. A series string, shown in Fig. 4.61, is made to produce the required voltage and the parallel connections to produce the required current. The precise circuitry is dependent on whether the distribution of solar flux on the cells is constant or highly variable, and on the requirement to make the array as fault tolerant as possible.

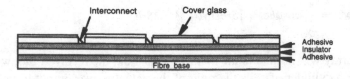

Fig. 4.61.   Series string of solar cells.

If we consider a rotating spacecraft with a fixed solar array, then some cells will be fully illuminated and some will be totally eclipsed. Fig. 4.62 (1) shows how series diodes can prevent the dark chain acting as a load to the illuminated chain. If only part of a series chain is illuminated a darkened cell can be reverse biased in excess of its Zener breakdown voltage Vz. In this case the power dissipated in the darkened cell VzIsc could raise its temperature sufficiently high to cause permanent damage. This can be overcome by the introduction of parallel diodes as shown in Fig. 4.62 (2).

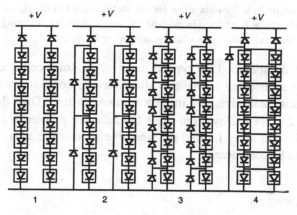

Fig. 4.62. Various solar cell interconnections.

An open circuit failure of a single cell causes the loss of a complete chain of the type shown in Fig. 4.62 (1), the loss of 4 cells if the configuration of Fig. 4.62 (2) is adopted but the loss of only the faulty cell if the layout of Fig. 4.62 (3) is chosen. However these modifications lead to a large increase in the number of interconnections between the cells and this leads to a reduction in the reliability of the whole array.

Cells can be constructed with the parallel diodes already fabricated on the wafer, but these are more difficult to produce and introduce an additional failure mode associated with each cell, and so are not generally used. Connecting several cells in parallel reduces the number of external diodes for a given number of cells but exacerbates the partial eclipse problems. A reasonable compromise is shown in Fig. 4.62 (4) which has pairs of cells in parallel and one protection diode per eight pairs of cells.

♦ *Example 4.3. Solar array area estimation*
Calculate the area of silicon solar cells required on the cylindrical surface of a spin stabilized spacecraft in a low Earth orbit with its spin axis normal to the Sun spacecraft vector.to produce 100 W of power continuously, Assuming the cells have an initial efficiency in converting solar power to electrical power of 15%, $100 / 0.15 = 667$ watts of solar power (at 1.35 kW/m^2) normal to a flat array of cells would be required which could be obtained from $667 / 1350 = 0.5$ m^2 of cells.

As only $1/\pi$ of the cells are effectively illuminated at any one time due to the cylindrical geometry of the spacecraft and allowing for degradation due to radiation (80%) and other inefficiencies such as the operating point on the characteristic curve (95 %), diode and harness losses (5%) and temperature of array at say 60°C (80%), this increases the required area to $0.5 \times \pi / (0.8 \times 0.95 \times 0.95 \times 0.8) = 2.7$ m^2. However this does not allow for the 30 minutes of eclipse every 90 minute orbit, which means that 50 Whr of energy are required to be produced when no solar power is available and this must be stored during the 1 hour sunlit portion of the orbit. Thus the actual power required is $100 + 50 = 150$ W which requires the final area to be $2.7 \times 1.5 = 4$ m^2 or about 5,000 cells of area 8 cm^2 each.

Even this result is very optimistic as it does not allow for the inefficiencies in the power storage system and control electronics, (to be considered in the next section), and we are assuming that the solar vector is normal to the spin axis at all times which is unlikely to be true.

Although we have already specified the typical mass of the solar cell and cover glass, when these are built into an array we have to allow for additional items such as the insulated backing material, adhesives and interconnection wiring. Thus the final figure will come to about 2 g for a 2 by 4 cm^2 cell for an array constructed on the surface of a spacecraft. For an array formed on steerable panels the mass of this structure would also have to be considered.

4.5.4   Storage cell specifications

Storage cells are required for peak loads and operation during solar eclipses. For a typical LEO at 550 km altitude, eclipses of the spacecraft occur for 30 minutes during every 90 minute orbit, the precise timing depending on the spacecraft's orbit relative to the equator. Typically there will be 5800 eclipses per year which implies 5800 charge–discharge cycles for the storage cells every year!

For geosynchronous orbits (an equatorial orbit with an altitude of 36,000 Km giving the spacecraft a period of 1 day and thus effectively remaining stationary over one part of the equator), the eclipse season occurs twice a year, the dark portion of the orbit varying from zero to over an hour, once per day, during a period of about 6 weeks, producing about 90 eclipses per year. These eclipses straddle the equinoxes when the Earth, Sun and spacecraft are almost in the same plane. Thus the batteries are only required for three months each year and even then their energy storage requirements are less than 10% of that required by the spacecraft during the sunlit part of its orbit.

The necessary properties of storage cells are: a robust structure so that they can withstand the mechanical rigours of the launch; hermeticity so that no loss of electrolyte occurs in the vacuum of space; the ability to operate at all attitudes; well matched $V I$ characteristics from cell to cell over the required operating temperatures; high energy storage per weight and volume; high reliability and the ability to withstand many cycles of charge and discharge without significant loss of efficiency (ratio of energy available to energy input) or capacity.

4.5.5   *Types of cell and their application*

The most commonly used electrode materials in storage cells are nickel & cadmium (NiCd), silver & cadmium (AgCd), silver & zinc (AgZn) and nickel & hydrogen (NiH_2), giving different storage energy to mass ratios; different numbers of charge – discharge cycles for a given storage efficiency and different costs. The NiH_2 cell has significantly better charging properties than the first three mentioned as it has almost unlimited overcharge capacity. This is because the gases produced during overcharge recombine to produce water. Hence no loss of electrolyte occurs and the degree of overcharge is easily determined by measurement of the pressure of hydrogen generated at the cathode. The NiH_2 cell stores more energy per unit weight than NiCd but not so much as AgCd or AgZn. Its disadvantages are the large volume required by the pressure vessel, per unit energy stored, between two and three times that required by the equivalent NiCd cell (Betz, [1982]) and an operating pressure of up to 5MPa. Thus any cracks in the pressure vessel would result in certain failure. Hence NiH_2 cells are only considered for large spacecraft requiring powers in the kilowatt range. They have been used on many communication satellites but not on LEO spacecraft yet.

AgZn produces the highest voltage and the greatest energy per unit weight for space qualified cells but its cycle life is significantly less than all the other types. Hence its use is limited to applications requiring high power at a low duty cycle. Spacecraft carrying very sensitive magnetic field detection systems must be designed so that self – generated magnetic fields are extremely small. This may preclude the use of NiCd cells due to the magnetic properties of Ni. In this case AgCd cells can be used but they are considerably more expensive than NiCd cells and have a much poorer cycle life.

With NiCd, AgZn or AgCd cells, although the number of charge – discharge cycles is the main factor governing their lifetime, the depth of charge and discharge is also a very important factor. A cell which has been charged close to its maximum energy storage level and only discharged by 25% may last for 10 to 100 times the number of cycles compared to a similar cell which has been discharged by 75% each cycle. Thus if weight permits, increasing the number of cells and hence maintaining a continuous high level of charge produces a very large increase in the cell's cycle life.

In LEO spacecraft whose storage cells undergo many thousands of charge – discharge cycles the most common compromise between battery size and cell cycle life results in a maximum Depth of Discharge (DoD) of around 25% and hence a battery capacity four times the spacecraft's nominal requirement. Table 4.4 shows how the predicted eclipse history has defined the type of storage cell chosen and the DoD allowed.

Table 4.4.   *Choice of battery type and DoD as a function of number of eclipses*

Spacecraft	%DoD	Launch Date	Lifetime Years	Battery Type	No of Eclipses
ESRO 2	15	1968	1	NiCd	6000
HEOS−A1	40	1968	1	AgCd	100
ERS1	20	1991	2	NiCd	10,520

Table 4.5 shows the general properties of the four main types of cell and how the DoD effects their cycle life.

Table 4.5.   *Properties of various storage cells*

Type of Cell	Nominal Voltage per cell	Energy Density WHr/kg	Cycle Life at Various Depths of Discharge			Temp Range °C
			25%	50%	75%	
NiCd	1.25	25−30	30000	3000	800	-10 to 40
AgCd	1.10	60−70	3500	750	100	0 to 40
AgZn	1.50	120−130	2000	400	75	10 to 40
NiH$_2$	1.30	40−80	>30000	>10000	>4000	-10 to 25

In the case of geostationary spacecraft which undergo less than 100 charge–discharge cycles per year, DoDs between 40 and 75% have been specified requiring the total cell capacity to be between 1.3 and 2.5 times the nominal.

To preserve battery life it is usual to specify an absolute maximum DoD and when that limit is reached, all non–essential loads are removed. Essential systems are normally considered to be the telemetry receiver, attitude sensing and control and the basic power and housekeeping circuitry.

Charging a cell to its maximum storage capability without significant overcharging is an important feature of the charge control electronics. Once fully charged, further charging causes the evolution of gas and although this may not cause catastrophic failure, the gas does not recombine at a significant rate, apart from the NiH$_2$ cell, and so represents a loss of electrolyte and hence of cell capacity. It is difficult to determine when the cell is fully charged as the cell voltage is dependent on charging current, state of charge and temperature. At a given charging current the voltage difference between a cell which has been charged to 75% of its maximum capacity and a fully charged cell may only be 70 mV in 1.4 V and this difference could also be caused by 10°C temperature change. Thus monitoring the cell's temperature and incorporating this information into its charging algorithm is essential.

Due to the temperature dependence of the cells' electrochemical reactions they are charged at a rate dependent on temperature. If C is the capacity in ampere hours, at 40°C the optimum charge rate is between $2C$ and $C/2$, at 20°C, between $C/2$ and $C/10$ and at 0°C between $C/10$ and $C/30$. At a given temperature the charging efficiency is maximum at the highest charge rate specified at that temperature. Typical voltage versus charging current curves for fully charged NiCd cells at different temperatures are shown in Fig. 4.63 (Thierfelder, [1982]).

These curves have to be determined empirically before launch for the specific set of batteries used. As overcharging leads to a more rapid increase in cell temperature than the normal charging process, temperature monitoring is vital to determine when the cell has become fully charged. As cells may become fully charged at different times it would be ideal to monitor the temperature of each cell individually. However this may not be practical and so monitoring is usually carried out in blocks of 4 or 8 or according to the geometry of their packaging.

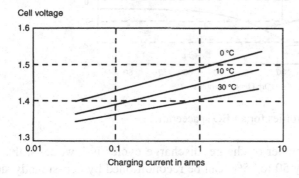

Fig. 4.63. VI curves for fully charged NiCd cell.

The charging current should be programmed so that initially it is constant at its maximum level for a given temperature, until 65% of the charge removed from the previous discharge has been replaced. Then the current is tapered down for the remaining 35% to maintain the voltage at a constant level. Thus charging and discharging profiles such as those shown in Fig. 4.64 are produced, with a final trickle charge of about 1% of the maximum charging rate when the cells are fully charged. The eclipse discharge current increases as the battery voltage falls if the power requirement of the spacecraft remains constant.

The cell charging profile will have to be varied as the cells age. This can be predicted from ground–based tests on similar cells or by noting the increased rate of rise of temperature as the cells become overcharged at their lower total capacity. For reliable operation the battery temperature should not fall below 0°C or rise above 40°C.

The overall efficiency of NiCd cells is about 75%. During discharge about 15% of the available power is lost in heat and 5% in pressure related effects. During charge a further 5% is lost in similar ways. This fact becomes very significant when the spacecraft power requirements are in the kW range as hundreds of watts of heat must be dissipated either into the spacecraft's structure or via radiator panels to ensure the battery temperature does not exceed 40°C. The difficulty of dissipating power from a single cell has limited the capacity range of NiCd cells for space applications to 50 Ah. For NiH cells capacities in excess of 100 Ah can be constructed. Within this limitation, the relatively small power dissipated during the charge of NiCd cells, compared to the discharge losses, simplifies the detection of the increased rate of rise of temperature if the cells are overcharged.

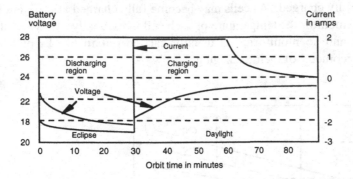

Fig. 4.64.   Battery profiles for a LEO spacecraft.

In an orbit where the number of charge–discharge cycles is low, cells, that have been regularly discharged by 50 to 75%, can be reconditioned by occasionally slowly discharging them by about 95% of their capacity and then fully recharging. For a geostationary spacecraft this process can be carried out twice a year, before each eclipse period commences, and this significantly improves the life of the cells. However for the shallowly discharged cells of a LEO system, this rejuvenation process is not effective and may actually reduce the cycle life of the cells. Also the length of time required to discharge the cells would be more than an orbit. This would necessitate the cells being partitioned so that the essential operations of the spacecraft could be maintained during the discharge period. Thus a battery reconditioning cycle is not normally included in a LEO mission.

♦ *Example 4.4.   Solar array area calulation allowing for storage battery and regulator efficiencies*

Continue the calculation for the solar cell array, started in the last section, The power required from the battery during the 30 minute eclipse is 100 W which implies the available energy must be $100 \times 30 / 60 = 50$ Wh. Allowing 80% for the efficiency of the discharge regulator and 80% for the battery efficiency on discharge, the energy required to be stored during the charge period is 50 $/ (0.8 \times 0.8) = 78$ Wh. As the charging efficiency of the battery is 95% and assuming the efficiency of the charge regulator is 80%, the input energy to the charge system must be 78 / (0.8 $\times 0.95) = 103$ Wh. In the simplest case this would require 103 watts for the 1 hr charging period. In reality a significantly higher power would be required during the first stage of the charging period to allow for the charging profile of the battery. Thus even in the simplest case the total power required from the solar cells is $100 + 103 = 203$ W requiring an array area of 5.4 m^2 which is more than double that required for sunlit operation only.

During charge the power losses are 20% (charge regulator) and 5% (battery) amounting to (103 - 103 $\times 0.8 \times 0.95) = 25$ W for an hour, and during discharge the losses are 20% (discharge regulator) and 20% (battery) amounting to (100 / 0.8 $\times 0.8 - 100) = 56$ W for 30 minutes.

The above example shows that for a regulated supply thermal design is governed principally by the discharge–cycle. The distribution of power during the charge discharge cycle is shown in Fig. 4.65. In these calculations we have assumed that all the solar array power is consumed during the charging period and so the shunt regulator would only come into operation during the tapering off period of the charge cycle.

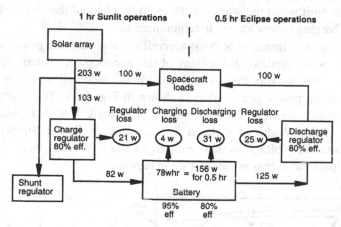

Fig. 4.65. Battery and regulator losses for a 100 W payload LEO spacecraft.

♦ *Example 4.5.   Sizing the storage battery*

Continuing with the above system, if we assume that the cell voltage degrades to 1.1 V per cell at the end of the mission, we have a 22 cell battery and the DoD, to maintain the required cycle life, is 25%, the capacity of the battery to produce the required 125 W for 30 minutes is 1 / 0.25 $\times 62.5$ = 250 Wh. As this is produced at a voltage of $22 \times 1.1 = 24$ V, the battery size is given by 250 / 24

= 10.4 Ah. Allowing for the capacity of the battery to deteriorate by 25% at the EOL of the mission, the initial capacity will have to be 10.4 / 0.75 = 14 Ah. With an initial voltage of 26 V, the Beginning of Life (BOL) energy storage capacity is 14 × 26 = 364 Wh and at an energy density of 30 Wh / kg this leads to a battery weight of about 12 kg. Volume densities of about 100 Wh / l indicate a volume requirement of about 3.6 l. Recent advances in battery design using lithium − carbon cells have improved energy densities and values up to 60 Wh / kg have been produced but these have yet to be proved in space applications.

4.5.6   Regulation

The design criterium for a spacecraft power system is to provide continuous power to the subsystems in an environment where the power supply (usually solar cells) may be routinely interrupted by the eclipse of the power source caused by the passage of the spacecraft through the Earth's shadow. This power has to be provided over a wide range of operating temperatures and maintained to the EOL of the spacecraft when the solar cells have been degraded by radiation and the efficiency of the electrical power control system has also deteriorated. Thus immediately after launch there is the problem of coping with excess power production. This is particularly true when the spacecraft has just come out of eclipse, is cold, and hence the solar cells are at their highest efficiency, although this problem is somewhat offset by the immediate charging requirements of the storage cells. Towards the end of the sunlit period of the orbit when the storage cells are fully charged the current that was available for the charging is now excess to requirements.

Whenever the load requirements of the spacecraft are less than the power available from the solar cells, some method of disposing of this excess power is required. At a given temperature and solar flux there is only one point on a cell's voltage–current curve where maximum power is available as shown in Fig. 4.66. If the array is not operated at this point, less electrical power is generated, but this requires the complexity of a Pulse Width Modulated Switched Mode Power Supply (PWM SMPS) to transform this voltage to that required by the spacecraft bus. A simpler solution is to dissipate the excess power in a shunt regulator.

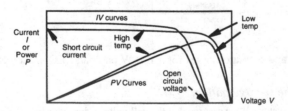

Fig. 4.66.   IV and PV curves for solar cells at two temperatures.

Hence there are two main types of power control systems, dissipative and non – dissipative, and each type can be associated with a regulated or unregulated power bus. Dissipative systems are usually limited to spacecraft requiring less than a kilowatt of electrical power due to the difficulty of dissipating large amounts of power in a shunt regulator without producing localized high temperatures and hence reducing the reliability of the dissipating circuitry. For spacecraft requiring powers in excess of a kilowatt, non – dissipative SMPS power control systems are more common. However for non – spinning spacecraft, the ability to divide the solar array into sections which can be switched on and off as required, reduces the power that has to be controlled by the regulators. In this case both the payload bus and charge regulators for high power spacecraft could be of the dissipative type.

Restricting sectioned arrays to non–spinning spacecraft is necessary to prevent the large power modulation that would arise from a spinner as the sections were regularly shadowed and illuminated.

4.5.7   Dissipative systems

Dissipative unregulated system

The simplest form of power conditioning circuit is shown in Fig. 4.67. During sunlit operations the solar cells supply power directly to the spacecraft bus and, via a current monitor, charge the battery. To control the battery charging current, the spacecraft bus voltage Vsb, which approximately equals the battery voltage Vb, is adjusted by dumping the excess current generated by the solar array in the shunt regulator. During eclipse the shunt regulator is switched off and the battery supplies the spacecraft loads. Vsb will not be stable during either part of the orbit but will follow the battery charge–discharge voltage characteristics shown in Fig. 4.64.

For an even simpler system, the shunt regulator could be omitted and the solar array used to supply the spacecraft loads and to charge the battery through an 'end of charge' isolating switch. However this would lead to a battery charging current entirely dependent on the characteristic curve of the solar array and the payload and this would not be conducive to optimum battery life.

Dissipative partially regulated system

The simplest form of bus voltage regulation maintains Vsb constant during the sunlit portion of the orbit. The current monitor of Fig. 4.67 is replaced by a linear charge regulator which can be used to define the charging profile for the battery irrespective of Vsb so long as $Vsb > Vb$. Thus Vsb will now only vary during the eclipse period.

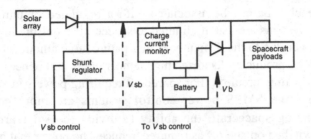

Fig. 4.67.   Dissipative unregulated system.

The design value for V_{sb} should be set a few volts above the voltage associated with a fully charged battery, where the 'few volts' are required to operate the charge regulator.   The solar array would have to produce at least V_{sb} at its EOL when operating at its maximum temperature.   Typical values might be 28 V for V_{sb} during sunlit operations, with the battery voltage and hence V_{sb} varying between 20 and 26 V during eclipse.   The charge regulator would have to provide the appropriate current with an input to output voltage range of 2 to 8 V.

Dissipative regulated system

To maintain V_{bs} constant throughout the entire orbit a discharge regulator has to be introduced as shown in Fig. 4.68.   This regulator has to be a switched mode type to boost the variable battery voltage V_b up to V_{sb} and so introduces the more severe problems associated with the design of high power fast switching circuitry.

The additional mass, power and complexity of a totally regulated system is significant, but the designs of the subsequent payloads' power subsystems are simplified as their input voltages are now constant.

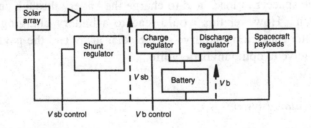

Fig. 4.68.   Dissipative regulated system.

For a non − spinning spacecraft an alternative technique is to divide the array into two sections with one section providing a higher voltage than the other. The low voltage section is used to power the sunlit payload and the other section charges the battery via a linear regulator. As the battery voltage can now be made higher than that required for the payload bus, another linear regulator can be used to maintain the bus constant during eclipse. Thus no complex switched mode power supplies are required.

If the above power subsystem is part of a LEO spacecraft with a proposed 10 year life, to allow for degradation of the solar array, the initial power output may be specified to be 50% more than that actually required. Thus for every 100 W of power required by the payload and additional 100 watts required for battery charging, a further 100 watts would be necessary to allow for the array degradation. When the batteries are fully charged the solar array still produces 300 W but only 100 W is required by the payload. Hence the shunt regulator has to be capable of dissipating 200 W. Thus we see from Fig. 4.65 that for every 100 W of payload power we have to install regulators which are capable of controlling about 400 W even though they are not used at this power level simultaneously. This is a feature that has to be considered when estimating masses and volumes for power subsystems.

4.5.8   *Non–dissipative systems*

Non–dissipative unregulated system

When the overall power requirements of a spacecraft rise above the maximum power that a shunt regulator can easily dissipate, power control can be accomplished by reducing the efficiency of the solar array and hence dissipating the unwanted power in the array. As the array is already designed to thermally dissipate about 85% of the power incident on it, dissipating another 5 or 10% more power is not a serious problem.

Fig. 4.66 shows how the electrical power generated by an array can be varied by the choice of the operating point on its *IV* curve. By allowing the array voltage to vary about the point associated with maximum power conversion, efficiency is reduced and hence the power available to the PWM SMPS can be controlled. As the load on the PWM SMPS varies, the electrical power generated by the solar array is matched to that required by the load. The simplest type of system is shown in Fig. 4.69.

During sunlit operations the SMPS is controlled so that Vsb is maintained at the voltage required to provide the correct charging current for the battery and during eclipse the battery supplies the payloads directly. Hence the payload voltage follows the charge–discharge profile of the battery. For a given power requirement, this type of power management system is similar in mass to the dissipative regulated system as they each require one switched mode power supply.

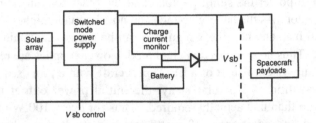

Fig. 4.69. Non–dissipative unregulated system.

Non – dissipative partially regulated system

If a second PWM SMPS is added so that the battery charging current and payload voltage can be controlled independently, Vsb can be stabilised during sunlit operations and we get a partially regulated system as shown in Fig. 4.70.

A linear charge regulator would not be appropriate here due to the high power dissipation at the charging currents associated with a high power spacecraft. The bus voltage is not regulated during eclipse, being dependent on the state of discharge of the batteries and their temperature.

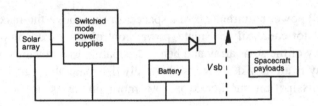

Fig. 4.70. Non–dissipative partially regulated system.

Non–dissipative regulated system

A completely regulated system, as shown in Fig. 4.71, requires a third regulator to maintain Vsb during eclipse.

Summarizing, for spin – stabilized spacecraft, dissipative regulators are electrically simple but complicate the thermal design and so are confined to the lower powered spacecraft, whereas the more complex non – dissipative systems are necessary where high powers are involved. For non– spinning high power spacecraft, the ability to sectionalize the solar array allows the use of dissipative shunt regulators to control the payload and battery charging systems but the discharge regulator, which has to supply the entire spacecraft power during eclipse, has to be a non – dissipative type.

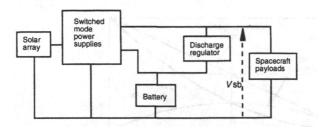

Fig. 4.71. Non–dissipative regulated system.

Regulation makes the system more complex but this results in the subsequent bus user's power supply being of a simpler design and so more power and mass efficient. This may well result in greater efficiency and less mass when all power regulators in the entire spacecraft are considered. A comparison of the various power system topologies proposed for ESA spacecraft is given by O'Sullivan, [1994].

As regulators are an important part of the power system let us consider these in more detail.

4.5.9   Regulators

Dissipative shunt regulators

These are linear regulators which dissipate the excess power in a series transistor–resistor network as shown in Fig. 4.72. The distribution of power between the transistor and resistor R is such that never more than 25% of the maximum dissipation occurs in the transistor, and at maximum dissipation the power dissipated by the transistor is practically zero. This is illustrated in Fig. 4.73.

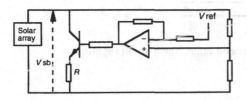

Fig. 4.72. Linear shunt regulator.

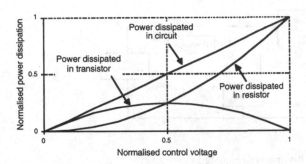

Fig. 4.73.   Linear shunt regulator power dissipation.

To increase the range of power dissipated, the circuitry can be extended to that shown in Fig. 4.74. In this system as Tr1 saturates Tr2 starts to conduct and so on as shown in Fig. 4.75. Thus only one transistor is actively dissipating power at any time, the others either being saturated or turned off. However power is still dissipated in semiconductor devices and when the overall power requirements exceed several hundred watts this becomes a serious thermal and reliability design problem.

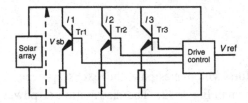

Fig. 4.74.   Extended range linear shunt regulator.

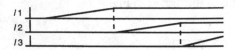

Fig. 4.75.   Operation of extended range linear shunt regulator.

Switched dissipative regulators

To minimize the power dissipated in the transistor, it can be used as a switch with power only being dissipated in the resistive element. If Vsb, in Fig. 4.76, exceeds a defined voltage the switch is closed and the combination of R and C is chosen so that Vsb will fall a few 100 mV in a few tens of microseconds. When Vsb has fallen by the required amount the switch is opened, the voltage rises and the process is repeated. The fraction of the time the switch is on is governed by the excess power to be dissipated. As *V*sb only fluctuates by a few 100 mV, the variation in array current, governed by the characteristic IV curve of the array, is relatively small compared with that extracted from the capacitor. Thus the EMC problems associated with pulsed currents flowing in the harness to the array and in the array itself are not severe. This can be reduced further with the appropriate filtering as shown.

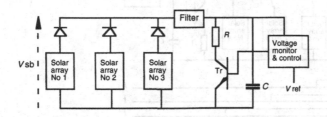

Fig. 4.76.   Pulsed digital shunt regulator.

Combined dissipative and non–dissipative regulators

To control power levels of up to several hundred watts generated by non–spinning spacecraft, both dissipative and non – dissipative systems can be combined in one system as shown in Fig. 4.77. Assume that the linear shunt formed by Tr and *R* can dissipate ultimately as much power as one section of the '*N*' arrays of solar cells (*W* watts). As soon as a reduction in the spacecraft's power requirements results in the linear shunt dissipating *W* watts, the linear shunt is turned off and array Solar Array No 1 is switched off. The linear shunt can then start increasing its power dissipation again until it reaches *W* watts when a second array will be removed and the process repeated. Thus excess power in the range zero to $N \times W$ watts can be linearly controlled with a maximum of only *W* watts being dissipated in the linear shunt, which would probably be of the extended range type. As the digitally controlled switches are only operated as required rather than repetitiously, the EMR is minimal.

The main disadvantage of this system are the series switches which present problems due to the high and variable voltage required for their control electrodes. The linear section also has to dissipate power within a semiconductor. The series switches can be replaced by shunt switches which short circuit the unwanted elements of the array and these have no control electrode problems as the switching voltages are now relative to spacecraft ground. However the short circuit current may increase magnetic cleanliness problems. The choice between series and shunt switches will generally depend on the magnitude of the array voltage with voltages in excess of 100 V probably requiring the shunt type.

For spacecraft requiring to control power in the kilowatt range, the use of linear dissipating networks is not thermally desirable and SMPSs are preferable, although their masses are a significant consideration, and being series elements their losses cannot be avoided.

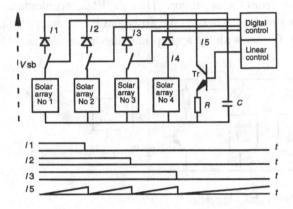

Fig. 4.77. Digital series, linear shunt regulator.

Series PWM SMPS

This functions by receiving power from the source through an inductor, in a regular series of pulses rather than continuously and as the power requirement varies the mark to space ratio of these pulses is adjusted accordingly. A basic series PWM SMPS designed as a battery charger from a solar array is shown in Fig. 4.78. Although ideally it would dissipate no power, due to the non–ideal properties of the constituent components, some power is lost, but efficiencies around 90% can be achieved.

The mode of operation is as follows. When the switch is on, the diode is reverse biased and energy from the solar array is built up in the inductor as well as supplying the battery for a time $\Delta T1$. As soon as $Vc2$ or $I2$ exceed the programmed value, the switch is opened, Vs changes polarity, forces the diode into conduction and the battery continues to be charged from the energy stored in the inductor for a time $\Delta T2$. When $Vc2$ or $I2$ fall below the design limit the switch is remade and the cycle repeats.

The capacitor $C1$ is necessary before the switch so that the current Isa from the solar array is maintained at the average value required by the regulator and not pulsed as the switch is operated. This is an important design requirement as current pulses in the long harnesses associated with the solar array would cause EMC problems. Although the current $I1$ is pulsed, as this is contained within the box containing the regulator its ability to generate interference can be minimised.

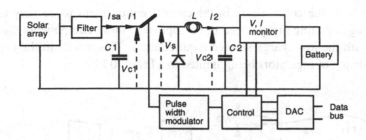

Fig. 4.78. Series charge regulator using a PWM SMPS.

From Fig. 4.78 we can see that when the switch is closed

$$Vc1 - Vc2 = L\frac{\Delta I2}{\Delta T1}$$

and when the switch is open

$$Vc2 = L\frac{\Delta I2}{\Delta T2}$$

as $\Delta I2$ is the same in both equations

$$(Vc1 - Vc2)\Delta T1 = Vc2\,\Delta T2$$

and hence

$$Vc2 = Vc1\frac{\Delta T1}{\Delta T1 + \Delta T2}$$

To minimize EMR, the system is designed to make $\Delta I2$ small and this results in the waveforms shown in Fig. 4.79. From the conservation of power, and assuming the design is for low ripple content

$$Vc1 \times Isa = Vc2 \times I2 \quad \text{or} \quad Isa = I2 \times Vc2 / Vc1$$

Thus the mean current from the solar array will be approximately that required by the battery multiplied by the ratio of the battery voltage to the solar array voltage. The DAC is driven from the micro-processor which supplies the programmed data to define the charging profile for the battery.

To minimize the size of the inductors and filters the switching frequency is made as high as possible. The maximum frequency is set by the losses in the diode and switch during the switch off transitions when the maximum reverse voltage develops across the device before its conduction has ceased. Although diodes and switches with very low stored charges are available, they tend to have low Zener breakdown voltages and the position of both the components in the circuit require them to be very reliable. Thus large voltage derating factors need to be applied to these components and this is incompatible with the low charge storage requirements. This results in the maximum switching frequency of regulators being limited to a few 100 kHz.

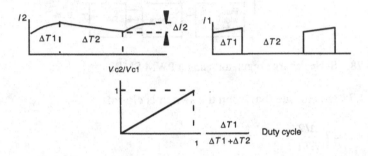

Fig. 4.79. Series regulator waveforms.

The PWM SMPS can either be used on its own, within the limitation that its output voltage will be less than its input voltage, or feed a separate symmetrical push–pull type of SMPS as shown in Fig. 4.80. Although this introduces more complexity into the system it does enable the output voltage to be set at any level below or above the supply voltage and enable the final voltage to be isolated from the original source of power as may be required to minimize EMC problems.

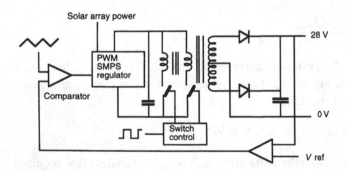

Fig. 4.80. PWM SMPS regulator & converter.

One of the advantages of the PWM SMPS with a solar array as its source of power, is that as the power required varies, it allows the solar array voltage to rise and fall on the array's *IV* curve and hence controls the power generated by the array. However the design of the PWM SMPS has to ensure that it operates on the negative slope of the *PV* curves shown in Fig. 4.66. If the spacecraft demanded more power than the peak available from the array, this would result in the PWM SMPS being forced onto the positive slope of this curve and this is not a stable situation, resulting in the array's voltage being pulled to zero. Thus a system to monitor the slope of the array's *PV* curve has to be incorporated into the design. The output of this monitor is fed into the spacecraft's power management's system in order to switch off non–essential loads and to change the operating characteristic of the PWM SMPS to a non–regulated, input current limited supply as required.

The PWM SMPS will normally only operate in the peak power region of the solar cells towards the EOL of the spacecraft, and so initially the PWM SMPS has to operate at high array input voltages. Within the PWM SMPS high currents have to be switched very rapidly. Thus by design, the PWM SMPS components have to function in circumstances which make it difficult to achieve the very high reliability required for this very important part of the system.

As PWM SMPS may regulate the entire spacecraft power, the masses of the associated cores in the inductors and/or transformers are not insignificant. Also as it is a series element in the power control system, it will always dissipate some power due to its own internal losses. Thus at the spacecraft's EOL when power is limited, the shunt regulator which can operate down to zero dissipation, would be the more desirable system.

Discharge (shunt or boost) regulators

Although the SMPS described above would be used as the discharge regulator for a high powered spacecraft, at lower powers, a simpler type of converter can be used. Such a system is shown in Fig. 4.81 and functions in the following manner. When the switch is on, energy is built up in the inductor and this is transferred to the load via the diode when the switch is opened. As the current in the inductor can exceed the load plus capacitor current, *Vsb* can be made significantly greater than *Vb* which cannot be achieved with the basic series SMPS. However this type of converter is not as efficient as the series type.

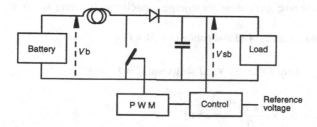

Fig. 4.81. Shunt regulator.

In the design of SMPS regulators there are several factors to be considered of which reliability and EMC are very important. As failure in a power supply can be catastrophic, all single point failure modes must be carefully investigated and, if possible, eliminated. Significant derating must be applied to all power supply components, and series parallel redundancy may be required for the filter components which will guard against a single short or open circuit as shown in Fig. 4.82.

The choice of electrolytic filter capacitors is critical as some types are specified with relatively low inrush currents and the values may be impossible to achieve in a power subsystem. However limiting the inrush current to some value is necessary to protect relay contacts which may be used to switch power to the various subsystems.

Regulator Filters

Adequate filtering to limit the current ripple in both the input and output harnesses is essential for EMC requirements and this must cope with both the basic switching frequency of the inverter and its associated high–frequency transients. A typical filtering network is shown in Fig. 4.82 (b).

♦ *Example 4.6.   Power supply filter design*

A regulator is designed to operate at 100 kHz with an average mark to space ratio for I3 of 1:1 and take an average current for $I1$ of 0.5 A. EMC requirements place an upper limit of 0.1 V peak to peak ripple voltage on the capacitor and a maximum of 1 mA peak to peak current ripple for $I1$, the current in the harness to the 28 V battery. Calculate the values of the filter components shown in Fig. 4.82.

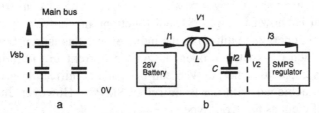

Fig. 4.82.   Power bus;   (a) redundant storage capacitors;   (b) filter network.

From the waveforms shown in Fig. 4.83 and noting that $I1 = I2 + I3$

$$I2 = C\frac{dV2}{dt} = C\frac{\Delta V2}{\Delta T} \text{ where } \Delta V2 = 0.1 \text{ V}, \Delta T = 5\mu s \text{ and } I2 = \pm 0.5 \text{ A}$$

$$\therefore C = I2\frac{\Delta T}{\Delta V2} = 0.5 \times \frac{5 \times 10^{-6}}{0.1} = 25 \text{ }\mu F$$

The ripple component of V2 ($\Delta V2$) is the same as the variable voltage V1 across L as we can assume the battery voltage is fixed. Thus in Fig. 4.84, the 28 V line also represents zero voltage across the inductor.

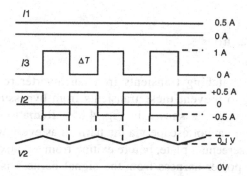

Fig. 4.83. Filter waveforms.

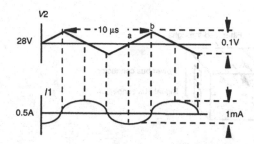

Fig. 4.84. Filter voltage and current ripple.

To determine the peak current in the inductor, we only need to consider the section 'ab' of $V2$.

$$V1 = L\frac{dI1}{dt} = \Delta V2 = \frac{dV2}{dt}t = \frac{0.1}{5 \times 10^{-6}} = 2 \times 10^{-4}t$$

$$\therefore \; dI1 = \frac{2 \times 10^{-4}}{L}t\,dt$$

Integrating this expression leads to

$$I1 = \left[\frac{2 \times 10^4 t^2}{2L}\right]_0^{\Delta T}$$

If ΔT is the duration of the 'ab' section of $V2$, $\Delta T = 2.5\,\mu s$

$$I1_{peak} = \frac{2 \times 10^4 \Delta T^2}{2L} = 0.5 \times 1mA = 0.5mA$$

$$5 \times 10^{-4} = \frac{2 \times 10^4 \times \left(2.5 \times 10^{-6}\right)^2}{2L}$$

$$\text{and} \quad L = \frac{10^4 \times \left(2 \times 10^{-6}\right)^2}{5 \times 10^{-4}} = 125 \ \mu\text{H}$$

Minimising the EMR from the switching transients from an inverter requires careful circuit layout and screening to prevent these transients from bypassing the filters by direct radiation. Such a system is shown in Fig. 4.85. Radiation from components in the screened enclosure cannot escape via the harness as every wire is individually filtered at the additional screening plate, before exiting from the box by a standard connector. Separating the power harness from the signal harness is good practice to minimise magnetic coupling from the power lines into the signal lines. If this is not possible then keeping the two sets of wires at opposite ends of a single connector is the next best solution.

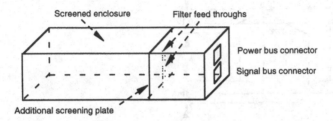

Fig. 4.85. Electrostatic screening.

De–centralised regulation

Most subsystems require several different voltage power lines such as ± 15 V for analogue and $+5$ V for digital circuits. These can be produced centrally very efficiently with only one or two (redundant) SMPSs. However this can present problems for the subsystems. Even if the voltage drop across the harness is insignificant, most subsystems will incorporate power line filters and protection networks, such as inrush current limiters and the voltage drop across these circuits could be significant, particularly for low voltage supplies. Thus apart from low powered spacecraft (less than 100 W) the main power bus, at typically 28 V, will generally be supplied to each subsystem which will provide its own isolating SMPS.

Multiple output regulation

With multiple output SMPSs which can be produced by having more than the single output winding shown in Fig. 4.80, only one output voltage can be used to control the regulation of the unit and so the other outputs will be less well regulated and may need additional linear regulation. This can result in a significant reduction in the efficiency of the power supply as shown in Fig. 4.86. Here the loss is caused by the linear regulator requiring its supply voltage $V_{us}1$ to be a few volts higher than its input voltage V_{rs}. This loss can be reduced considerably by producing a second lower supply voltage $V_{us}2$ from the regulator which powers the output power transistor only. In this case the power taken from $V_{us}1$ is only that required by the amplifier, and the power wasted from $V_{us}2$ is now relatively small as $V_{us}2$ need only be a few hundred millivolts greater than V_{rs}.

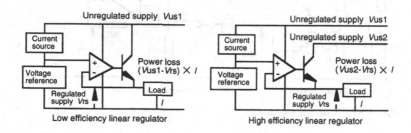

Fig. 4.86. Linear regulator efficiency.

The power from the main spacecraft bus to each subsystem's regulator will normally be switched using a latching electromagnetic relay with current limiting circuitry or possibly fuses.

Regulator mass estimates

Depending on the type and size of the main power subsystem, from the simple unregulated to the regulated multiple PWM designs, the mass to power ratio can vary from 2 to 10 kg kW^{-1}. Within a typical subsystem containing its own isolating SMPS, supplying power up to a few tens of watts, the weight to power ratio will usually be in the range of 10 to 40 g W^{-1}.

4.5.10   Power supply monitoring

Monitoring the voltage of the bus and the regulated supplies is essential to verify the performance of the regulators. Similarly a detailed knowledge of the currents taken by the subsystems provides very useful information on the health of their circuitry. However producing an accurate indication of the current taken from a variable bus is not as simple as it first appears. Fig. 4.87 shows the circuit of a basic current monitor, where the current to be monitored I develops a small voltage V_r across the resistor r.

As an amplifier will usually require its positive supply to be a few volts above its inputs, this is carried out with the four R resistors. r will normally be a fraction of an ohm with R being tens of kohms and so r can be considered negligible with respect to R. Applying the principle of superposition to the equivalent circuit, the output voltage Vo is the sum of the outputs due to $Vb / 2$ (Vo^+) and ($Vb - Vr$)$/ 2$ (Vo^-).

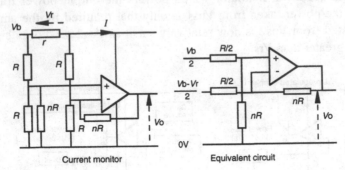

Fig. 4.87.   Current monitor.

Consider the $Vb / 2$ term first

$$Vo^+ = \frac{Vb}{2}\frac{nR}{(nR + R/2)}\frac{(nR + R/2)}{R/2} = nVb$$

and the ($Vb - Vr$)$/2$ term gives us

$$Vo^- = \frac{(Vb - Vr)}{2}\frac{nR}{R/2} = n(Vr - Vb)$$

$$\therefore Vo = Vo^+ + Vo^- = nVb + n(Vr - Vb) = nVr \neq f(Vb)$$

♦ *Example 4.7.   The problems of current monitoring*
Consider the result of an error ΔR occurring in the upper left hand resistor R of Fig. 4.87. For simplicity assume I is zero so that Vr is zero, Vb is 28 ± 5 V and $\Delta = \Delta R / R$. The only two terms which are affected by ΔR are $Vb / 2$ and $R / 2$ associated with the + input.

$$Vb/2 \rightarrow Vb\frac{R}{R + R(1 + \Delta)} = Vb\frac{1}{2 + \Delta} = \frac{Vb}{2}(1 + \Delta/2)^{-1} = \frac{Vb}{2}(1 - \Delta/2)$$

$$R/2 \rightarrow \frac{1}{1/R + 1/R(1 + \Delta)} = \frac{R(1 + \Delta)}{2 + \Delta} = \frac{R(1 + \Delta)}{2(1 + \Delta/2)} = \frac{R}{2}(1 + \Delta)(1 - \Delta/2)$$

$$= \frac{R}{2}\left(1 - \frac{\Delta}{2} + \Delta - \frac{\Delta^2}{4}\right) = \frac{R}{2}\left(1 + \frac{\Delta}{2}\right)$$

Thus Vo now becomes the sum of $Vo^{+'}$ and Vo^- where $Vo^{+'}$ incorporates the changes in Vo^+ due to ΔR.

$$V_o = \frac{Vb}{2}\left(1-\frac{\Delta}{2}\right)\left(\frac{nR}{nR+(1+\Delta/2)R/2}\right)\left(\frac{nR+R/2}{R/2}\right)+n(Vr-Vb)$$

$$= \frac{Vb}{2}\left(1-\frac{\Delta}{2}\right)\left(\frac{n}{n+1/2+\Delta/4}\right)\left(\frac{n+1/2}{1/2}\right)-nVb \text{ as } Vr = 0$$

and assuming that $\Delta/4 \ll n$

$$Vo = Vb\left(1-\frac{\Delta}{2}\right)n-nVb = -\frac{n\Delta Vb}{2}$$

As we have specified that Vr is zero, Vo should also be zero and so the above value is an error due to ΔR. If Vb and Δ were constant this error voltage could be removed, during set–up procedures before launch, however this is not the case. Inserting values for n and the variation in Vb gives the maximum unknown error in Vo of

$$\pm 50 \times 5 \times \Delta/2 = 125\,\Delta$$

If we specify that Vr must not exceed 0.1 V, $Imax = 1$ A and Vo has a full scale of 5 V, then $n = 50$ and $r = 0.1\ \Omega$. If an allowable measurement error is 1% then the maximum error at Vo is 50 mV. Therefore

$$125\Delta \le 50 \times 10^{-3} \quad \text{or} \quad \Delta \le 4 \times 10^{-4}.$$

Thus the resistor values must be matched and stable within this error. If the operating temperature range is 80°C, this requires the temperature coefficient of the resistors to be better than $4 \times 10^{-4}/80$ or 5 parts per million per °C. Hence this circuit requires components of very high quality and stability and detailed attention to circuit layout.

4.5.11   Noise reduction

As the photon or particle sensor will usually produce a low level signal at a high impedance, some FEE will be required to condition the signal so that it can be transferred to the ME at a reasonable S/N. This is illustrated in Fig. 4.88. The current $I1$ in the ground line between the FEE and ME will be relatively quiet as only low level analogue signals are being processed in the FEE. However $I2$ will probably be noisy as the ME will contain the high level digital electronics with its associated fast pulsed waveforms. The noise content of this current should be minimised by adequate capacitive decoupling and filtering of the noise–producing circuits, but the residual noise current will generate a noise voltage v_n across the impedance of the signal ground line between the ME and the power system's ground. Thus v_n will appear between the FEE and its structure and so couple into the system at its most sensitive area.

Noise can also be generated by magnetic radiation coupling into the loop formed by the signal ground and chassis and become apparent at the highest impedance part of the circuit which is again the sensor. These problems are not simply solved by shorting the structure to the signal ground at the FEE, as the circulating noise current due to magnetic radiation will increase due to the reduction of the loop impedance and will appear across the inductance of the signal ground wire. This will also allow noise current to flow in the chassis which is highly undesirable.

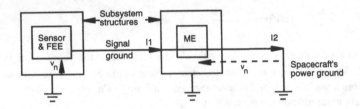

Fig. 4.88.   A source of ground induced noise.

These problems can be overcome if the main electronics system's power is supplied by its own isolating SMPS. This enables the signal ground to structure connection to be made at the sensor but broken at the power supply, which ensures that the noise voltage now appears at a part of the circuit which is fairly insensitive to noise. The problem of interfacing such a system with the telemetry without reintroducing ground loops can be accomplished as shown in Fig. 4.89 with the aid of an opto coupler, or Fig. 4.90 with the aid of differential coupling.

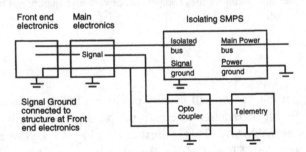

Fig. 4.89.   Isolating SMPS with opto−coupled signal.

Opto couplers present problems due to their power requirements if required to operate at high speed, and their degradation with age and radiation. Although differential coupling does not give such good isolation, it has the advantage of good common mode rejection. The probability of any induced noise being added equally to both signal lines (common mode) is made high by twisting them together and ensuring the combined impedance of the driving and receiving circuitry is the same for each line.

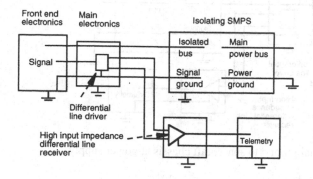

Fig. 4.90. Isolating SMPS with differential signal coupling.

An internal screen connected to signal ground but isolated from the spacecraft's structure, as shown in Fig. 4.91, permits direct coupling of signals to the telemetry system without generating ground loops, but it does require extra volume and weight at the FEE of the subsystem.

If the design of the SMPS is such that the signal ground is connected to the structure at the power supply, this can be accommodated with the system shown in Fig. 4.92.

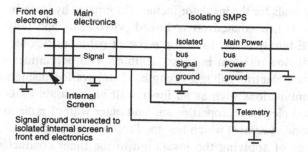

Fig. 4.91. Isolating SMPS with isolated screen in FEE.

Ensuring that a system's grounding philosophy does not generate loops is a common EMC requirement. This is often specified in terms of a minimum resistance and a maximum capacitance between the various ground lines and chassis ground. Typical values are 10 MΩ for the resistive component and 10 nF for the capacitive component.

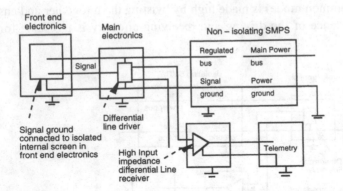

Fig. 4.92.   Signal ground to spacecraft chassis at power supply.

4.6   Harnesses, connectors and EMC

4.6.1   *Harnesses*

Harnesses are composed of two fundamental materials, the conductor and the dielectric or insulator. The ideal requirements for the dielectric are high insulation resistance, the ability to withstand a high electric field over a wide temperature range, very low vapour pressure, flexibility but good mechanical strength and the ability to withstand photon and particle radiation without degradation.

The recommended materials for the insulator include fluorinated hydrocarbons such as TFE (tetrafluoroethylene), irradiated cross linked polyolefins and polyimides. Within this range the TFE family is less radiation resistant but more flexible and the polyolefins are more radiation resistant but less flexible. The polyimides such as Kapton have a very high dielectric voltage strength. Dual walled insulations are sometimes used with combinations such as an inner wall of radiation cross linked polyalkene which has good electrical properties and and outer wall of radiation cross linked polyvinylidene fluoride (Kynar) which has good mechanical properties.

There are two methods of applying the insulation to the inner conductor, tape wrapping and extrusion. Although wrapping is the simpler construction, under constant flexing the possibility of exposing the inner conductor or allowing the ingress of contaminants leads to the preference of the extruded type for space use. Due to the vibration environment during launch it is most important to ensure that cables do not pass over any sharp edges or corners under tension. This is particularly for the TFE family as they tend to 'cold flow' even under static conditions. Adequate cable strapping, the use of protective sleeving or removal of the sharp edge is essential.

Due to the requirement to minimise the weight and volume of any spacecraft component, the gauge of wires used is generally minimised. For conveying power the size will be defined by an acceptable voltage drop across the cable. However for signal wiring the minimum acceptable gauge is about 26 in the American Wire Gauge (AWG) due to the requirement to maintain adequate mechanical strength particularly at the termination of the wire. The conductor is usually silver plated stranded copper wiring with at least ten conductors.

Certain circumstances require wires of very much smaller gauges. This may be to minimise thermal conduction through the wiring. In these cases brass or stainless steel conductors can be used. Stainless steel presents termination problems and crimping is usually the only solution.

4.6.2   Harnesses and EMC

As harnesses act as antennae to EMR, methods must be found to minimise the results of this effect. Due to the physical size of the spacecraft, except at very high frequencies, the EMR generated internally is 'near' field radiation. A generally accepted criterion for near field conditions is that the separation of the 'culprit' wire sourcing the EMR and the 'victim' wire receiving it is $< \lambda / 2\pi$, where λ is the wavelength.

This leads to the near field condition occurring in most spacecraft for frequencies less than about 100 MHz where $\lambda / 2\pi$ is about 0.5 m.

Magnetic field coupling

If we have a current I flowing along a wire it will generate a magnetic field **B** around the wire defined by Ampere's Law where

$$\oint \mathbf{B}.d\mathbf{l} = \mu_0 I$$

and μ_0 is $4\pi 10^{-7}$ Hm^{-1}. For a circular path distant r from the wire, B is defined by the expression shown in Fig. 4.93.

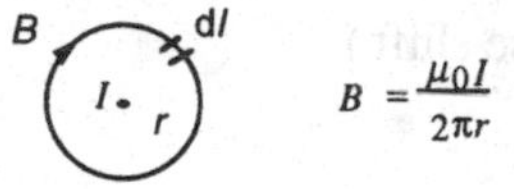

$$B = \frac{\mu_0 I}{2\pi r}$$

Fig. 4.93.   Magnetic field due to current I.

If we have another wire ce in the vicinity of the wire ab, and ab is parallel to and coplanar with ce as shown in Fig. 4.94,

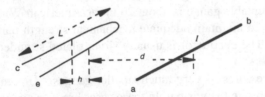

Fig. 4.94. Magnetic coupling.

the magnetic flux ϕ passing through the loop formed by ce, of area A, is given by

$$\phi = \int_{d}^{d+h} BdA = \int_{d}^{d+h} BLdr$$

where L is the length of the loop, d is the distance between ab and ce and h is the separation of the two halves of the ce loop.

$$\phi = \frac{\mu_0 IL}{2\pi} \int_{d}^{d+h} \frac{dr}{r} = \frac{\mu_0 IL}{2\pi} [\ln(r)]_{d}^{d+h} = \frac{\mu_0 IL}{2\pi} \ln\left(\frac{d+h}{d}\right)$$

From Faraday's law we know that the voltage V induced in a single loop of wire linking a magnetic flux ϕ is given by

$$V = -\frac{d\phi}{dt}$$

$$= -\frac{\mu_0 L}{2\pi} \frac{dI}{dt} \ln\left(\frac{d+h}{d}\right)$$

If a component of the current I varies in a sinusoidal manner,

$$I = I_0 \sin(wt) = I_0 \sin(2\pi ft)$$

$$\frac{dI}{dt} = I_0 2\pi f \cos(2\pi ft) = 2\pi fI \quad (\text{ignoring } \pi/2 \text{ phase shift})$$

$$V = \mu_0 ILf \ln\left(\frac{d+h}{d}\right)$$

If we define the wire ab as the culprit and the loop ce as the victim, to reduce the coupling between the two, the loop area of ce (Lh) should be made as small as possible by ensuring the victim's signal and its return wire are as close together as possible, and d should be made as large as possible by separating the high current radiating harnesses from those containing the susceptible wiring.

♦ *Example 4.8. Magnetic coupling between harnesses*
In a system where the culprit harness has a current of 100 mA flowing at a frequency of 10 MHz separated from the victim harness by 0.1 m, and the victim harness is 1 m long with a 1 mm separation between its signal and return wires, $L = 1$ m, $f = 10$ MHz, $I = 100$ mA, $d = 0.1$ m, $h = 1$ mm, the induced voltage in the victim harness will be 12.5 mV.

When the frequency is high, the self inductance of the wire has to be considered relative to the impedances in the circuit. As an example 24 AWG wire has an inductance of about 1uH per metre. Thus for the circuit shown in Fig. 4.95 where Z_i is the input impedance of the susceptible amplifier, Z_s is the impedance of the source driving the amplifier, L_v is the inductance of the wire connecting Z_s to Z_i, and V_v is the voltage induced in the victim harness at an angular frequency w, the induced voltage appearing at the input of the susceptible amplifier V_i is smaller than V_v, as indicated below.

$$V_i = V_v \frac{Z_i}{Z_i + jwL_v + Z_s} \rightarrow V_v \frac{Z_i}{j2\pi fL_v} \quad as \quad f \rightarrow \infty$$

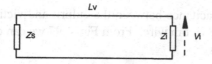

Fig. 4.95. Self–inductance of victim wire.

Thus the self–inductance of the victim circuit limits the high–frequency coupling which would otherwise rise indefinitely with frequency. The self–inductance of the culprit circuit limits the maximum current causing the problem.

The most common mechanism for reducing magnetically generated EMR is shown in Fig. 4.96. Here the area into which the EMR is induced is minimised by tightly twisting the signal and its return (2 twists per cm). Furthermore as the wires change place every half twist the induced voltages tend to cancel. It is important to note that the connection to the ground plane is only at one end of the harness, otherwise the signal return wire and the chassis ground form a large area ground loop allowing any magnetic flux present, to couple a significant current into this loop. Although the signal wire and the chassis ground would also form a similar loop under these conditions, the impedance of this loop would be much larger than the signal return ground loop due to the impedance of the signal circuitry. Thus the two loop currents would not be the same and so the interference caused would not be cancelled by any common mode rejection (CMR) circuitry .

It is difficult to specify whether the twisted pair or the coaxial cable is better at reducing magnetic interference problems. With coaxial cable, the outer sheath is the return path for the signal current. Relative to the culprit wire, the loop formed by the central wire and near side of the sheath is not coincident with the area formed by the central wire and the far side of the sheath. Hence if the magnetic field varies significantly across the diameter of the cable some residual flux will be coupled into the harness. This would make the twisted pair slightly better than the coaxial cable. However if the twisted pair is not uniformly or tightly twisted it may be inferior.

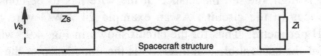

Fig. 4.96. Twisted pair connection, sensor grounded.

If two wires carrying the same current but in the opposite sense are placed very close to each other it is clear that the two fields will cancel. Hence harnesses that are designed to minimise susceptibility to EMR, also minimise emissions.

Electric field coupling

This is basically due to the capacitance between the culprit and victim wires (Cs) and the impedance to ground of the victim wire. From Fig. 4.97 we can calculate the cross coupling between the two wires.

$$\frac{Vv}{Vc} = \frac{Zv}{Zv + Zcs}$$

Zv includes the impedances at both ends of the victim harness (Zvs and Zvi) and the impedance to ground of the harness due to Cv.

♦ *Example 4.9. Electric coupling between two harnesses*

Using the configuration shown in Fig. 4.97, if Cv = 100 pF, Cs = 5 pF, Zs = Zl = 10 kΩ, as $1/wCv$, at 10 MHz, is much less than both Zvs and Zvi, the voltage on the victim wire due to the culprit wire is given by

$$\frac{Vv}{Vc} = \frac{Cs}{Cv} = \frac{1}{20}$$

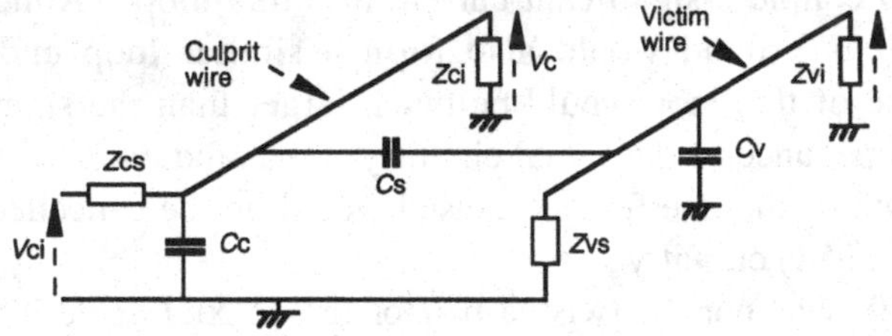

Fig. 4.97. Electric coupling.

4.6.3   Shielding techniques

Assuming that the proximity of the victim and culprit harnesses is fixed, the only way of reducing the coupling is by interposing an electric shield between the two wires. This is usually accomplished by using screened wires. From Fig. 4.98 (a) we can see that although Cs is increased due to the larger diameter of the shield, this capacitance is irrelevant as it is between two nominally grounded shields.

The assumption that the impedances of the shields to ground are zero may not be true. Even a very small inductive impedance, as shown by $L1$ and $L2$ in Fig. 4.98 (b), can reduce the effectiveness of the screen severely at high frequencies. With an inductive bond $L1$ of only 10 nH, a capacitance Cc between the inner and outer of the screened cable of 100 pF and a source impedance Zcs of 50 Ω, the signal on the screen of the sourcing cable, at 100 MHz, is reduced by less than 30 dB relative to its inner conductor.

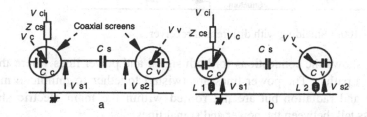

Fig. 4.98. Electric screening; (a) ideal ; (b) reduced screening due to inductive ground straps.

If the coupling capacitance Cs between the two shields is 10 pF, the attenuation of the signal at the second shield $Vs2$ relative to $Vs1$ will be less than 40 dB. The coupling between the shield and inner wire of the second cable may be close to unity if the impedances of the circuit are high. Thus the overall attenuation is limited to less than 70 dB by an inductive bond as low as 10 nH. To reduce the impedance of the bond strap, thick copper braid can be used which is equivalent to several wires and hence inductors, in parallel.

The question of where the screen should be grounded to maximise the Faraday cage effect is important. The screen length must not become a significant fraction of the wavelength λ of any likely source of EMR. The worst case arises when the screen length equals $\lambda/4$ as it becomes an effective $1/4$ wave dipole which can produce a high impedance at the non−grounded end. Although there is no absolute criterion for grounding screens, a reasonable guideline is to ground every $l/5$ where l is the shortest wavelength that may interfere with the receiving circuitry. It is also important to ensure that the Faraday cage does not negate the harness configuration designed to combat the magnetically coupled interference. Grounding configurations which provide both electric and magnetic shielding, in differing situations, are shown in Figs. 4.99 and 4.100. Although the electric shield and chassis do form a magnetic loop this is not significant as it does not contain any signal paths or returns.

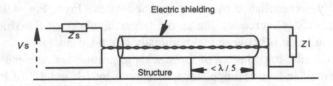

Fig. 4.99.   Shielding with receiver isolated from structure.

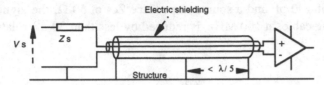

Fig. 4.100.   Shielding with differential receiver.

Fig. 4.101 shows the connections for both signal and power lines where there are multiple signal paths. The power lines are twisted together to minimize magnetic susceptibility and radiation but are not routed within the main electric shield to minimise cross talk between the power and signal lines.

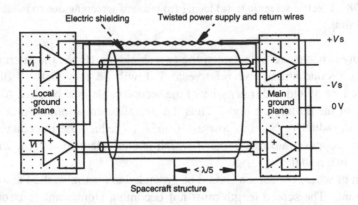

Fig. 4.101.   Shielding with multiple signal paths.

As the power supply return is the only connection between the two ground planes, its impedance must be made as low as possible to minimise the common mode noise caused by currents flowing along this wire. Where this is a problem, reducing the impedance between the two ground planes by connecting both ends of the signal screens to their respective ground planes, even though this introduces ground loop noise, may actually reduce the total noise, if the CMR capability of the second stage of amplification is inadequate.

Within a ground plane the return paths of differing parts of a subsystem may need to be kept separate. If this is not done the results illustrated by the ADC circuit in Fig. 4.102 may arise. Here the common impedance of the ground plane introduces a noise voltage Vn into the signal path due to the noisy current I, and this is added onto the input signal before it reaches the ADC.

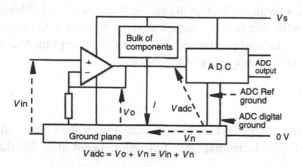

$Vadc = Vo + Vn = Vin + Vn$

Fig. 4.102.   Single ground plane.

This problem may be overcome by the use of split ground planes as shown in Fig. 4.103. In this case, the input buffer amplifier and the ADC's analogue ground reference are taken to the signal ground plane which has negligible currents flowing in it and hence Vns is very much less than Vnp.

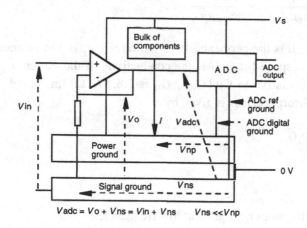

$Vadc = Vo + Vns = Vin + Vns$ $Vns \ll Vnp$

Fig. 4.103.   Power and signal ground planes.

4.6.4   Shielding efficiency

Due to the EMC problems that can occur in spacecraft, it is important to calculate the degree of shielding that is necessary to preserve the integrity of the signal that is to be measured and to ensure that the associated circuits do not emit radiation that would violate the spacecraft's EMC specifications.

When an EM wave strikes a shield there is partial reflection at the surface, partial absorption in the shield and a further reflection at the exit from the shield although this is usually negligible. The reflection loss is complex as it depends on both the structure of the shield and the type of radiation involved, whereas the absorption loss is only dependent on the structure and the frequency of the radiation.

Absorption losses

The absorption loss A, expressed in dB, is deduced from the reciprocal of the exponential attenuation of the electromagnetic field as it penetrates the shield or screen.

$$A = 20 \log \left(\left(e^{-t/\delta} \right)^{-1} \right) = 20 \log \left(e^{t/\delta} \right) = 20t/\delta \log e = 8.7t/\delta$$

where t = thickness of shield and δ is the skin depth of the radiation in the material.
From electromagnetic theory it can be shown that the skin depth is defined by

$$\delta = \frac{1}{\sqrt{\pi f \mu G}} = \frac{1}{\sqrt{\pi f \mu_0 \mu_r G_0 G_r}} = \frac{1}{15 \sqrt{f \mu_r G_r}}$$

where f = frequency in Hz, μ is the permeability of the screen, μ_r is the permeability of the screen relative to free space (μ_0), G is the conductivity of the screen, G_r is the conductivity of the screen relative to Cu (G_0), G_0 is $5.8.10^7$ $\Omega^{-1}m^{-1}$ and μ_0 is $4\pi 10^{-7}$ Hm^{-1}. Thus the absorption loss is given by

$$A = 8.7t \times 15 \times \sqrt{f \mu_r G_r}$$

$$= 130t \sqrt{f \mu_r G_r} \, \text{dB}$$

$$= 130t \sqrt{f} \, \text{dB for a Cu screen}$$

As $A \propto \sqrt{f}$, the absorption becomes negligible at low frequencies.

♦ *Example 4.10.   Absorption of EMR by copper screen*
For a copper screen 0.1 mm thick the absorption loss, using the above equation, at 100 MHz is 130 dB,, at 10 MHz is 40dB and at 10 kHz is only 1.3 dB.

Reflection losses

The reflection loss R is calculated from the ratio of the wave impedance of the EMR Z_w to the surface impedance Z_s of the shield material.   Thus $R = f(Z_w/Z_s)$. The value of Z_w is shown in the near and far field regions for both electric and magnetic fields in Fig. 4.104.

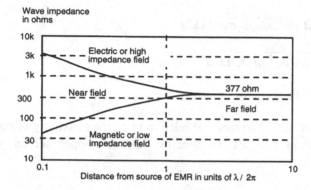

Fig. 4.104.   Wave impedance of EMR.

If r is the separation of the victim and culprit systems, Z_w varies from $Z_0 = \sqrt{\mu_0/\varepsilon_0} = 377\ \Omega$ (ε_0 is the permittivity of free space and equals 8.85×10^{-12} Fm^{-1}) in the far field region, to $Z_0\lambda/2\pi r$ in the near electric field region and to $Z_0 2\pi r/\lambda$ in the near magnetic field region.

In the near field EMR, we can see that at low frequencies, for electric fields, reflection loss increases, whereas for magnetic fields, reflection loss decreases.

In the usual situation where $Z_w \gg Z_s$, the worst case reflection loss, at normal incidence, is given by

$$R = 20\log\left(\frac{Z_w}{4Z_s}\right)\text{dB}$$

Intrinsic surface impedance of a conductor

From general electromagnetic theory it can be shown that if the skin depth δ of the screen is much less than its thickness t, its intrinsic

surface impedance Z_S is given by $\sqrt{\dfrac{j\omega\mu}{G + j\omega\varepsilon}}$, where $\varepsilon = \varepsilon_0\,\varepsilon_r$,

$\varepsilon_0 = 8.85 \times 10^{-12}$ Fm^{-1} and ε_r is its relative permittivity, σ is its resistivity in Ωm ($G = 1/\sigma$) and ω is the angular frequency in radians/s. As $G \gg \omega\varepsilon$ at all radio frequencies for metals, the intrinsic surface impedance of EMR shielding material is given by

$$Z_s = \sqrt{j\omega f\mu\sigma} = (1+j)\sqrt{\pi f\mu\sigma} \ \ \Omega/\square$$

$$|Z_s| = \sqrt{2\pi f\mu\sigma} \ \Omega/\square = \sqrt{2\pi f\mu_0\mu_r\sigma_0\sigma_r} \ \ \Omega/\square$$

$$= \sqrt{2\pi 4\pi 10^{-7} \times 1.72 \times 10^{-8} f\mu_r\sigma_r} \ \ \Omega/\square$$

$$= 3.7 \times 10^{-7}\sqrt{f\mu_r\sigma_r} \ \ \Omega/\square$$

where σ_r is the resistivity of the shield material relative to σ_0 the resistivity of copper. However δ is not less than t at all frequencies. For a copper screen where $G_r = \mu_r = 1$

$$\delta = \frac{1}{15\sqrt{f}} \text{m} \quad \text{or} \quad \frac{66}{\sqrt{f}} \text{mm}$$

Thus at 1 MHz the skin depth is 0.066 mm and at 10 kHz this has risen to 0.66 mm. Hence for a 0.1 mm thick copper screen, the formula deduced for Z_s does not apply at 10 kHz. If we assume an exponential variation of the surface impedance Z_s' with depth, then

$$Z_s' = \frac{Z_s}{1-e^{-t/\delta}} = \frac{Z_s}{k} \text{ where } k = 1-e^{-t/\delta} = 1-e^{15t\sqrt{f\mu_r G_r}} \quad (= Z_s \text{ when } \delta \ll t)$$

If $\delta \gg t$, $k = 1 - e^{-t/\delta} = 1 - (1 - t/\delta) = t/\delta$

and $Z_s' = Z_s / k = Z_s \delta / t = \sqrt{\dfrac{2\pi f\mu\sigma}{\pi f\mu G} \dfrac{1}{t}} = \sqrt{2\sigma/t} \ \ \Omega/\square$.

Where δ is comparable with t, the k term should be included in near field reflection loss calculations.

Far field reflection losses

If we are operating at a high frequency where far field conditions apply, $Z_w = Z_0 = 120\pi = 377 \ \Omega$ and $\delta \ll t$, the reflection losses R are given by

$$R = 20 \log\left(\frac{Z_w}{4Z_s}\right) = 20 \log\left(\frac{94}{3.7 \times 10^{-7} \sqrt{\mu_r \sigma_r f}}\right)$$

$$= 20 \log 256 \times 10^6 - 20 \log \sqrt{\mu_r \sigma_r f}$$

$$= 168 - 10 \log (\mu_r \sigma_r f)) \quad \text{dB} = 168 - 10 \log f \text{ for Cu}$$

♦ *Example 4.11.   Far field reflection loss for copper screen*
The reflection loss for copper, using the above equation, at 100 MHz is 88 dB.

Near field electric reflection losses

The near field electric wave impedance $Z_w = Z_0 \lambda / 2\pi r = 120\pi \lambda / 2\pi r = 60 \lambda / r$. If we substitute c/f for λ, where c is the velocity of light and f is frequency, the electric reflection loss in dB is given by

$$R = 20 \log\frac{kZ_w}{4Z_s} = 20 \log\frac{k60\lambda}{4Z_s} = 20 \log\frac{15kc}{Z_s rf}$$

$$= 20 \log\frac{15\left(1 - e^{-15t\sqrt{f\mu_r G_r}}\right) \times 3 \times 10^8}{3.7 \times 10^{-7} \sqrt{\mu_r \sigma_r f} \times rf}$$

$$= 322 - 10 \log (\mu_r \sigma_r r^2 f^3) + 20 \log\left(1 - e^{-15t\sqrt{f\mu_r G_r}}\right)$$

$$= 322 - 10 \log (r^2 f^3) + 20 \log\left(1 - e^{-15t\sqrt{f}}\right) \text{ dB for Cu}$$

♦ *Example 4.12.   Near field electric reflection loss of copper screen*
Using the same copper screen as before and assuming the separation r of the source of EMR from the screen is 0.5 metre, using the above equation, the electric reflection loss at 10 MHz is 120 dB and at 10 kHz is 191 dB.

Near field magnetic reflection losses

The near field magnetic wave impedance $Z_w = Z_0 2\pi r / \lambda = 120\pi 2\pi r / \lambda = 240\pi^2 r / \lambda$, and hence the magnetic reflection loss in dB, for $Z_w \gg Z_s$, is given by

$$R = 20 \log\frac{kZ_w}{4Z_s} = 20 \log\frac{k240\pi^2 r}{4Z_s \lambda} = 20 \log\frac{k60\pi^2 rf}{Z_s c}$$

$$= 20 \log \frac{\left(1 - e^{-15t\sqrt{\mu_r G_r f}}\right) 60 \pi^2 r f}{3.7 \times 10^{-7} \sqrt{f \mu_r \sigma_r} \times 3 \times 10^8}$$

$$= 14.5 + 10 \log \frac{r^2 f}{\mu_r \sigma_r} + 20 \log \left(1 - e^{-15t\sqrt{\mu_r G_r f}}\right) \text{ dB}$$

$$= 14.5 + 10 \log(fr^2) + 20 \log \left(1 - e^{-15t\sqrt{f}}\right) \text{ dB for Cu}$$

♦ *Example 4.13. Near field magnetic reflection loss of copper screen*

Using the same specifications as before the magnetic reflection loss at 10 MHz is 78.5 dB and at 10 kHz is 31 dB. The above equation is not valid below a few hundred Hz due to the falling value of Z_w at low frequencies and the requirement for $Z_w \gg Z_s$.

Summarizing the absorption and reflection losses in dB for the 0.1 mm thick copper screen 0.5 m away from the culprit radiator produces the data shown in Table 4.6.

Table 4.6.   *Screening losses of a copper shield at various frequencies*

Frequency in Hz		10 k	10 M	100 M	
Near Magnetic Field	Absorption	1	40		dB
	Reflection	31	78		dB
	Total	**32**	**118**		**dB**
Near Electric Field	Absorption	1	40		dB
	Reflection	191	120		dB
	Total	**192**	**160**		**dB**
Far Field E and M	Absorption			130	dB
	Reflection			90	dB
	Total			**220**	**dB**

This shows that magnetic screening at low frequencies is the most severe problem.

For all these calculations it has been assumed that the screening is adequately grounded so that it is not resonant at any significant frequency. If this is not the case then the screening can become completely ineffective as previously described.

In EMC calculations associated with harnesses, the probability of reaching an answer that is within an order of magnitude of the actual case is not good. This is due to the complexity of the fields within the spacecraft, the lack of detailed knowledge of the culprit sources and the quality of the twisting and screening and local capacity to ground of the victim wires. Hence if EMC is likely to be a problem then all the reduction techniques discussed here will probably need to be invoked.

It is important to note that high–frequency EMR can be rectified by non–linearities in low–frequency amplifiers. Hence any low–frequency modulation of the EMR can be demodulated into the pass band of low–frequency amplifiers which initially might be thought to be immune to the interference. Thus harnesses and circuitry associated with low–frequency amplifiers should still be designed to minimise their susceptibility to high–frequency EMR.

Although most spacecraft will have their own EMC specifications, US MIL – STD – 461 is frequently the basis from which they are developed, with US MIL – STD – 462 specifying the measurement procedures.

4.6.5   *Outgassing requirements and EMC*

It is usually necessary to provide holes in the screens of electronic subsystems to allow trapped gases to escape during the launch. However these holes can provide a path for EMR and this must be prevented. One technique is to design the holes as a waveguide with a cut – off frequency above the highest frequency which might lead to problems.

For a rectangular waveguide, if λ_g is the guided wavelength and λ_0 the wavelength in vacuum,

$$\frac{1}{\lambda_g^2} = \frac{1}{\lambda_0^2} - \left[\left(\frac{l}{2a}\right)^2 + \left(\frac{m}{2b}\right)^2\right] = \frac{1}{\lambda_0^2} - \frac{1}{\lambda_c^2}$$

where a and b are the cross–sectional dimensions of the waveguide, l and m are the propagation modes and λ_c is a critical wavelength defined

by
$$\frac{1}{\lambda_c^2} = \left[\left(\frac{l}{2a}\right)^2 + \left(\frac{m}{2b}\right)^2\right].$$

If $\lambda_0 > \lambda_c$, $1/\lambda_g^2$ is negative and thus cannot propagate in the guide. Thus any frequencies less than f_c, where $f_c = c/\lambda_c$, will be highly attenuated. If we let $l = 1$ and $m = 0$, $1/\lambda_c = 1/2a$ or $\lambda_c = 2a$, and this defines the lowest possible cut–off frequency. For all other values of l and m, $\lambda_c < 2a$. Thus the waveguide must be designed so that its smallest cross–sectional dimension is less than $\lambda_c/2$. This gives us a cut–off frequency f_c defined by $c/2a$ Hz for a rectangular waveguide. If the waveguide is circular with diameter a expressed in mm, the cut–off frequency is given by $176/a$ GHz.

For the venting system to function as a waveguide the ratio of the length of the tube to its diameter must be at least 10, and this usually limits the hole size to about a mm.

As this will probably not provide sufficient area for outgassing requirements, several holes will be needed.

Details of the shielding properties of a variety of materials can be found in White, [1980].

4.6.6   Connectors

Spacecraft connectors can be broadly divided into two main groups. Those which interface between boxes or subsystems and those within a subsystem which interface between PCBs. There are also the more specialized connectors for coaxial cables and high voltage wiring.

Low voltage harness connectors

Connectors have to provide good electrical contact and be mechanically stable. The latter implies that the mated pair should be locked together so that demating cannot occur during the launch vibration. Multiway connectors are usually made in one of two geometries, circular or rectangular. For high packing density the rectangular style typified by the 'sub miniature D type' is often used. These have two or three rows of pins on a 0.1 inch pitch.

Crimping harness wires to removable connector pins is preferable to the use of fixed solder bucket type pins as crimping produces a more standardized result and there are no flux residues to remove. For a given gauge of wire, size of pin, pin locater in the crimping tool and crimp diameter, the strength of the crimped joint is well defined and samples can be tested before inserting into the connector to verify the crimping technique. Hence reliability is improved. Typical strengths of crimped joints, using stranded silver plated copper wire, are shown in Table 4.7.

Table 4.7.   *Crimped wire strengths*.

Wire Gauge AWG	Crimp Strength N
20	90
22	50
24	40
26	20
28	12

Fig. 4.105 (a) shows how the crimped pin should appear, with the gap between the end of the wire's insulation and the start of the pin specified by $0 < s < d$ so that there is no possibility of the crimp occurring on the insulation. A very thin heat shrink sleeve over the crimped part of the pin, the exposed wire and the first few mm of the insulation can be used to relieve the strain at the top of the pin. Although mechanical strippers for insulation removal are acceptable for wires thicker than 20 AWG, smaller gauge wires should be stripped with thermal strippers. Apart from the possibility of mechanical strippers creating nicks in the wire, the mechanical force exerted on the wire can move it relative to its insulation and when it creeps back, this will upset the specification for 's'.

If screened wires have to be used with normal connectors, a small length of insulated 24 AWG wire must be soldered onto the screen and the pin crimped to that wire as shown in Fig. 4.105 (b). The lack of screening between pins in the connector can be accommodated by grounding intermediate ones as shown in Fig. 4.105 (c).

For very high packing densities a micro miniature version with a 0.05 inch pin pitch is available. These also use crimped pins but they are permanently installed during the manufacturing process. Thus the precise harness length has to be specified to the manufacturer if micro Ds are to be used both ends and this information may not be available early enough to meet the overall spacecraft schedules.

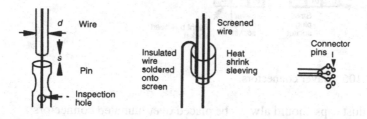

Fig. 4.105. Terminating wires; (a) Crimped pin; (b) Screened wire;
(c) Screened wire at multipin conector.

The dielectric contained by the connector must be acceptable to the spacecraft's standards for low vapour pressure materials and be mechanically sound. Di–Allyl Phthalate or DAP is commonly used.

The connector shells and pins must also comply with various specifications such as very low ferromagnetic content. Cadmium or zinc plating is generally not acceptable due to their relatively high vapour pressure and a tendency to form whiskers in vacuum. Various alternative finishes are available with gold over nickel preferred, but hard anodized aluminium and electroless nickel plate may be acceptable.

Circular connectors tend to be more robust than the D type and are used where the packing density is not a severe problem. They are generally locked using some form of bayonet arrangement. A variety called 'scoop proof' is designed so that before the pins and sockets mate, the two connector shells must be perfectly aligned thus preventing any deformation of the pins or sockets. This is an obvious good design but, of course, further increases their size.

The possibility of bending pins on D type connectors is high unless care is taken to ensure they are mated and demated correctly. For large connectors which have high insertion and withdrawal forces, it is good practice to provide special tools to ensure this problem does not occur.

It is best to arrange power and signal wires in different connectors for EMC reasons, but if this is not feasible then maintaining them at opposite ends of the connector is the next best arrangement. It is also sensible to choose a connector size that gives a few spare pins to allow for unexpected developments.

To maintain the reliability of connectors, the number of insertions should be noted and limited to about 50. As this number can easily be exceeded in a complex test programme the technique of using connector savers is invoked. This consists of back – to – back plugs and sockets inserted into the harness and bulkhead connectors, as shown in Fig. 4.106. Thus the constant mating and demating only occurs at the saver interface and eventually the saver connectors are removed and discarded.

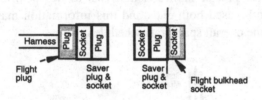

Fig. 4.106. Saver connectors.

Protective dust caps should always be placed over unmated connectors.

Locking techniques

Bulkhead D type connectors are fastened to the chassis using tapped posts with studs. The harness connectors are then fastened to the tapped post with screwlocks. If brass posts are required (non – magnetic) it is essential that the maximum torque recommended for the post is not exceeded so as not to shear the small diameter stud. However it is necessary that they are adequately torqued so that inserting the screwlock does not loosen the post. Before flight all screwlocks should be locked in place with a small amount of an acceptable epoxy adhesive.

Circular connectors may be designed with a bayonet or similar type of locking device or require wire locking through special holes in the plug and socket. No system which can be demated by tension in the harness should be employed. The use of back shells to clamp the wires at the back of D type connectors is good mechanical practice but is often not carried out due to lack of space. However if good EM screening is required then a metallic backshell will be required with the cable screen forming an RF tight bond to the backshell. This is more easily carried out with circular connectors.

Connectors are available with built in RF filters covering the frequency range 10 kHz to 10 GHz. This is a very volume efficient method of filtering rather than individual line feed–through filters mounted on the bulkhead. The filter section should be of the Π type as this attenuates both the received RF and any RF that may be emitted by the box. Care must be taken to ensure that the filter capacitors, which may be as large as 1 μF, do not adversely affect the required signals in terms of rise time or the stability of the driving circuitry. Optimum filtering would require different filter characteristics for the signal and the power lines.

When specifying connectors it is important to note whether the male or female pins are more exposed. Thus connectors on harnesses, which are more likely to be mechanically damaged, should be of the more protected type. This argument takes priority over specifying that pins supplying power should be of the protected type.

High voltage connectors

High voltage connectors are particularly difficult to construct due to the problems of corona breakdown occurring in the appropriate pressure region. High voltages should be conducted along screened wires. If no grounded screen is used the outside of the insulator will eventually charge up to the full voltage on the inner conductor. This can then discharge causing noise problems both to the high voltage user and other subsystems. Thus high voltage connectors will usually be of the coaxial type. To ensure that any trapped gases within the connector cannot cause problems, a type of connector can be used which presents a high impedance path between the pin and the outer screen so that the gases cannot acquire sufficient energy to become ionized. This is carried out by means of an O ring seal as shown in Fig. 4.107.

The O ring does not need to provide an airtight joint but is to provide sufficient impedance that most of the electric potential appears across the ring rather than across any trapped gases. These types of connectors have been successfully flown on many spacecraft.

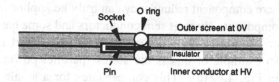

Fig. 4.107.   High voltage connector.

PCB connectors

PCB connectors should be of the two part type and conform to the material specifications imposed on harness connectors. They should always be fastened to the PCB before the pins are soldered to minimize any stress on the pins. If harnesses have to be soldered directly to PCBs, it is vital that stress relief is provided to prevent flexing of the wires, during vibration, occurring at the wire–solder interface which eventually results in a breakage. Stress relief can usually be accomplished by providing tie–down holes in the PCB within a cm or two of the joint.

4.7   Reliability

4.7.1   Introduction

To produce a reliable electronic subsystem, one has to consider the reliability of the design, the quality of the components and the techniques used in the production and assembly of the circuit boards and their interfaces. As an error in any of these stages reduces the reliability of the whole subsystem, it is reasonable to state that;

Overall reliability = Design reliability × Component reliability × Fabrication reliability

4.7.2   Design techniques

All components have sets of limiting operational and storage parameters associated with them. A good design will ensure that these limitations are never exceeded and should only be approached with great caution. Advantage should not be taken of extreme properties as these will probably degrade to the same absolute values as those of average components and hence will actually degrade by a much larger factor. It is essential to read the manufacturer's specifications very carefully and to query any ambiguous statements.

There are few instances where component redundancy can truly be applied and this is usually limited to passive components, discreet semiconductors and some basic logic gates. Component redundancy should be able to cope with either a short or open circuit of the component. An example of this is the series–parallel power supply decoupling network shown in Fig. 4.82 (a). This compensates for a single failure, either short or open circuit, though not without a loss of overall capacitance. Each capacitor must be rated as if it were the sole component. The problem with such techniques is that failures before launch are not easily detectable.

Power to each PCB should be current limited within the PCB but one must ensure that the faulty board cannot be powered by another board to which it has signal interfaces. This can occur with Complimentary MOS (CMOS) devices through their input protection diodes and hence special interfaces are needed between such boards.

4.7.3   Heat dissipation

When considering the derating factors to be applied to a subsystem, power dissipation must be viewed with much more concern than when convection cooling is available. Even with sealed units containing gas, convection still does not take place due to the lack of gravity, and forced motion of the gas is required.

Components on a simple PCB without a ground plane will lose most of their heat by radiation. If the PCB is in a stack of similar PCBs, such losses may be minimal due to radiation being received from adjacent cards. A large improvement can be made using PCBs with a continuous ground plane which is adequately thermally connected to the spacecraft chassis or a radiator panel. In this case normal deratings will generally apply. However components dissipating more than a few watts on a standard Eurocard may need thermal straps from the top of the components to an adequate heat sink. Thus the thermal design of the card mounting system is of great importance and the position on the card of high power dissipating components must be considered from the thermal viewpoint rather than the signal flow requirements.

Components dissipating large powers such as transistors in regulators, should be mounted onto the subsystem's chassis which should also be well thermally grounded to the spacecraft's chassis or other heat dissipating network. As the device will need to be electrically isolated from the chassis for EMC requirements, some form of insulating washers will be required. Brittle washers made from beryllium oxide, which is a good electrical insulator and good thermal conductor, are not considered reliable due to their likelihood of cracking from mechanical stress, vibration or thermal expansion. Flexible washers with low outgassing properties should be used.

4.7.4   Latch-up

This is a problem caused by spurious on–chip PN junctions, normally benign, being made active by some unusual condition. This may be a burst of radiation or the device being driven outside its specified operating range. Typically a PNPN element between the supply voltage and the substrate of the IC, which is normally inactive, is turned on, allowing excessive current to flow. This may be sufficient to blow the bond wire or damage the chip and is an example where current limiting is essential. A simple solution to this problem is shown in Fig. 4.108 where current limiting to a few 100 mA is required.

This also illustrates a form of component redundancy which compensates for an open circuit transistor but does not compensate if a short circuit occurs. T3 is a foldback network to prevent the current limiting transistors T1 and T2 from over dissipating if the over current is sustained for a long period. The delay network associated with $R2$, $R3$ and C is incorporated to allow any filter capacitors on the PCB to become fully charged without tripping the foldback network. The limiting current is approximately given by $\beta V/R1$ and this applies at all times. $R5$ and $R6$ ensure that the current is divided approximately equally between T1 and T2. Paralleling the pass transistors minimises the potential drop across the system under normal operating conditions compared with using a higher powered device. This circuit requires resetting by removal of the input power. Automatic resetting after a specific time could be introduced, but power dropouts should always be monitored.

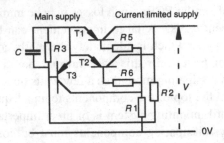

Fig. 4.108. Simple current limit.

With all current limiting systems, care must be taken to ensure that any power supply decoupling capacitors cannot supply sufficient energy to negate the effect of the limiting circuit. 300 mA for 0.5 ms will fuse an on chip wire bond and at 5 V this corresponds to the energy stored in a 30 uF capacitor In digital circuits a few nF across each IC is much better than a few μF at one location. The use of distributed capacitance between a power plane and a ground plane in a multi–layer board is by far the most effective decoupling technique for removing fast transients and reducing ground bounce.

4.7.5   *Interfaces and single point failures*

Interface components on PCBs are more likely to be damaged than those in other areas. Damage can be caused by induced electrostatic charge, when the board is not plugged into its usual load during test procedures or by surges induced into interface wiring due to problems occurring in other systems. One of the simplest precautions that can be taken in data communication links, is not to let the chip come into direct contact with the outside world. This can be achieved using the circuits shown in Fig. 4.109.

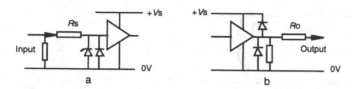

Fig. 4.109.   Interface protection; (a) input; (b) output.

The resistors to ground prevent terminals floating to an indeterminate voltage. Rs and Ro ensure nearly all the power from a spurious electrical surge is dissipated in the resistors rather than in the ICs. The diodes will prevent excessive negative or positive voltage swings from occurring at the input or output pins. Although protection diodes are built into most CMOS ICs, due to their small size they will be rated at a much lower power dissipation than larger external diodes. The Zener diode at the input replaces a diode connected to the positive supply which would allow a signal voltage applied to an input, to power up the power supply line of an unpowered PCB. The double diode protection of an output circuit is acceptable as only surges would be expected here.

When choosing the value for the input resistor Rs, allowance has to be made for the integrating effect it has on the rise time of the input signal both due to the input capacity of the chip and the PCB tracks. A 10 kΩ resistor will typically limit the rise time of a signal applied to single CMOS gate to 100 ns. This feature may be used to advantage to prevent the interface responding to unwanted spikes. The value of Ro may have to be limited to a few tens of ohms, if required to drive a low impedance network, but even at this level it will still offer some protection.

Subsystems containing latches or other bistable components should always be designed so that when power is initially applied, automatic resets are generated to ensure that circuits always start up in the required mode. All hazardous circuits (High Voltage Units, Control mechanisms etc) should be designed to come on in their safe mode.

When using inductors in filters, relays, motors etc, the inductive energy stored during the normal operating mode should be calculated so that the appropriate diode and damping resistors networks can be added to reduce switch off transients to an acceptable level. As the time constant of an Lr circuit is L/r, if only a simple diode is used, as shown in Fig. 4.110 (a), the time constant may be excessively long as the resistive component r, of the inductor may be quite small. Although the addition of R in Fig 4.110 (b) will increase the transient voltage to $Vs + IR$, where I is the inductive current when the switch is closed, the recovery time is reduced to $L/(R+r)$ and this may be a significant feature.

When electrical isolation between two circuits relies on the high impedance due to the gate of a FET, an additional resistance in series with the gate, ensures that if the gate fails the isolation is maintained.

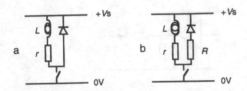

Fig. 4.110.   Dissipating inductive energy.

The removal of single point failure mechanisms, where possible, at interfaces used by more than one subsystem, is a prime design requirement. This is why simple current limiting, of a type that cannot short the supply, is sensible at each PCB.

Switching from a faulty system to its backup will normally be carried out by ground control to allow the fault to be analysed before corrective action is taken.

Failures may not be catastrophic but degrade performance by reduction of mission life, available power, or pointing accuracy. For example loss of data storage would still allow real time operation. As redundant systems are expensive in weight, volume and cost, it is important to ascertain if the failure of a system is catastrophic or acceptable.

The use of redundant units in power supplies, telemetry receiver and command decoder subsystems is desirable. Automatic switching between redundant units in the latter two subsystems is sensible as failure in one of these systems could be catastrophic. The highest reliability components would be required for both systems and switching arrangements. Solar cell array pointing systems may also qualify for redundancy although some time would usually be available, due to on–board batteries, for corrective action to be taken before loss of power became catastrophic.

If a battery regulator fails open circuit, the battery cannot charge but the system can operate in daylight. If it fails short circuit, the battery could be permanently connected to the solar cells which could destroy the batteries with disastrous results. Thus it would be sensible to have series redundancy in the switching circuitry between the regulator and the battery. Evaluating the consequences of a failure is an important design consideration because of the costs of the preventative and backup measures.

4.7.6   Housekeeping

The values of many parameters (e.g. temperature, supply voltages and currents) are not required for normal operation but do give a useful indication of the health and performance of a subsystem. As a large number of these parameters may need to be monitored, they are generally multiplexed at a very low rate in the analogue state and then applied to an ADC for subsequent transmission to the telemetry system.

Digital signals that may need to be monitored at a low rate are the positions of latched relays and the status or positions of mechanisms. An important feature of monitoring is that the monitoring circuit should be as close to the feature being investigated as possible and be extremely reliable. As an example, monitor both the state of a relay switching power to a network, using a redundant pair of contacts, and the power rail itself, as loss of power may be due to a faulty relay or a faulty power bus.

It is very wise to provide a simple, easily accessible (to external test equipment) monitor of the final analog output voltage of a subsystem before it has been digitized. This can supply vital information when attempting to diagnose a fault during a test programme when accessibility to the hardware is extremely limited. A typical problem might be an an increase in the noise threshold, which after digitization, will not initially be apparent because the noise amplitude is below the voltage of the lower level discriminator (see section 4.3.4). This will only become noticeable, via the normal data processing scheme, when a real signal plus this noise triggers the lower level discriminator but determining the nature of the problem from this combined data may be difficult. However if the basic analogue channel can be observed, diagnosis becomes much simpler.

The analogue monitor should be designed with EMC, and the possibility of cross talk, in mind and so only a low impedance attenuated version of the final analogue signal should be made available at the monitor point.

It is clear that to ascertain a problem has occurred may require the presence of a real signal and this may not easily be generated (may require the subsystem to be at very low pressure). Hence it is sensible to include in the subsystem design the ability to generate electronically a simulated signal. This should be an on–board pulse generator commandable on/off by the normal data processing system, rather than an external connector as this would necessitate additional wiring to the sensitive front end of the system and be prone to EMC problems.

Temperature monitors can use thermistors which require only one or two additional resistive components to produce results accurate to about a degree. Voltage monitors can be constructed using resistors only to produce voltages in the range required by the ADCs. When monitoring relay contacts and switches, ensure the voltage supply to the switch is current limited. Typical monitoring networks are shown in Fig. 4.111. The voltages from these monitors V_a, V_b, V_c and V_d will normally be multiplexed to an ADC either within the subsystem or within the spacecraft's data handling system.

Monitoring hot spots such as on transistors in power converters can give early warning of possible failures. This is particularly useful in the design phases.

Monitoring circuits may use the subsystem's own power but it is sometimes necessary to monitor the state of various parameters before this power is applied. This can occur if there are constraints on the operation of a unit outside a given temperature or pressure range. Such monitoring can be accomplished by the use of an independent power line supplied to the subsystem purely for monitoring purposes.

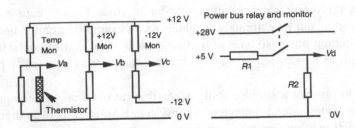

Fig. 4.111. Housekeeping monitors.

Precision monitoring of high voltage supplies demands high value resistive potential dividers with carefully matched voltage coefficients of resistance if the supplies are variable.

Current monitors need to be included before the final regulation to prevent the voltage drop across the current sensing device degrading the regulation.

The wiring to monitoring circuits should be in twisted pairs or adequately screened and filtered to minimise EMC problems, particularly if supplied from an external power source, as the monitors may well be close to sensitive circuits.

4.7.7   Component specification

Due to the cost of any major spacecraft project the reliability of its component subsystems must be made as high as is reasonably possible. As subsystems are made from individual components the first requisite must be to ensure that these are reliable.

Items made under a particular generic code can have vastly different electrical properties depending on the manufacturer, and even between different batches made by the same manufacturer. A typical example is the ubiquitous 741 Operational Amplifier whose specifications are still those which applied when they were first produced. Thus you could develop a circuit which works satisfactorily on an old 741 whose parameters might be 100 times worse in some respects than its modern counterpart. However during the production stage a good modern device but still within the same specification, is used and problems arise because it responds to glitches to which the older one does not. Other types of problem can occur because manufacturers have different methods of testing and specifying the results of their tests.

Thus ESA and the US NASA (United States National Aeronautics & Space Administration) have each developed a precise set of manufacturing, testing and component parameter specifications to overcome problems of this nature and to produce components of a defined failure rate.

The US Military Standard 883 was produced in 1968 to produce an economically feasible standardized IC manufacturing screening procedure which would achieve an in–equipment failure rate of less than 0.08% per 1000 hours for their class B specification and 0.004% per 1000 hours for their class A specification (class A is now called S for space worthy). The standard establishes uniform methods and procedures for testing micro–electronic devices, including basic environmental tests to determine their ability to withstand the more rigorous conditions met in military and space vehicles. However it did not provide the detailed electrical parameter specification associated with specific devices, nor did it contain a mechanism which could be used to enforce the correct interpretation of the standard. These loopholes were closed by the US MIL – M – 38510 (recently replaced by US MIL – I – 38535) or JAN (Joint Army and Navy) IC program. Thus ICs procured to US MIL – M – 38510 have been manufactured and tested to US MIL – S – 883 and have a well defined set of operational parameters (measured in a defined manner), specified on device 'slash' sheets. The 38510 / XXXXXXX designates a specific IC, its slash sheet, the screening level, the package and lead type.

US MIL – S – 883 now establishes three standards of product assurance, level S, B and C. Level S is intended for critical applications such as spacecraft where repairs are almost impossible. Level B is for slightly less critical such as airborne equipment and level C is for reliable ground–based equipment subject to relatively easy access.

Components produced to level C require 100% visual inspection during manufacturing processes, extensive environmental testing, but no extended 'burn – in' is required and parameters are measured at ambient temperature.

Class B components receive all the class C tests plus 160 hours burn–in and 100% electrical tests at high and low temperatures. Also some of the parameter specifications are tighter.

Class S components are produced to a more stringent set of class B test, plus particle impact noise detection, non–destructive bond–pull tests and post–capping radiographic inspection. However the most important distinction between class S and the rest is the requirement for component serialization. This allows full traceability back to the original wafer.

US MIL – S – 19500 is a detailed specification for discrete semi–conductor components such as diodes and transistors, with US MIL – STD – 750 specifying the manufacturing and testing program. Thus US MIL – STD – 750 is the test program for discrete components comparable to US MIL – STD – 883 for ICs, with US MIL – S – 19500 being the discrete component specifications equivalent to the US MIL – M – 38510 for ICs.

Within US MIL – S – 19500 there are 4 levels of reliability called JAN, JANTX, JANTXV and JANS, JANS being the highest. Completed JAN devices are qualified on a sample basis. Completed JANTX devices require 100% testing but not pre–cap inspection. JANTXV and JANS devices require pre–cap inspection. JANS devices also require wafer inspection, particle impact noise detection, radiographic and external visual inspection and serialization with traceability back to the original wafer. They also require device specific slash sheets.

Manufacturers meeting the US MIL – M – 38510 requirements for ICs and the US MIL – S – 19500 requirements for discreet semiconductors, qualify to have these components included in the NASA Goddard Space Flight Centre (NASA / GSFC) Qualified Parts Lists (QPL). These are the US MIL – STD – 975 NASA Standard Electrical, Electronic and Electromechanical Parts List, QPL – 19500 for semiconductors and QPL – 38510 for microcircuits. It enables users to ascertain which devices have been made to this standard and the associated manufacturers, but the task of updating and distributing this literature is such that many devices which meet the specifications do not appear in the QPLs for some considerable time.

Both ESA and NASA produce documents called Preferred Parts Lists (PPLs) ESA PSS – 01 – 603, [1995]; US MIL – STD – 975M, [1994]; PPL20, [1993] which specify sets of parameters associated with a particular type of component, but not all these parts have necessarily been manufactured.

In the UK we have BS 9000 which is the General Requirements for Electronic Components of Assessed Quality and is similar to US MIL – M – 38510 and US MIL – S – 19500 giving a QPL which is specified as BS9002. Subclasses of BS9000 such as BS9400 give general test specifications for ICs and further subdivisions extend to specific ICs as in the slash sheets. Screening procedures specified in BS9400 are shown in Fig. 4.112. Note the importance that is attached to the removal of 'infant mortality' failures in all high reliability testing by the use of high – temperature bakes and burn – ins.

ESA have a similar system developed by their Space Components Coordination (SCC) group and they produce a rigorous set of manufacturing and testing procedures for electronic components. Parameters of devices made to these specifications have to comply to an SCC slash sheet. The ESA specification similar to level S is SCC level B.

Devices which are not made to US MIL – STD – 883 specifications can be 'upscreened' by carrying out the tests specified in the post manufacturing stages. However as the pre – cap (before the lids are sealed to the bases) tests and inspections have not been performed on a 100 % basis they can not be considered to be as good as the fully qualified devices. Nevertheless this procedure has to used when the electronic task can not be carried out with the limited types of qualified devices available.

Pre – cap inspections are carried out to verify such items as uniformity of lead bonds and quality of metallization. However these can be carried out after capping on a destructive sample basis, but this does require that all the ICs come from the same manufacturing batch to ensure the test samples are representative. An acoustic process which can probe the mechanical properties of semiconductor chips has recently been developed (Bonati *et al.*, [1993]) and should provide significant additional information on both pre – capped and capped devices.

Tests such as a 48 hour bake at 150°C, thermal shock 0 to 100°C, constant acceleration and hermeticity with full parameter measurements before and after, can be performed on a batch basis. However certain tests specify a maximum change in some parameters rather than just specifying limits. This requires each device to be serialized. Traceability from the original silicon wafer, through all manufacturing and test processes, to the PCB, ensures the highest reliability in the final product. If a component fails in equipment and analysis shows the failure is due to some manufacturing fault, all devices from that batch can be replaced.

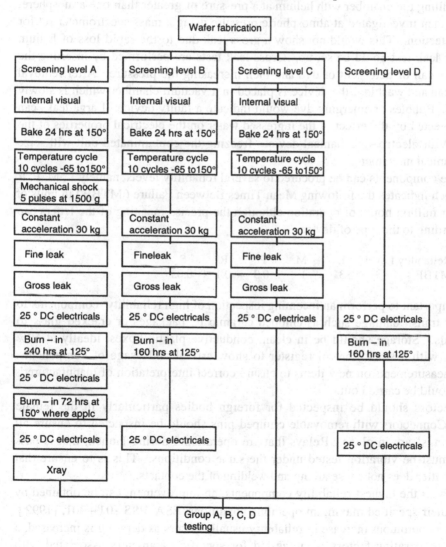

Fig. 4.112. Screening levels BS9400.

The reliability of most electrical and electro–mechanical components is very significantly improved if they are hermetically sealed in a benign atmosphere. This prevents corrosion, the ingress of water and other contaminants, the loss of surface oxide layers permitting cold welding to occur and the reduction in the dielectric strength of gases resulting in corona and arcs as the external pressure reduces. Thus the need for checking the integrity of the seal is most important. A single test may not be able to detect both large and small leaks.

Fine leak tests involve inserting the device in a vacuum chamber, pumping down, and then filling the chamber with helium at a pressure of greater than one atmosphere. Leaks are then investigated at atmospheric pressure using a mass spectrometer set for helium detection. This would not show a gross leak due to the rapid loss of helium before the leak test could be set up. Gross leak tests are performed by inserting the device in a can of detergent or dye and then increasing the pressure. After removal from the can and washing, the device is placed in a vacuum chamber which is slowly evacuated. Bubbles or emerging dye would indicate a faulty device. Large leaks can also be detected by the effect of the ingressed water on the electrical properties of the device. With electro–mechanical devices freezing the contaminated unit will seize the mechanical movement.

Passive components can be procured to several reliability levels, namely L, M, P, R and S which indicates the following Mean Times Between Failure (MTBF) in units of failure per million hours of operation although the precise meaning of the letters can vary according to the type of device.

Reliability Level	L	M	P	R	S
MTBF	3	1	0.3	0.1	0.03

It is important to perform an incoming inspection of high reliability components to check for transit damage such as chipped ceramic IC packages or cracked glass to metal seals. Storage should be in clean, conductive plastic boxes, ideally at 50% humidity, with a stock control register to show arrival and usage of components. Sample measurements on new items to ensure correct interpretation of manufacturer's coding should be carried out.

Connectors should be inspected for foreign bodies particularly in the female portion. Connectors with removable crimped pins should be inspected to ensure all 'napkin rings' are in place. Relays that are operational during launch, that is pass current, must be vibration tested under the same conditions. This is to ensure that contact chatter does not cause arcing and welding of the contacts.

Even with the highest reliability components, an improvement can be obtained by derating their specified maximum operating conditions ESA PSS–01–301, [1992]. Although a continuous increase in reliability usually occurs as derating is increased, a fixed set of derating factors is suggested for specific parameters associated with various devices. All the 'per cent' values quoted below refer to the factor that the recommended maximum operating specification should be multiplied by.

Resistors: Voltage 75%; Power 50%. Chip resistors should be additionally derated above 20°C to zero power at 100°C. This also applies to high precision resistors.

Capacitors: Voltage 50%; Maximum operating temperature 85°C; Inrush current 75%.

Solid Tantalum capacitors with an effective series circuit impedance of Rs: For $0.1 < Rs < 3$ ohms per volt, Voltage 40%; Maximum operating temperature 50°C.

Connectors: pin / pin and pin / shell voltage 50%; Current 50%; Mates plus demates 50; Reduce maximum operating temperature by 30°C.

Wires: Voltage 50%; Maximum operating temperature 85°C;

Current rating in amps for various wire gauges

AWG	28	26	24	22	20	18
Currents	1.8	2.5	3.3	4.5	6.5	9.2

In bundles these need to be further derated to 50% for more than 15 wires. However, efficient distribution of power may call for even further derating. For example a 1 A current supplied along 3 metres of wire requiring a voltage drop of less than 0.1 Volt indicates a maximum wire resistance of 0.1 Ω.This can only be met with 20 AWG wire (at 33 Ω per 1000 m) rather than the 28 AWG indicated.

Fuses: Avoid where possible but if unavoidable, use at least two in parallel each at the rated blow current.

Diodes: PIV 65%; Current 60% at 25°C derating to zero at 115°C.

Transistors: Vcb and Vce 65%; Ic and Ib 75%; Power 60% at 25°C derating to zero at 115°C within safe operating area; Maximum operating temperature 115°C.

Linear ICs: 80% Max supply voltage; 70% maximum voltage input; Maximum operating temperature 85°C.

LEDs: Current limited to 20% of specified maximum.

Digital ICs: Maximum operating temperature 85°C, Supply voltage greater than the minimum specified but less than 80% of the maximum.

For missions lasting several years, allow for parametric changes due to aging.

Transistors and diodes: Design for leakage current 5 times nominal maximum at the specified temperature. Allow for an increase in forward voltage of PN junction of 10%.

Transistors: Allow for a 25% decrease in current gain and a 10% increase in $Vsat$.

Linear ICs: Allow for a change of 1mV in Vos, a 25% change in supply current and similar changes in leakage currents as for transistors. Devices which have minute input offset currents (Ios) but significant bias currents (Ib) may change their Ios by orders of magnitude as Ib changes.

Linear IC regulators: 0.25% change in output voltage.

All the above deratings are highly desirable but may not be practicable, particularly in terms of the maximum temperatures caused by the sum of the ambient temperature and that due to power dissipation. In this case there is a simple empirical rule that states that the lifetime of a semiconductor device decreases by a factor of 2 for every 10°C rise in temperature. For metal film resistors the doubling factor is for a 30°C rise, and for ceramic and tantalum capacitors it is for a 20°C rise.

For tantalum capacitors the voltage acceleration factor is very dependent on the circuits series impedance Rs. For an Rs of 0.1 Ω per volt of operating voltage, a voltage doubling within the rated voltage, reduces lifetime by a factor of 10 whereas for an Rs of 3 Ω/V, voltage doubling hardly affects lifetime. Details of these failure rates are shown in the ESA publication ESA PSS − 01 − 301, [1992].

In addition to the above, changes due to radiation damage have to be considered in terms of total dose and are specific to device type.

The choice of materials associated with electronic subsystems for space use is assisted by the following publications; ESA PSS − 01 − 701, [1990]; NASA Materials Selection Guide, [1990] and the Handbook for the Selection of Electrical, Electronic and Electromechanical Parts for GSFC Spaceflight Applications, (US 311 − HNBK − 001, [1990]).

4.7.8   Failure rates

The overall failure rate of a subsystem can be calculated using the failure rates associated with the individual components as indicated in US MIL−HDBK−217F, [1991]. The standard formula is

$$R = e^{-\lambda t}$$

where λ is the failure rate and R is the reliability of operating for a time t. The total failure rate for n similar components $= n\lambda$.

♦ *Example 4.14.   Failure rate calculations*
Metal film resistors with a base failure rate of 0.45×10^{-9} hr^{-1} are procured to level S reliability, improving the failure rate to 0.0045×10^{-9} hr^{-1}. If derating the power and voltage give a further improvement in reliability by a factor of 5 and there are 10 resistors in the circuit, the failure rate due to the resistors is $0.0045 \times 10^{-9} \times 10/5 = 0.009 \times 10^{-9}$ hr^{-1}. If the circuit also contains one amplifier, procured to level B, with a failure rate of 12×10^{-9} hr^{-1}, and its operating parameters are derated to produce an improvement in reliability by a factor of 4, the failure rate due to the IC is given by $12 \times 10^{-9}/4 = 3 \times 10^{-9}$ hr^{-1}. Thus the total failure rate $= 3.009 \times 10^{-9}$ hr^{-1}. Hence the reliability for a three year mission is given by

$$e^{-3.009 \times 24 \times 365 \times 3 \times 10^{-9}} = 0.99992.$$

4.7.9   Outgassing

Low outgassing materials are required for two main reasons. Gases may condense on optical or other sensitive surfaces and degrade their performance. Spin–stabilized spacecraft are very carefully spin balanced before launch. Any loss of mass asymmetrically will cause the spacecraft to nutate or wobble about its spin axis and so lose its pointing accuracy.

A generally accepted criterion for materials is that they must not lose more than 1% of their total mass (Total Mass Loss TML) and they must not produce more than 0.1% of condensable materials (Collected Volatile Condensable Materials CVCM). These specifications must be considered in conjunction with the total quantities of the material used in the spacecraft. Also materials used close to, or in hermetically sealed enclosures with optical components or surfaces would require specifications several times smaller than the general limits.

The type of electronic materials that need to be considered are cable insulation, heat shrink and other sleeving, conformal coatings, PCBs and other insulating sheet materials, lacing cords, potting compounds, adhesives, dielectric materials in connectors, grommets, card guides, thermal blankets, marking materials and paints.

The outgassing performance of many of these materials can be obtained from ESA or NASA publications: Campbell et al., [1993]; ESA RD – 01, [1992]. Also see section 2.7.3 for more details on this topic.

4.7.10   Fabrication

There are many detailed specifications for the fabrication of high reliability electronic subsystems ESA PSS– 01 – 708, [1985]; ESA PSS – 01 – 710, [1985]; ESA PSS – 01 – 726, [1990]; ESA PSS – 01 – 728, [1991]; US MIL – HDBK – 338, [1984]. However one feature that cannot be stressed too strongly is the regular independent inspection of production processes and the discussion of any inconsistencies between the design, production and management teams. Lack of this activity can prove to be very costly! Although this chapter cannot detail all the techniques described within the above publications, there are a number of processes which are worth emphasising.

Leads emanating from glass to metal seals should not be bent at the metal glass interface as this can easily chip or crack the glass. This is particularly important with hermetically sealed, gas filled relays, where vacuum operation could cause cold welding to occur.

Wire to post connections on PCBs should not be mechanically sound before soldering to ensure that it is the solder that provides both the electrical and mechanical strength. Thus a wire wrap of between 180 and 270 degrees around the post is preferred. This also makes rework much simpler. Joints should show the profile of the wire around post and not be an 'all hiding' blob. Two wires can be wrapped around one post but they should not be twisted.

Component leads must always be bent with round edged tools to prevent nicks occurring and when cutting leads, the shock should always be directed away from the component. Leads should not be bent closer than two lead diameters from the component and the bend radius should not be less than the lead diameter.

Components with radial lead spacing of 0.1 inch or less should not be fitted flush to the PCB as solder rising through the plated through holes, by capillary attraction, can spread out under the component, invisibly bridging adjacent tracks. The use of components with small lugs under them (or the insertion of some equivalent material) will prevent this happening.

If a component weighs more than a few grams, note should be taken of the height of its centre of gravity above the PCB and the size and number of leads. If this indicates inadequate support, additional support such as lacing tape, clips, staking material or other mechanical structures should be employed.

Pure rosin, halide free, fluxes should be used for soldering. Component leads should be pretinned and any gold plating removed in this process. Flux should be washed away as soon as possible after completion of the unit and in any case on a daily basis. Acceptable solvents are ethyl alcohol and iso propyl alcohol but care must be taken due to their flammability.

The use of ultrasonic cleaning systems for PCBs is unacceptable due to the possible mechanical damage to the components. Once cleaned a conformal coat of an approved material may be applied to the PCB. This provides electrical protection against floating conducting debris that might exist in the spacecraft (wire cuttings), protection against atmospheric corrosion, and as additional mechanical support to the components. The type of material should be resilient to allow for differential expansion of coating and components, conform to the outgassing requirements and allow rework to be carried out without too much difficulty. A common material used is Solithane 113 with catalyst C113–300 used in the ratio 100 to 73 which provide a TML of 0.4% and a CVCM of 0.03% Its only drawback is the noxious gases produced on rework at soldering iron temperatures. This necessitates such rework being carried out in a fume cupboard. Other materials are available but they all tend to have noxious decomposition products.

Harnesses should be manufactured away from the spacecraft on a model. This reduces the likelihood of wire clippings entering the spacecraft and enables the harness wiring to proceed while other work is being carried out on the structure. Captive harnesses should be avoided where possible.

4.7.11   Radiation

All semiconductors start to undergo measurable changes in their properties when the accumulated total dosage of radiation (gamma, electron, proton, Xray etc.) exceeds a certain level (Holmes−Siedle, [1993]). This may be as low as 500 rad or in excess of several Mrad for devices specially made to be radiation hard.

The definition of a rad is that amount of radiation of any type that will deposit 100 ergs (10^{-5} J) per gram of material. As the ability to absorb energy from radiation varies with the type of material, there is a more precise definition for semiconductor devices, the rad (Si) where the material in which the radiation is deposited is defined as being Silicon. This is the most common term used by IC manufacturers to present the radiation tolerance of their products. The term 'gray' is also used and is a dose of 100 rad, or the amount of radiation that will deposit 1 J per kg of material.

To relate the radiation dosage to the particle / photon fluxes expected in various orbits, the following units are also used. The particle flux f or number of particles per square cm per second and the particle fluence ft or total number of particles per square cm where t is the lifetime of the spacecraft.

The stopping power s of a material is the ability it has to remove energy from radiation per unit density per unit path length. The dimensions 'per unit density per unit length' is equivalent to 'area per unit mass' which is how s is typically specified. The total dose, fst is the energy absorbed per unit density per unit volume which is equivalent to energy absorbed per unit mass.

♦ *Example 4.15. Radiation total dose calculation*
    If the flux f is 10^4 1 Mev electrons cm^{-2} s^{-1}, the duration of the radiation t is 3 years = 10^8 s, the stopping power s for 1 Mev electrons in Si is 3 Mev cm^2 g^{-1}

Dose = $10^4 \times 10^8 \times 3 = 3 \times 10^{12}$ Mev g^{-1}

1 Mev = 1.6×10^{-6} erg

Dose = $3 \times 10^{12} \times 1.6 \times 10^{-6} = 5 \times 10^6$ erg g^{-1}

    =50 krad (500 gray).

This is a total dosage which would significantly effect most semiconductors.

When calculating the total dosage, the flux is normally integrated over one orbit and an average dose rate specified. Typical fluxes of electrons and protons as a function of spacecraft orbit are available in ESA PSS−01−609, [1993]. If we assume that most electronic circuits will be housed in an enclosed metal structure, if only for EMC reasons, then it is likely that a thickness of at least 1 mm aluminium will be available for radiation screening. If the package is close to the centre of the spacecraft then this thickness will be much greater and even if the package is close to the skin, the area facing inwards will have the entire spacecraft as a shield. Thus a 1 mm screen is a reasonable basis for estimating total dosage.

For low Earth orbits in the altitude range 200 to 1000 km at low inclinations (less than 30°) dose rates up to several krads per year should be considered. At 90° the dose rate may be a factor several times worse. From 1000 to 4000 km altitude the dose rate can rise up to about a Mrad per year. At geosynchronous orbits the dose rate is in the range 100 to 1000 krads per year. The energies of the trapped electrons are in the range 10 kev to 10 Mev and up to 200 Mev for trapped protons.

The effects of radiation can be decreased by locally shielding the Si with high energy absorbing material. A good material is Tantalum but as this is only 20% more efficient than Cu for a given mass, Cu or Bronze is more convenient if thickness is no problem. The absorbing material tends to produce secondary emission particles when bombarded with high energy primary particles and hence in low Earth orbits once the shielding is greater than several mm Al, which will have removed most of the low energy particles, further shielding is not very effective in terms of the additional mass required.

A common material used as a source of radiation for testing is Cobalt 60 which is a gamma ray emitter with energies in the range 1.17 to 1.33 Mev. These induce secondary electrons from scattering and absorption. The effects on semiconductors are similar to those caused by electrons and protons from the space environment.

Effect of radiation

When a material is subject to either particle or photon radiation, the loss in energy of the incident radiation is transferred to the material and can produce a number of different results. The two most significant to semiconductors are ionisation (the production of ion–electron pairs) and displacement damage of the crystal lattice. These effects can be transient or relatively permanent.

Displacement damage

This occurs when sufficient energy is imparted to a lattice atom, by a high energy ion or proton, to enable it to move far enough away from its original position to produce a stable vacancy site. These act as recombination centres for hole–electron pairs and alter such parameters as 'input offset voltages' V_{os}, 'input offset currents' I_{os}, gain A and leakage currents in transistors, diodes and operational amplifiers. This type of damage can be considered permanent in space although in the laboratory it can be cured by annealing. If we specify a failure for linear ICs as $DV_{os} > 1$ mV, $DI_{os} > 1$ nA or $DA > 10$ dB, for standard devices such as an LM108 or an LM101, this starts to occur at a total dose of about 10 to 20 krad.

Ionisation damage

The production of hole–electron pairs in the dielectric forming the gate of a MOSFET will result in a positive charge forming there as the more mobile electrons will drift away under the influence of the gate source field. Thus in an N type device a voltage which enhances the conduction of the channel will be created, and due to the high insulation resistance of the material, this can be considered permanent at room temperature. Eventually the MOSFET which is 'off' at zero gate to source voltage, turns on and unless the design can cope, renders the device useless. This type of failure is significantly reduced when the voltage is removed from the device as the hole–electron pairs can recombine before drifting apart. Thus on entering a part of the orbit in which high radiation fluxes can be expected, turning the system off has significant advantages.

In bulk CMOS fabrication, parasitic PNPN junctions occur which do not turn on as the current gain from one junction to the next is too low to be self–sustaining. See Fig. 4.113. If $\beta 1 \times \beta 2$ is less than 1, the device cannot switch on.

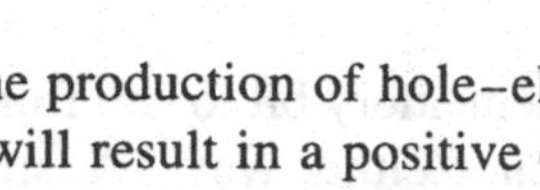

Fig. 4.113. Parasitic PNPN junction.

However if a sufficient large quantity of charge is generated at the right place, this can increase the current gains of both transistors to the point where they both switch on. As the PNPN network is effectively across the power supply, the increase in current is only limited by the ohmic resistance in this path, which could be catastrophic for both the unit affected and other items interfaced to it. This problem can be cured by altering the basic architecture of the MOS device using isolation techniques to ensure the spurious PNPN cannot form, but this involves a more complex IC manufacturing process, increases the area of silicon used per element and is much more costly. Manufacturers producing radiation tolerant devices typically qualify their components in four total doses; 3, 10, 100 and 1000 krad and these are given the symbols M, D, R and H in the US MIL–M–38510 and US MIL–S–19500 classifications. Safety margins of a factor of 2 should always be included when estimating the degree of radiation tolerance required.

Transient effects

The introduction of charge can also cause a memory bit to flip from '1' to '0' or from '0' to '1'. There appears to be no preference either way. Apart from data errors, this can cause a processor to crash, if the contents of its stack or other vital parameters are changed. Although this is not catastrophic to the unit involved, the system it controls could be switched to a disastrous mode. The term for these events is 'Single Event Upset' or SEU and although the amount of energy needed to cause this result is dependent on the cell size, it is typically about 100 Mev and comes from high energy ions or protons, that is 'cosmic rays'. Thus one can predict from the orbit the probability of a device encountering such a particle and from the type of device the probability of the particle causing an SEU (Pickel *et al*., [1980]).

Designing for radiation protection

To overcome processor crashes, a 'watch–dog timer' can be designed into the circuit. This is implemented by programming the processor to regularly read a byte of data stored in ROM and compare this with a preset value. If the comparison is true the comparator inhibits the generation of a processor reset pulse which would normally be triggered about once a second. A byte of code rather than a bit is required because under crash conditions the processor might run in a loop which could still output data. If the processor crashes, the subroutine which is used to generate the code will not be called and so the reset pulse will be generated. Depending on the processor it may be necessary to interrupt its power supply rather than generate a reset pulse as some systems can lock into a mode that a reset pulse will not control.

Even though the processor can be automatically reset, this may not be adequate if the processor is included in a command generation subsystem. Between the crash occurring and the reset pulse generation, spurious commands could be generated with catastrophic results. This can be prevented by ensuring the system will not respond to commands in less than the time $T2$ required to generate the reset pulse after the last trigger pulse has occurred.

Fig. 4.114 illustrates a watch–dog timer where the time delay $T3$ between the issue of a command pulse, spurious or genuine, and its execution is greater than the time $T2$, the period of a retriggerable monostable. As the 'delayed command' enabling pulse $T4$ is reset by the watch–dog timer $T2$, the activate pulse is not generated.

The latch – up current problem can be solved by current limiting to around 100 mA with a foldback to 10 mA within about 1 ms. In this time not enough energy would be available to cause permanent damage. At currents of less than 10 mA the device will usually unlatch, but total power removal may be necessary.

Due to the diode protection network in most CMOS devices, the interfaces between any sections with separate current limiting systems should be carefully examined to ensure the affected device could not remain powered through its input. Thus buffers without input protection diodes or with current limiting resistors should be included to solve this problem.

The design of the current limit should not introduce voltage drops at normal currents of greater than 0.1 V or if this is not possible, the voltage rail for these supplies should be made higher to accommodate this problem. The limiter must have a delay network built in to prevent the initial charging of the power line decoupling capacitors from operating it. Although the circuit shown in Fig. 4.108 is current gain and temperature dependent, these effects and their dependence on the expected radiation dose can be allowed for in the initial design.

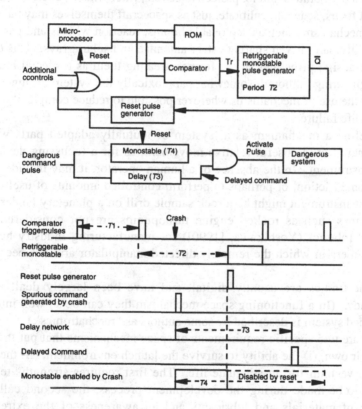

Fig. 4.114. Watch–dog timer.

5

Mechanism design and actuation

Scientific instruments carried aboard spacecraft often have to be equipped with mechanisms to operate shutters, protective covers, filter wheels, aperture changers and devices that focus, scan and calibrate, just as spacecraft themselves may have to be equipped with mechanisms such as deployable booms, reaction wheels and gas valves for manœuvrability, and driven shafts to steer antennae and solar arrays. This chapter focusses on the design principles of the former, treating them as a special branch of mechanical engineering, although somewhat paradoxically the system designer's first duty is to avoid the use of mechanisms wherever possible to reduce complexity and the risk of end–of–life failure.

We define a mechanism as a 'system of mutually adapted parts working together'. It may provide useful relative movement, as for a focussing device in a photographic instrument. In the absence of a human operator, it may include a drive motor to overcome friction, or perhaps to perform controlled amounts of useful work. For example, the instrument might be a rock sample drill on a planetary lander. High powered machines such as rocket–engine turbopumps are not considered. The development of robotics (Yoshikawa, [1990]) for manufacturing industry has been widespread in an era in which the remotely directed manipulator arm has been useful in space.

The electronics are usually digital, and have been largely dealt with in Chapter 4 already. (In a functioning space mechanism they can be so well integrated that the combined system is described by some authors as ' mechatronics'.)

There are two specific requirements of space mechanisms that put them in a category of their own, (i) the ability to survive the launch environment, (ii) the ability to operate in a vacuum for an adequate life. The first requires strengths tests and vibration tests to be made during the development process, the second calls for a careful choice of materials and lubricants and an awareness of the extremes of temperature to be met with and the possibility of a damaging radiation environment.

We discuss the requirements under two broad headings, design consideration and actuation mechanisms, with some limited knowledge of the engineer's repertoire of mechanisms assumed.

5.1   Design considerations

5.1.1   *Kinematics*

In the design of a mechanism there has to be some study of the relative motion of parts. For formal analysis, this is the province of kinematics.

A mechanism can be regarded as a chain of rigid elements (or bars or links); between contacting pairs of these relative motion is possible at idealized points. This motion may be turning, sliding, or screw–like, depending on the design details. One of the bars is regarded as fixed (to a frame such as the spacecraft structure). If the bars lie in a plane, the kinematic chain is made into a rigid frame by fixing any two bars to each other.

The methods of kinematic analysis may be found in textbooks (Erdman & Sandor, [1991]). Computer graphics are making possible 'animated' displays of motion (Cable, [1989]).

5.1.2   *Constraints and kinematic design*

The principles of kinematic constraint underly much good mechanical design. Their applicability to precision mechanisms is discussed here, and to optics in Chapter 6.

Essential assumptions are:

    i)   any unconstrained rigid body has 6 degrees of freedom;
    ii)   contact between bodies occurs at points and is frictionless;
    iii)   a contact with another body removes a degree of freedom;
    iv)   the number of contacts with a second rigid body is the number of constraints on the first body;
    v)   the contacts must be well chosen and independent; some configurations of contact points are degenerate singularities and ineffective.

Hence, to locate a body, 6 constraints are required, and 6 contacts (Fig. 5.1(a)), each maintained by a minimal force, are necessary and sufficient. The force closure may derive from some component of a preload, or, as for an instrument standing in a laboratory on the ground, from the body's own weight. In principle, 6 equations of static equilibrium can be solved to determine the contact forces. A classic illustration of kinematic location is the 3–groove type of Kelvin clamp (Fig. 5.1(b)), long used as a standard interface for portable surveying instruments. Each of 3 ball feet makes 2 contacts with the paired plane surfaces offered by the grooves. These features are very effective for the accurate replacement of demountable alignment mirrors, as discussed in a later chapter. Note that the real contact is more than the idealized point, which is but the centre of a small circular (Herzian) area of each contacting surface, bearing a finite stress; also that some friction may have to be overcome to bring the contacts to their resting places.

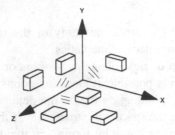

Fig. 5.1(a). Kinematic position–defining surfaces concept, 3 + 2 + 1, to locate a cuboid (omitted).

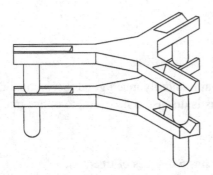

Fig. 5.1(b). Kinematic location by Kelvin's 3–groove arrangement.

To permit a single degree of freedom such as a rotation or translation, kinematic design dictates that 5 appropriate contacts are sufficient. It follows that 4 contacts leave 2 freedoms; 3 leave 3. An example is the 3–legged stool or trivet on a frictionless floor, free to slide in perpendicular directions and to rotate about a vertical axis.

Overconstraint results from too many contacts, and may lead to uncertainty of location or imprecise motion. Unstable location can occur when 2 contacts are alternately redundant and functional, as illustrated by a 4–legged stool with a short leg. Where design departs from the kinematic ideal into overconstraint, good reasons can often be found, but there may be penalties of higher cost due to tighter manufacturing tolerances. (The common example is a shaft rotating on ball bearings; see below.)

'Semi–kinematic' designs are modelled on kinematic ideals, but contacts are expanded to pads large enough to bear loads at acceptable stresses. For the rocket–launched instrument, this allows preloads to be high enough to maintain contact during severe vibration. Further, the contacting bodies are recognized as elastic, and this fact can be exploited to accommodate small manufacturing errors, by 'elastic averaging'.

Fig. 5.2. Fabry–Perot interchange for Infrared Space Observatory, before addition of wiring harness. (MSSL photo).

The 5 – bolt mounting for an interferometer interchange mechanism (ISO LWS, Fig. 5.3) is a semi–kinematic design. It is robust, yet allowed repeated mating and accurate realignment to the spectrometer. Although the principle of Fig. 5.1(a) asks for a 6th connection, there is no bolt parallel to the shaft (and optical) axis, in which the interferometers would be free to float in the bolt clearance until immobilised by friction from bolt preload. Each bolt passes through an annular contact pad, so that the stress due to preload is localised. An incremental tightening procedure ensured contact at all pads, particularly the pair at the side, before all bolts were finally torqued.

5.1.3   *Bearings, their design and lubrication*

The traditions of instrument–making and fine mechanisms have been part of a technology using oils or grease to reduce friction; maintenance is mainly a matter of putting back the oil lost by evaporation and migration, but this is generally impossible in space. Further, the knife–edge bearing is not robust for a rocket launch. Bearings, largely but not entirely, must be designed and built with dry lubricants. These are represented by sputter–coated molybdenum disulphide (MoS_2), ion–plated lead, and PTFE (in a transfer film). There are oils of very low vapour pressure, such as perfluoralkylethers. All are expensive, and do not eliminate the risk that they could contaminate cold optics. We shall discuss lubricants more widely after surveying the field of applicable bearing designs, beginning with ball bearings and ball screws. Plain bearings are cheaper but less precise.

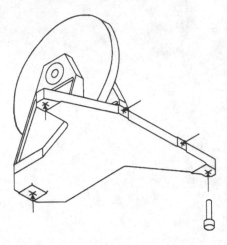

Fig. 5.3. Five–bolt semi–kinematic location and fixing for the ISO
Long–Wavelength Spectrometer interchange mechanism. Compare Fig. 5.1(a).

The basic type of rolling–element bearing is the deep–groove ball bearing (Fig. 5.4 (a)). Often called the radial bearing (to distinguish it from the pure thrust bearing), it is also fairly strong in the axial direction. Its rings and balls are made, with notably high precision, of a hard material which is typically steel, AISI 440 C. This is a rust resistant steel with 17% Cr, 1% C. The common alternative is the steel SAE 52100 (1.45% Cr), which is slightly prone to corrosion if not stored in low humidity. The number of balls that can be assembled is limited (e.g. to 7), and a ball cage, typically crown–shaped to snap in from one side, keeps the balls at even spacing. In the proprietary instrument bearing called BarTemp, the cage is made of a composite of glass–fibre–reinforced PTFE with added MoS_2. This is called a sacrificial cage; as the balls roll round the rings, they slide on the cage, transferring to both balls and rings a solid lubricant film of PTFE. Other deep–groove bearings have cages of bronze, steel, or phenolic composite etc.

The angular–contact bearing (Fig. 5.4(b)) is probably the ball bearing most commonly selected for space mechanisms. In its separable form, this can have a dry lubricant (MoS_2 or lead) applied to rings and balls, and the cage can be made of Duroid or other glass / PTFE / MoS_2 composite. Preloaded, there is no need for working clearance, and stiffness is high axially and radially, with a well defined load–dependent torque. The preload can be set either soft, by using axial disc springs (Belville washers, wavy washers, diaphragms), or hard, by mounting a duplex pair with a small offset of rings; Fig. 5.5 shows two styles of back–to–back hard preloaded duplex pairs. Soft preload is more tolerant of temperature change, but one bearing ring must be able to slide.

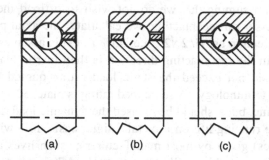

Fig. 5.4. Ball bearing grooves: (a) deep groove; (b) angular contact, separable; (c) 4–point contact.

Thin section bearings are offered by manufacturers and can be used to save weight and bulk. Four–point contact (gothic–arch section, Fig. 5.4(c)) bearings also offer a weight advantage, but preload can only be adjusted by changing ball size and the snap–over type of cage limits the application of dry lubricant.

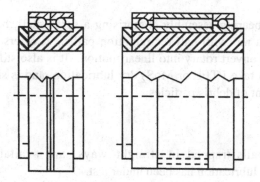

Fig. 5.5. Duplex pairs of angular–contact bearings, made to dimensions which ensure preload on assembly.

Good performance requires high precision. Manufacturers offer different qualities, graded according with dimensional tolerances, to a standardised code. ABEC 7 is commonly specified.

All–ceramic ballbearings have not fully emerged from development, but tungsten carbide and silicon nitride balls are available. Due to differential temperature expansion, soft preload is advised.

Ball–bearing assemblies are offered in a range of sizes. The small bearing will be preferred for its lightness and compactness, so long as it has adequate strength and life expectancy. A minimum load rating is calculated as follows. Beginning from a conservative vibration–case acceleration, and the mass of its shaft plus its load, the product gives a vibration inertia load F. Taking a back–to–back angular–contact duplex pair as an example, and assuming that we do not wish to offload the balls because of the high impact stresses on recontacting, we calculate the axial preload required in the disconnecting direction as $F/2\sqrt{2}$, i.e. 35% F. (This assumes a 45° contact angle.) The limit load in the recontacting direction is then inertia load plus preload, i.e. 1.35 F. This should not exceed the static load rating quoted by the bearing manufacturer. Note the terminology. A static load rating is matched against a load that is clearly in part dynamic, but it should be realised the dynamic load rating is a figure to use in computing the running life on a metal fatigue basis, following the formulae (from Weibull statistics) given by most manufacturers; such lives are not necessarily lower than the revolutions which will exhaust any dry lubricant. Here we meet a space technology conundrum, for if a sacrificial cage or coating is tested on the ground, the cage weight will increase the support load reacting on the cage, and shorten the life, compared with what might be expected in orbit. But at least ground tests on qualification prototypes may be expected to offer conservative results, allowing for orbital weightlessness.

Ball screw bearings

The ball screw is a ball–bearing assembly comprising a screwshaft, balls or rollers in helical raceways, and a nut with ball recirculation paths. It offers lower friction than the sliding screw to convert rotary into linear motion. It is also stiffer, is preloadable against backlash, can take PTFE or MoS_2 dry lubrication, and is slow to wear, although ultimately intolerant of debris particles.

Other bearings

Linear bearings, with balls trapped in straight ways, are available as manufactured assemblies, and dry lubrication has been under test.

Our survey of various types of ball bearing ought not to leave the impression that plain bearings are always inferior for space mechanisms. When a little working clearance, and shorter life, can be allowed, the simplicity, lightness and compactness are advantages. Self–lubricating shells and thrust washers, made of sintered bronze loaded with lead and PTFE, are available as commercial items (Glacier DU). Clearance of about 10 μm is necessary. Spherical and rod–end bearings are also available. Alternatively, bearing bushes and shells can be manufactured from Vespel or Duroid composite to any size required.

Lubrication of bearings

To return to the bearing lubrication problem, it will be seen from the foregoing that a wide range of space lubricant needs are answered by self-lubricating solids such as PTFE and polyimide composites, and by MoS_2 sputter coatings. MoS_2 coatings should be run-in and tested under vacuum, then stored in a dry atmosphere or inert gas. The ion-plated lead coating, when used with a lead-bronze ball cage, offers a longer dry -lubricated life (but not in air). ESTL tests (Rowntree, [1994]) show that, compared with sputter-coated MoS_2, friction torque is higher, so there is a price to pay for the durability. Another coating, which has been applied to 440 C balls is titanium carbide, noted for its smoothness, but not low friction.

The case for liquid lubricants, such as low vapour pressure oils and greases based on them, is not easily made for a space instrument. If contamination is a serious concern, they are best avoided. Tiny quantities may be admissible. For continuously running shafts, such as sealed momentum wheels, despin bearings and scanning mechanisms, oil has often been a successful choice. Parker gives a review of the various requirements and the candidate lubricants available (Rowntree, [1994]). A shortlist has been given in section 2.7.13.

5.1.4   Flexures and flexure hinges

The elastic properties of metals are very reliable, more so than their friction properties. Hence, when the motion between two parts is small enough that the parts can be joined by a bending elastic member fixed to both, a flexure bearing becomes very attractive. The only friction effect is that due to internal hysteresis, which is very small. No lubricant is required. There will be an elastic self-centring effect, not always a disadvantage. By making the flexure from a hard material, such as beryllium copper, which exhibits a high strain before onset of yield, the restrictions to motion of this type of design can be minimised.

Fig. 5.6 shows the form of a parallel-motion flexure, in this case monolithic, rather than built up from bolt-fastened parts. Each of the two blades can bend elastically to a contra-flexural curve as indicated when some small force is applied.

Fig. 5.7 is a cross-flexure hinge, in which the applied hinge bending moment bends the short, unclamped lengths of spring blade to a circular arc form so that the rotated part appears to turn around the axis at the initial intersection of the neutral planes of the blade. Both types of flexure have been widely used in laboratory applications as well as space instruments. Smith and Chetwynd, [1992] give a modern survey of flexure design and analysis which includes formulae for stiffness and stress of notch–type hinges. The latter allow monolithic designs relying on precision machining, as opposed to leaf–spring flexures where the spring blades must be clamped at their ends or else brazed to plates. The leaf–spring assembly allows larger rotations without overstress, but the possibility of Euler buckling under compression by vibration should be considered.

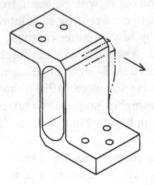

Fig. 5.6.   Parallel–motion flexure (machined from solid).

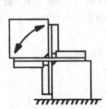
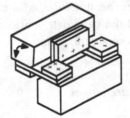

Fig. 5.7.   Cross–flexure hinge.

5.1.5   *Materials for space mechanisms*

The choice of materials for space mechanisms is guided by the same principles that apply to primary and other secondary structures in a spacecraft (as discussed in section 2.8), with the particular exception of parts interacting with relative motion. It follows that aluminium alloys are frequently chosen for both fixed and moving parts, and austenitic steels for shafts and fasteners. Honeycomb panels have been used for ultra–light baseplates.

5.1.6   *Finishes for instrument and mechanism materials*

Thermal considerations may require a paint finish, probably black inside the spacecraft, white outside. Some etch pre–treatment of aluminium alloys is required.

Phosphate–chromate passivation has the advantage of giving a surface layer that is a reliable electrical conductor whereas an anodic film is an insulator.

5.2   Actuation of space mechanisms

Remote operation of instrument mechanisms in space is performed with electrically driven motors, step motors, and pyrotechnic or phase–changing devices. The motors are constructed from materials, bearings and lubricants suitable for space application. Conductive heat dissipation may have to be arranged within the thermal design (see Chapter 3). Control may be open loop, or closed with feedback. Gearing may be incorporated to increase accuracy and torque. These topics are reviewed below.

5.2.1   DC and stepping motors

Brushless DC motors

The conventional commutated direct–current motor is suitable and efficient as a servomotor in feedback control devices, and can be operated in space vacuum, although with reduced life because of brush wear. The brushless DC motor uses solid–state switches instead of commutator and brushes, and so has a longer life, limited only by its bearings. Whereas the conventional DC motor has an armature (rotating coil) inside fixed magnets, the brushless motor has fixed coils and a permanent magnet rotor. This construction is similar to that of the synchronous AC motor, with the addition of Hall effect components or optical sensors to detect the magnet poles in the rotor (Kenjo, [1991]).

Stepping motors

The stepping motor, or stepmotor or stepper is a kind of DC motor, often small, designed to rotate discontinuously. It will move to an angular position, or series of positions, in response to a pulse, or sequence of pulses. Moving rapidly in discrete steps, it will go forward or back to any position commanded, with an accuracy of a few per cent of its step angle. It is therefore a digital actuator, easily matched to the electronics of digital systems. It shares with the brushless DC motor the advantage of non–rotating (stator) coils, and it can be simple and relatively cheap. Its coils and bearings are adaptable to the special requirements of space mechanisms.

The two broad classes of stepmotors are permanent magnet (or PM) stepmotors and variable reluctance (or VR) stepmotors. The first class is subdivided into 'simple' and 'hybrid' types.

Permanent magnet stepmotors

The PM stepmotor has a permanent magnet rotor carried transversely between soft iron stator poles wound with coils. A simple 4–phase (i.e. 4–coil) stepmotor is shown diagrammatically (Fig. 5.8). It is like a brushless DC motor, but without sensor components. When the stator coils are unenergised the rotor aligns itself with a pair of stator poles, so minimising the reluctance of the magnetic circuit. Such an aligned position, called a *detent* position, is stable to small disturbing torques and is a power–off characteristic of all PM stepmotors. When the aligned stator coil is energised with polarity such as to increase the magnetic flux, the position stability is stiffened. If, instead, the neighbouring rather than the aligned stator coil is energised, the rotor turns to its next position, again minimising the reluctance of the magnetic circuit; there is some overshoot followed by a damped oscillation, resulting in a short settling time. Sequential switching of the coils results in a progressive stepping and the rotor turns under *open–loop* control. The limitations on the possible stepping rate are discussed below.

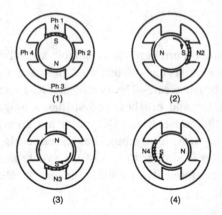

Fig. 5.8. Stepwise operation of 4–phase permanent magnet stepmotor.

For any PM stepmotor, the stable stiffness near a detent position is associated with a useful holding torque, which is the peak value of the stabilising torque with the appropriate coil energised. This torque will be greater than the power–off detent torque. Note that the detent action can be detected by turning a PM stepper's shaft against reluctance torque, and that some intermediate shaft angles may appear to be stable, or metastable, due to bearing friction. The friction, though small, will arise from lubricant (solid or fluid) and preload.

Variable reluctance stepmotors

In contrast to the permanent magnet in the PM motor, the VR stepmotor has a soft–iron rotor into which are machined salient pole–pieces or *teeth*. Only when the stator coils are energised do these rotor teeth become magnetised. Many design variations are possible; we show a simple 3–phase example (Fig. 5.9) in which a 30°–step *counter–clockwise* motion results from sequential *clockwise* switching.

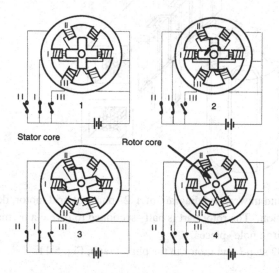

Fig. 5.9.   Simple 3–phase variable reluctance step motor.

More typically, a 4–phase VR stepmotor with 12 rotor teeth and 16 stator teeth gives 7.7° steps, while a 6– phase motor with 50 teeth delivers 1.2° steps. Such a VR motor yields a good step–angle precision, and it can be run fast. Having no permanent rotor magnetisation, however, a VR motor exhibits no power–off detent.

Hybrid stepmotors

The PM hybrid class of stepmotors combines the small angle and high speed of the VR motor with a magnetic detent property, effected by building a multi–tooth rotor around an axial permanent magnet. Figs. 5.10 and 5.11 illustrate a simple 2–phase hybrid developed for a cryogenic space instrument.

Half–step drives

Stepmotors can be 'half–stepped' by devising pulse sequences which alternately energise one coil and its adjacent coil. However such sequences cannot be stably halted, at an intermediate detent angle, after an odd number of half–steps. The truth tables for the variety of stepping sequences are usually given by manufacturers of stepmotors and motor controllers (Kenjo, [1984, 1991]).

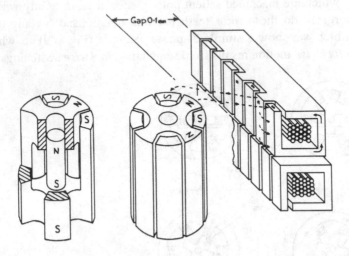

Fig. 5.10. Separated and unwrapped diagram of a 2–phase hybrid motor, designed for 16 steps per revolution. The rotor part is half–sectioned to reveal a permanent magnet between soft–iron pole–pieces.

See parts of a similar 32 step / rev. motor in the photograph, Fig. 5.11.

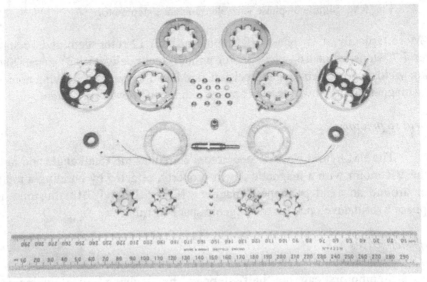

Fig. 5.11. Parts of 2–coil cryogenic stepmotor developed for ISO LWS; compare Fig. 5.10.   (MSSL photo).

Some stepmotors, particularly hybrids, are amenable to so–called microstep (or ministep) drives, for which one natural step is subdivided into many small steps, and, instead of square wave pulses, a stepped approximation to sinusoidal current variation is delivered. Kenjo, [1984] explains further how this can improve resolution in some applications.

Torque and speed

The torque–speed characteristics of a typical stepping system – as distinct from a stepmotor alone – are shown in Fig. 5.12(a). They differ because the load inertia as well as the rotor inertia plays a part in determining these characteristics. The available torque is limited by the maximum current amplitude and in PM motors by the saturation magnetic flux and by the need to avoid fields strong enough to demagnetise the rotor. These three parameters, total inertia, current amplitude and effective magnetic flux, determine the speed at which the stepping system advances. Current rise–time, determined by the electrical parameters of the drive circuit, is less important.

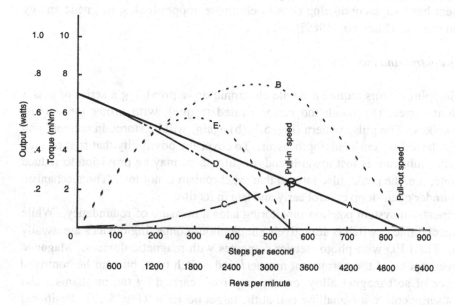

Fig. 5.12.(a). Torque vs. speed characteristic for typical stepping motor.

Any stepping system has natural upper limits to its stepping rate and shaft speed beyond which it cannot be driven in step with the pulse rate. This limit, the *pull–out* rate, is reduced rapidly by the addition of friction, thereby defining the torque–speed pull–out curve (Fig. 5.12(a)). There is also a lower *pull–in* rate below which the system can be started, and run, in synchronism with the drive train. Between these two rates a speed is said to be in the slew range. It follows that the slew range can only be accessed by starting below it and then accelerating up to it.

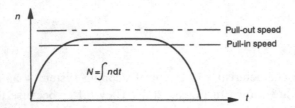

Fig. 5.12.(b).   Pulse repetition rate vs. time for stepmotor drive.

The relative simplicity of construction of stepping motors is illustrated in Fig. 5.11, which shows the parts of a 2–coil hybrid developed by the Mullard Space Science Laboratory. It operates at 4K and drives a 2 – interferometer interchange within the spectrometer in the Long Wavelength Spectrometer of the Infrared Space Observatory (Davis *et al.*, [1991]).   A set of VR steppers operating in the same low–temperature environment have superconducting coils to eliminate copper losses; magnetic energy dissipation remains (Luciano, [1989]).

Stepmotor control and monitoring

Stepping motors require a simple electronic drive providing a series of pulses of sufficient current (to match the motor's rated torque), with protection against inductive spikes. The pulse pattern (Fig. 5.12(b) again) will be stored in memory and applied via the subsystem's microprocessor. To cover the possibility that friction may increase after lubricant is lost towards end–of–life, there may be provision to reduce the slew rate, i.e. the peak pulse rate, so that synchronism is not lost. The mechanism is thereby under open–loop control early and late in its life.

Some means of system position monitoring adds a measure of redundancy. While the commercial microswitch is attractively simple, non–contacting devices are usually preferred. The LED with photo–sensor competes with magnetic devices. Magnetic sensors need not be of the permanent magnet/reed switch type, but can be contrived with a piece of soft magnet alloy, or 'moving iron', carried by the mechanism and coupling a sense coil to a signalling coil at the target position (Fig. 5.13). Positional resolution can be increased with inductive devices, wired to phase discriminating circuits. Even greater accuracy is available from an appropriate shaft encoder. To conserve power, any monitoring device can be switched off after the mechanism has reached its required position.

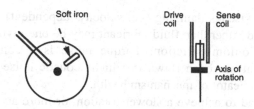

Fig. 5.13. Magnetic position sensor.

5.2.2   Linear actuators

The following devices used for linear actuation require no special comment:

*   nuts and ball–nuts on feedscrews controlled by steppers
*   solenoids with moving iron or moving coil, possibly mounted on a
    leaf–spring flexure (Smith & Chetwynd, [1992])
*   piezoelectric ceramics, giving a high force with low power
    dissipation in return for a high operating voltage over a short stroke
*   electrically heated memory–alloy actuators
*   electrically heated phase–change wax capsules

Pyrotechnic devices are reviewed in section 5.2.6.

5.2.3   Gear transmissions

Ratio considerations

Gears are a complication to be avoided if possible, a direct motor drive being cheaper to make and free from extra problems of lubrication and additional wearing parts. The stepping motor, especially in hybrid versions with small step angle, has made it possible to transfer some of the complication to electronics, with gain in system reliability and life. But a train of geared wheels may be the only way to match a proven and available motor to the output motion requirements specified for a new mechanism.

For consideration of the dynamics of the load in a mechanism, the parameters are its inertia, some friction, and possibly an elastic reaction. If the required motion of the load is rotation, these become a moment of inertia and a torsional elastic constant (both calculable) and a friction term. As in vibration analysis, frictional dissipation may be modelled as an equivalent viscous damping (i.e. velocity–dependent) term. However, the real situation with solid rather than fluid lubricant may be one involving static adhesion (stiction) and Coulomb friction. Torque noise is a common observation; at least one experiment has been flown in which acoustic noise from bearings has been monitored as an indicator of mechanism health.

Usually gears are introduced to achieve a slower rotation, or more accurate positioning, of the load. The reduction ratio is the number of teeth on the driven wheel divided by those on the motorised wheel. If there is a train of gears on a succession of shafts, the overall ratio is evidently the product of the successive ratios.

The torque needed at the motor to overcome a given friction plus elastic torque at the load is load torque divided by gear ratio (k). The system moment of inertia to be accelerated by motor torque is the sum of the motor inertia (I_m) and (I_1 / k^2), where I_1 symbolises load inertia.

It is shown in servomechanism theory that the fastest acceleration of a given load by a given motor occurs when

$$I_1 / k^2 = I_m \qquad \text{or} \qquad k^2 = I_1 / I_m$$

This is usually taken as the condition for a load to be matched to its motor, but is not critical, and need only be used as a rough guide to the gear ratio required. Other factors, such as tooth size and volume available to house the mechanism must be considered; unless, that is, a spaceworthy commercial gearbox, or motor–gearhead unit, provides a ready–made solution, complete with lubricant.

Types and sizes of gears

The standard profile for a gear tooth is based on the involute curve, which has the properties that (i) as the point (or line) of contact of the driving tooth slides over the driven tooth, the angular velocities of the geared wheels remain constant or proportional; and   (ii) this constancy is not sensitive to small errors in the between–centres distance of the wheel shafts. This allows backlash to be adjusted if means are provided.

Standard sizes for gear teeth result from the adoption of a range of values for the 'module' D/N, where N is the number of teeth and D the pitch circle diameter; the latter is the nominal diameter of an imagined wheel rotating by friction (without slip) through contact at its rim with a corresponding mating wheel. A related (but obsolescent) term is DP (for 'diametral pitch'), N/D, where D is measured in inches. The pitch of teeth if measured round the pitch circle is $\pi D/N$, i.e. $\pi \times$ module. See table below for some gear sizes commonly used in small mechanisms.

Table 5.1. *Some gear sizes used in mechanisms*

Module / mm	(.2117)	.2500	(.3175)	.4000	.6350
Diametral pitch / in $^{-1}$	120	(101.6)	80	(63.5)	40

The external spur gear is the common choice, but the involute profile may also be applied to internal teeth, as found in epicyclic gears for planetary gearboxes and gearheads, giving access to a wide range of overall ratios. The harmonic drive has found many applications for large ratios; this is a commercially available assembly exploiting a flexing toothed ring, distorted by an elliptic wave–generator with rolling–element bearing, inside a circular gear or spline having 2 more teeth than the flexspline. Ratios up to 320, with little backlash, are available; lubrication has usually been grease (Braycote 601); see Rowntree, [1994]. High reductions could also be achieved with a worm and worm–wheel, whose lower efficiency prevents backdriving. Bevel gears are used on occasion.

For further discussion of gear design principles see texts such as Drago, [1988] or Shigley, [1989].

Materials for gears in space vacuum

Driving pinions are commonly made in steel; the authors have had satisfactory experience with stainless austenitic steel (400C) at low stress. Rowntree, [1994] reports that in NASA's Shuttle manipulator epicyclic gearboxes plasma–nitrided, precipitation hardened steel is used. Luciano, (ESA, [1989]) reports the use of the polyimide/MoS_2 composite Vespel SP3; like other self–lubricating polymers such as Duroid 5813 and polyacetal, its low elastic modulus recommends it because for any given tooth load/width ratio the contact area is increased.

Driven gearwheels have been made of aluminium, titanium, steel and bronze. Aluminium alloy 6082, anodised (chromic/sulphuric, thin), run against a self–lubricating polymer and lightly stressed, is expected to have a long life. The same material, when age hardened, hard anodised and MoS_2 sputter–coated, has demonstrated 2.4×10^6 revolutions at 119 MPa tooth–contact stress (Rowntree, [1994]). Aluminium 6061, age hardened, hard anodised and MoS_2 coated, has been qualified at 4 K (Davis *et al.,*[1991]). The hard anodising treatment gives typically a 75 μm–thick, Al_2O_3 skin.

Lubrication for gears

Self–lubricating polymers such as Duroid benefit from a vacuum run–in of about 10^3 revolutions. Sputter–coated MoS_2, about 1 μm thick, has been found good for tooth encounters up to 10^6. Low vapour–pressure grease is typified by Braycote 601. Creep barriers are recommended. This and other perfluorinated oils may eventually degrade with aluminium and titanium alloys.

5.2.4   Fine motions

The conversion of a macromotion, such as a stepmotor increment, into a micromotion, such as would be required for a focus or tilt adjustment, needs either gearing or some other lever principle. The use of feedscrews, wobble pins and cascades of spring elements are treated by Smith and Chetwynd, [1992]. A motion attenuator can be effected by the use of opposing cantilevers, or other spring elements, of different stiffnesses. In such systems a large part of the input motion is absorbed by the compliant element while a small, but consistent and exact, proportion appears at the output platform.

5.2.5   Ribbon and belt drives

Ribbon drives as used in computer magnetic disk drives, are discussed by Smith and Chetwynd, [1992]. Thin beryllium, or hard steel would be a suitable material. Preloading is necessary to ensure driving friction, as for plain belts. The toothed flexible belt, as widely used in mechanical engineering, would have to be moulded into a spaceworthy, low–outgassing, composite elastomer.

5.2.6   Pyrotechnic actuators

The remote operation of many vital functions on a spacecraft requires more or less powerful 'one–shot' actuators. Pyrotechnic actuators are electrically initiated chemical energy 'fireworks' which are appropriate for boom, antenna, or solar array release, unlocking of launch locks, opening of protective shutters over optics apertures, and release of clampbands at separation.

A typical 'pyro' burns a small quantity of solid (propellant) chemical and evolves gas, at high temperature and pressure, to do work; but the gas is kept sealed in a ductile case which expands. Ignition is by an electrically heated 'bridge' initiator, which is a standardised component, also used for the pyrotechnic flame generators which ignite rocket motors, or for shock–producing hermetic detonation cord for kicking–off a possibly recalcitrant clampband. There is a large energy release for low installed mass. For redundancy, at least two pyro actuators (sometimes called 'squibs') are installed, and the pyro subsystem may be wired with series duplication as well as parallel redundancy.

The range of available well–proven devices has expanded to include bellows actuators (see Fig. 5.14), pin pullers or pushers, cable shearers, release nuts (replacing explosive bolts), and pyrotechnic gas valves.

Fig. 5.14.   Bellows actuator.

The verification of correct operation prior to flight presents a difficult problem; the actuator cannot be tested and re–used. Continuity of initiator wires can be checked at a current well below the firing level. The pyro design will have been very thoroughly tested through a wide range of temperatures etc. The installed devices will have been taken from a production batch of which a significant proportion will have been fired to prove their reliability. Such precautions are costly, but total failures in flight are costlier.

Precautions against accidental firing of the pyro must also be taken. The screening of the firing harness against EMR is a necessity. To ensure there can be no power in the firing circuit until operation is imminent, the physical removal of an arming plug is required. Its socket will be fitted with a safeing plug, to short the initiator wires. The arming plug is replaced just before launch (see Fig. 5.15). An unregulated bus may be supplied to fire the pyro directly, or a capacitor storage system may be used as illustrated. In the first case the current limiting resistor R is necessary in case the heater wires fuse short circuit instead of open. Fusistors can be used here i.e. current limiting fuses. The capacitor storage firing system is preferable as the firing current is contained within the subsystem, rather than being distributed back to one of the spacecraft power supplies, but the energy required to fire the pyro may require a prohibitively large capacitor.

The typical small bellows actuator illustrated is 15 mm long and 8 mm in diameter. It provides an actuating thrust of 20 N through an extension of 10 mm. Its initiator has a resistance of a few ohms with a design firing current of about one ampere and a maximum no–fire current of 10 mA.

New devices, such as phasechange wax capsules and memory metal actuators (both types electrically heated), overcome the single–shot handicap of pyro devices, but will be used for less powerful applications, since no chemical energy is released. Another one–shot device, not to be despised for its simplicity, is the spring–loaded pin, released by electric heating of a fusible link.

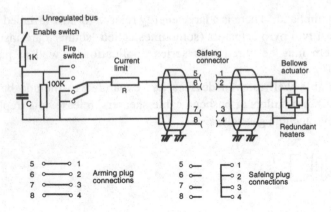

Fig. 5.15.   Actuator firing circuit.

5.2.7   *Space tribology and mechanism life*

In the pioneering years of artificial satellites there was a widespread awareness of severe problems of friction and wear which had been observed in laboratory vacuum experiments. Fundamental experiments in friction and surface physics, in which clean, unlubricated (and unoxidised) metal surfaces were rubbed together, showed tendencies for high frictions and adhesion by cold–welding. Tribology, which is the engineering science of interacting surfaces in relative motion, began at about the same time. Within a few years substantial parts of the fruits of tribology research were applicable to vacuum and space mechanisms. The techniques have been outlined in section 2.7.13.

At the beginning the choice was between attempting to seal a grease or oil lubricant in the mechanism, or launching it without lubricant. Either way worked, but with limited life in orbit, especially if loads were not light. By the decade of the seventies, however, techniques including solid lubrication were well advanced.

Life testing in vacuum has played its part. In the absence of oil to carry away particles produced by rubbing of mating surfaces, they stay on the surfaces. Transfer films, as of PTFE from composite ball–bearing cages to balls and tracks, reduce friction. One mechanism of bearing failure is seizure by packing of clearances with wear debris or piled–up lubricant. We should bear in mind that debris that falls off in a ground test may not come away in orbit.

With the best of space lubricants, carefully chosen and used, lives of many millions of rotations are achievable.

6

Space optics technology

One of the early motivations of space science was the opportunity for astronomy to use hitherto inaccessible wavelengths, from gamma–and X–rays to infrared, and visible–light astronomy, from orbiting telescopes, was allowed better seeing, free of atmospheric contamination. Meteorological observations and earth remote sensing required orbiting cameras and infrared radiometers. All needed applications of optics.

It will be clear from foregoing chapters that, for space instrumentation, optics should (i) have qualities of rugged mechanical design, (ii) be built of lightweight non–contaminating materials, and (iii) survive years of unattended use in orbit. This chapter offers an introductory account of materials and opto–mechanical design techniques which have been serving these ends. Optical design as such, and physics of sensors, are beyond our scope. The steady improvement of sensor and detector systems, often of great sensitivity, has fostered parallel development of computation and suppression of stray light. As in all space endeavours, pre–launch qualification testing should be carefully and thoroughly conducted. Operation and adjustment in space requires mechanisms whose life may be limited, but in–orbit repair or replacement is either impossible or very costly; hence trade–off decisions may be difficult to make.

6.1   Materials for optics

As remarked in paragraph 2.7.12, the concern with glasses and ceramics is their brittleness while exposed to the launch environment. The chart (Fig. 2.31) of fracture toughness versus strength for the diversity of materials shows optical glasses, as a class, to have high strength but low toughness.

Together with the (even stronger) engineering ceramics, the critical length of crack may be 0.1 mm or less and so too short to be observable. The implication is that, due to stress concentration at the sharp tip of a crack, quite low forces may generate the high stress to propagate fracture from any one of the myriad microcracks at a ground edge. When possible, flame sealing of edges may reduce these sources of weakness. This treatment might be particularly applicable for materials such as fused silica 'quartz' glass whose strength is sensitive to exposure time under stress in laboratory air. Even normal humidity appears to be mildly weakening; one could say 'corrosive'.

Because of brittle fractures, the tensile and bending strengths of optical glasses and ceramics are lower, sometimes much lower, than their compressive strengths. Bending strength is relatively easily checked in a 3–point bending test, in which a bar of the material bridges two simple supports while sustaining a centrally applied load; the result is often directly relevant to design problems, and is often reported as the 'modulus of rupture'. Scatter of results between different samples reflects the spectrum of crack sizes and the probability of the worst crack being close to the zone of highest tensile stress during bending; see Fig. 6.1 for the mirror substrate material Zerodur. This is a stress / probability / time diagram. While it is not wrong to interpret the right hand sloping line (Weibull modulus 2.5) as showing a mean strength of 108 MPa ± 25% under moderately rapid loading, the suppliers recommend an allowable stress of 10 MPa where the load duration may be some tens of years. This implies a factor of safety over 10, and such factors have been advised for fused silica glass too. Indeed, work has been done to relate design factor of safety to Weibull modulus, higher values of which imply greater consistency and less scatter. This can be achieved via the breakage probability. The message is that, in the absence of copious test data about a glass or ceramic, only a large safety factor is comfortable.

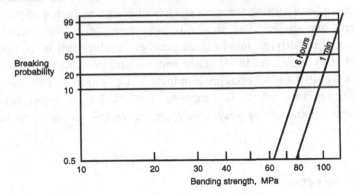

Fig. 6.1. Stress / probability / time diagram for a glass ceramic.

(Zerodur ; by courtesy of Messrs Spindler & Hoyer and the makers.)

For further information data on the selection of materials for optics, see Yoder, [1992] or Barnes, [1993].

6.2   Materials for mountings and structures

The desired qualities of lightness, strength, stiffness, low–outgassing and resistance to radiation have been reviewed in the foregoing chapter. Dimensional stability and vibration damping properties have been touched on.

The traditional material for lens mountings was brass, especially naval brass. Its virtues include strength, forgiving ductility, machinability. But it is a zinc–bearing alloy of copper, with, we fear, a propensity to sublimate in space vacuum. Also because of its density, its specific strength is inferior to aluminium alloys such as 6061 and 6082 which already share the other desired properties.

Neither the aluminium or copper alloys have low CTE (coefficient of thermal expansion), however. Sometimes this is a disadvantage. Invar is a low CTE alloy of iron and nickel (36%). Because it is ferromagnetic, of inferior specific strength, and loses its low thermal expansion outside the range 4–38 °C, it is rarely used for space instruments. Carbon fibre composites can have lower CTE and are much lighter, with high specific stiffness.

A well–designed carbon fibre / epoxy composite offers low CTE with outstanding specific stiffness and strength. The blemish is the small degree of moisture absorption, which may be eliminated by using PEEK (polyetheretherketone) as the matrix resin. But while this and other composites may be competitive for the frame structure of a large camera, spectrometer, telescope, or other reflecting optics assembly, they are not so attractive for lens, prism and filter mountings on the smaller scale. It is clear that epoxies and other polymers outgas and are eroded by atomic oxygen if exposed in low earth orbit, increasing the risk of condensing and polymerising contaminants on cooler optical surfaces.

6.3   Kinematic principles of precise mountings

We begin by returning to the principles of kinematic constraint, introduced in the foregoing chapter as an aspect of precise mechanism. Usually the intention in a mounting is that the mounted lens, mirror prism or other components shall have an unambiguous position and orientation, over a range of temperature. This implies that all 6 degrees of freedom between component and mounting frame shall be constrained. Therefore, ideally, 6 contacts are required, each clamped with preload to effect a force closure. Whereas the force can be provided by gravity when (for instance) a surveyor's theodolite is returned to its tripod or triangulation station, there can be no gravity closure when in orbiting free fall. Enough spring force should be provided so that accelerations during test or launch cannot separate any of the 6 pairs of contact surfaces.

Departures from the ideal are frequent and, with discretion, allowable. Thus in the base of the Fabry–Perot assembly for the Long Wavelength Spectrometer for the Infrared Space Observatory (ISO), there are 5 bolt holes through 5 pads (see Fig. 5.3). Each bolt effectively clamps a pair of narrow annular surfaces, each approximately, with sufficient accuracy, a point contact. The 5 conceptual points are arranged like the straight prismatic vee slide (Fig. 5.1), but the unconstrained freedom is limited to the assembly clearance averaged over the bolts, and is trivial because it is parallel to the path of infrared rays through the Fabry–Perot etalons (to about 1 arc minute). If it had been of value, this sixth limited freedom could have been turned into a firm constraint by providing, through a lug from the case, a sixth bolt drawing a pad on the baseplate of the Fabry–Perot assembly towards itself. To accommodate tolerances on the positions of the other 5 bolts, a gap between pad and lug could have been packed with shims. Note that all 5 (or 6) attachment bolts are preloaded when they are tightened to a torque standardised for their size, as in Table 6.1.

Table 6.1.   *Table of bolt torques standardised at MSSL for the tightening.of instrument—size bolts of austenitic stainless steel, to ensure consistent assembly and avoid overstraining*

Size	Torque Nm	Major dia. mm	
			Torques fit formula
0 - 80 UNF	0.11	1.51	
2 - 56 UNC	0.34	2.17	$\dfrac{\sigma}{13\text{GPa}}\left(\dfrac{D}{\text{mm}}\right)^{4}\text{Nm}$
4 - 40 UNC	0.90	2.82	
6 - 32 UNC	2.0	3.48	
8 - 32 UNC	4.4	4.14	where σ is
10 - 32 UNF	8.4	4.80	allowable stress

Effective designs can be realised by employing elastic flexures to provide freedoms that in the laboratory would be effected by ball feet sliding in grooves with gravity force closure. Thus the X–ray telescope, operated on the OAO –3 Copernicus spacecraft (launched 1972), was attached to the spacecraft through 3 equispaced inverted–V composite flexures which were stiff tangentially but compliant radially (Fig. 6.2). Each pair of V–legs was equivalent to a pair of constraining contacts, so there were effectively 6 constraints. The flexures were poor thermal conductors by design, and their radial compliances allowed for temperature differences between telescope package and spacecraft without capricious joint slip. Another style of 3–leg flexure was adopted for each of the instruments on SMM, and other styles have supported telescope primary mirrors.

The examples quoted so far are for whole instruments, but the principles can guide the design of hardware down to the last small optical component within the assembly. While an underconstrained part or subassembly is detectable by its looseness within the freedom left to it, overconstraint is less obvious. Indeed, one of the difficulties that taxes the designer's skill is how to extend a conceptual constraint, to keep stress down, by exploiting elastic averaging. This we consider below.

Close examination of the ball–on–plane constraint reveals limitations. The classic studies by Herz of the elasticity of such contacts established that the stiffness is non – linear, initially very low. Low–frequency vibration problems arise if force closure is inadequate. Further, the real surface departs from the idealised conception because of its roughness and friction. Recent work in nanotechnology reveals hysteresis and other non–conservative effects at the nanometre level. These observations counsel caution, rather than kinematic doctrine, where ultraprecision is the goal (Smith & Chetwynd, [1992]).

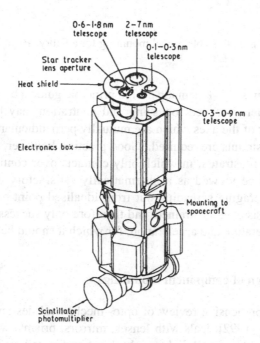

Fig. 6.2. Mounting of early X–ray telescope, on OAO–3 Copernicus, employing 3 inverted–V glass–composite flexures.

The idea of elastic averaging is that, although a real contact consists microscopically of a large number of asperities joining from both sides of an interface, they can be conceived as one ideal point contact of infinitely strong material. The challenge to the designer is to expand the contact area around the point, with sacrifice of a small measure of precision, yet remaining within the overall specification. A typical example is the clamping of annular pads around fixing bolts, with shims introduced to adjust alignment, for a Fabry–Perot etalon assembly (Fig. 6.3). The contact pad has an outside diameter of order one–tenth the distance between pad centres, and unless no shim is fitted there need to be local rotations, accommodated by elastic bending deformation, of about 3 arc minutes.

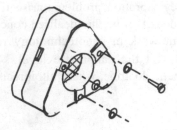

Fig. 6.3.   Shim–adjustable 3–pad mounting for a Fabry–Perot interferometer etalon.

To what extent can semi–kinematic design principle guide a basic problem such as the mounting of a lens? Since focussing and centration may be as important as rotation round either of the axes which are mutually perpendicular to the optical axis, it appears that 5 constraints are required. Good practice, however, can offer a solution like the lens mount illustrated, in which only contacts over continuous surfaces are provided. This can be viewed as a kinematically satisfactory design with a high degree of elastic averaging spreading out from idealised point contacts. But it can equally be regarded as overconstrained, and therefore only successful within limits of environmental temperature and acceleration, for which it should be tested.

6.4   Detail design of component mountings

In his comprehensive review of opto–mechanical design for a full range of components, Yoder, [1992] deals with lenses, mirrors, prisms, windows and filters; mirrors are categorised as small, lightweight, large and metallic. The purpose here is to illustrate the problem of mechanical design for spacecraft environmental testing by reference to a rugged lens mounting. The professional reader will wish to follow up his special interests in the referenced literature which includes a bibliography annotated by O'Shea, [1986].

In reviewing the launch environment we saw (Fig. 2.10) that response accelerations of small masses (< 1 kg) are typically above 50 g. A target for stiffness might be a natural frequency above 200 Hz. We consider the mounting of a plano–convex lens, and assume a temperature range –15 to + 40 °C. A bevelled ring, with male thread, closes on the spherical convex surface with tangency at the glass–metal interface; see Fig. 6.4. As it is screwed into the mounting tube, it can be tightened with a measurable torque to preload the lens to 50 × lens weight.

Kowalstine's empirical formula is:

tightening torque = 0.2 × (required preload) × thread pitch circle dia.

The interface load in the dynamic case for a 50 g inertia force will be, conservatively,

force = 100 × lens weight

but this will be reduced by friction and by that small share of load carried by elastic shearing of the RTV bonding elastomer. The preload at low temperature will increase due to differential contraction, but there will be no accompanying dynamic load. The tangent interface ensures that the compressive stresses in the glass are of the same order, front and back. The stress–raising effect of a sharp corner interface is avoided.

Attractive features described by Yoder for a mounting of this type are (i) spherical grinding of the rim of the lens to avoid damage on assembly, for which less than 0.02 mm clearance may be provided; (ii) injection of RTV silicone elastomer to seal the lens in the mount and bond the mounting ring against loosening under vibration. On the other hand, the bevel nut (which provides force closure to maintain the axial constraint of the lens) is stiff, the lens is overconstrained, and the mounting preload is sensitive to temperature if designed in aluminium alloy. The effect is reduced if tube and nut are titanium, or if an elastomer O–ring or other flexible element is introduced.

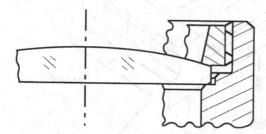

Fig. 6.4. Typical lens mounting designed for a 50 g response to environmental testing.

6.5   Alignment, and its adjustment

Although alignment will be most frequently interpreted as angular orientation, we begin by considering a problem in bringing apertures to a common line.

An X–ray collimator was built for the Solar Max Mission (Acton *et al.*,[1980]), soft X–ray polychromator, bent crystal spectrometer. Nine nickel grids, each pierced with some 50,000 apertures 0.35 mm square, were set out on a magnesium alloy 'optical bench' (Fig. 6.5). Each grid was an identical electroform, attached to a frame for 3–point semi–kinematic connection to the bench. The kinematic principle ensured that, each grid being properly positioned on its frame, each aperture was aligned with the 8 corresponding apertures in the other grids. No adjustment was required, and the intended 6 × 6 arc minute field of view was achieved. The robust design withstood routine vibration testing, and also the subsequent launch.

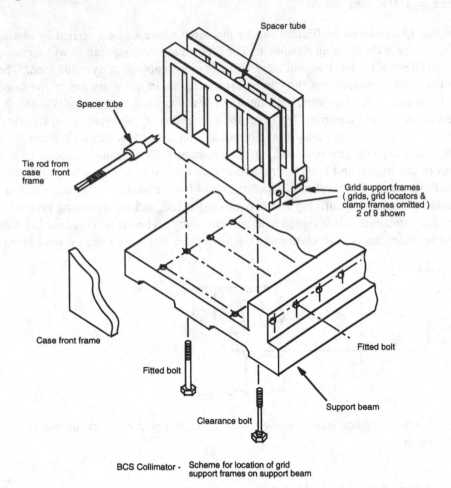

Fig. 6.5. Rugged optical bench construction of collimator, in dimensionally stable magnesium alloy, for precise location of 9 grid mountings (SMM X–ray spectrometer).

The angular alignment of the same spectrometer illustrates a typical problem: to bring a suite of instruments to a common alignment. At the outset, each instrument was prescribed to have 3 compliant legs which were shim–adjustable for height, with a resolution of about 5 arc seconds with a suitable combination of shims. The pointing direction of the optical axis of the spectrometer was defined by the normal to an optical reference flat mounted on a demountable 3–legged accessory (or trivet); this was accurately remountable to the collimator's optical bench by the contacting of 3 ball–ended feet in 3 grooves, in a Kelvin clamp arrangement. The functional success of the spectrometer also required the correct orientation of each of 8 diffracting crystals, fixed to mounts, and each of these had a shim–adjustable 3 –screw connection to the stiff frame of the spectrometer. Laboratory preparation required the use of alignment tools which substituted a preset optical flat for each crystal; shims were introduced at the crystal mounts to bring these into co–alignment with the optical reference flat, with a precision of a few arc seconds. This involved an autocollimating telescope, traversing vertically and horizontally on stands joined on the Kelvin clamp principle, and adjusted to a reference direction defined by a large optical–flat mirror. The shim arrangements were found, as expected, to be demountable and remountable, maintaining consistent alignment through successive vibration tests in different axes.

Elastic relaxation in orbit has been mentioned above and can be computed by elastic analysis to determine if critical alignments are affected.

Dimensional instability misalignment is a more difficult problem because it involves that kind of long–term non–elastic deformation of materials which, at least at high material temperature, is called creep. The plastic strains that are identified with creep are driven by stress at extremely low rates if temperatures are not high. The stresses are more likely to exist as an accident of heat treatment or other processing than because of some preloading feature in a design. In either case there is a self–balancing force system within structural parts, leading to creep strains which might be 1 μm per metre per year. Creep for carbon fibre composite, due to moisture desorption in vacuum, may be 10^{-4}. Only for the most critical assemblies will this be a concern and it can be mitigated by appropriate disposition of structure – for instance, by avoiding reliance on a beam form of construction.

♦ *Example 6.1. Creep misalignment estimate*

The secondary mirror of a telescope stands off from the primary by 1 m and is supported by a tube of diameter 0.4 m.

To estimate the misalignment after a period in which opposite generators of the tube's cylinder expand and shrink 1 μm.

The rotation of radii at the base of the secondary, relative to their initial directions parallel to lines in the base of the primary, is

1 μm/($^1/_2$ × 0.4 m) = 5 microradians or 1 arc second.

For other aspects of alignment and optical testing, adaptations of Hartmann and 'star' tests are applicable. See Malacara, [1978 & 1993].

6.6   Focussing

At bottom, the achievement of correct focus for a lens or mirror system may be as simple as assembling parts, made correctly within manufacturing limits, then prudently testing that the optical design specification has been achieved. If not achieved, then the addition of shims or removal of metal, at points dictated by a kinematic view of the opto–mechanical design, should bring it within specification. In some designs there may be screw–threaded components to unlock, adjust and relock.

More sophisticated designs will incorporate focussing mechanisms to permit in–orbit refocussing as instrument dimensions are affected by change of temperature. The risk of loss of optical performance, if and when the mechanism fails, could be reckoned small compared with the gain in operating time by using otherwise useless periods of the spaceflight such as eclipses.

6.7   Pointing and scanning

An orbiting telescope can be directed by commands to the spacecraft attitude control system, which will function through its inertia wheels or gyros, probably with the intervention of star trackers or fine error sensors. Sun–pointing instruments may require independent mechanisms to raster–scan the solar disc, such as that for the X–ray polychromator on SMM (Turner and Firth, [1979]).

6.8   Stray light

In an optical instrument any light which does not follow the intended path from object to detector is known as stray light. Stray light, from whatever source or by whatever optical path it reaches the detector, contributes to an unwanted background or 'veiling glare' which reduces the contrast and degrades the image quality.

A space optical instrument is likely to be even more sensitive to the effects of stray light at its detector than an earthbound device, because of the wider spectrum of radiations entering it. High sensitivity is usually a design goal. For simplicity, the term 'stray light' is used here; in practice, all radiations within the spectral sensitivity band of the detector must be considered in the analysis, as a result of which the instrument will include a number of apertures and baffles designed to minimise the effect of stray light at the detector.

Any surfaces or edges within the instrument, e.g. the tube wall, any apertures baffles or vanes, struts, mounts for optical components, even the optical surfaces themselves as well as contamination on them, can act as secondary sources of stray light if themselves illuminated. The illumination may be direct, from an object inside or outside the intended field of view, or as a result of diffraction, or of specular or scattered reflection from another surface which receives illumination. All possible ray paths must be considered.

All surfaces such as baffle faces and edges will be finished with a suitable highly absorbent black finish, as will all lens and mirror mounts etc. The spectral scattering reflection characteristics of each surface will need to be included in a full analysis. The most critical surfaces are those that are visible from any point on the detector and hence the most damaging sources of stray light are those which illuminate these surfaces. The most significant stray light paths and the power transmitted along each to the detector can be computed. The aim will be to provide baffles as required at positions, and of a size, tilt, and surface form and finish to block these stray light paths.

6.9   Contamination of optical surfaces

The use of clean rooms, to which only highly filtered air is supplied, has long been standard practice for the assembly of spacecraft and their critical instrumentation. The procedures originated with aviation gyroscope engineering, if not with hospital operating theatre practice before that. A great advance was to arrange the necessary changes of air by bulk movement at laminar flow speed, i.e. about 0.2 m/s. This requires the provision of blocks of particle filtering porous material, of size in the order of cubic metres, often built as a complete wall (for horizontal laminar flow down the length of the assembly room) or ceiling (downflow units). Less expensive apparatus, in the form of laminar downflow clean benches, is also used, often within an already clean filtered–air facility.

The standards generally adopted were laid down in the USA as Federal Standard 209A. A high standard in wide use is Class 100, which requires no more than 100 particles larger than 5 µm sieve size to be suspended in each cubic foot. (Particles fall under gravity, but very slowly, about 5 µm/s.) See Fig. 6.6.

It is important to realise that an empty clean room, even if operating at class 100, still allows uncovered surfaces to get 'dusty', by particle fallout. The rate is approximately 10 particles > 5 µm per square decimetre (90 per square foot) each 24 hours. Further, the arrival and activity of personnel will bring more contaminating particles (e.g. from human skin and breathing) and stir up previously settled particles. These will then drift downstream. Upstream, the cleanest conditions prevail.

Finger grease is likely to be deposited from the touch of an ungloved human hand. Often special handling tools must be devised. Such mechanical ground support equipment (MGSE) must be cleaned by swabbing with iso–propyl alcohol periodically.

Outgassing, vacuum pump oils, slow chemical processes, harness connectors, and lubricants from fasteners or vibrator slip tables may all give rise to filmy condensates. These are molecular, often hydrocarbons, rather than particulate. The hazard to optics is that they may be redistributed in orbit by convection from hot and outgassing surfaces to the much cooler surfaces of space–facing lenses and mirrors.

Vacuum bake–out is a preparation technique from ultra–high vacuum practice to reduce outgassing in test and orbit. The bake–out temperature must be low enough to avoid modification to the heat–treatment condition of metal alloys, or creep effects, especially of polymers. Since outgassing may be an activation energy phenomenon, there is a trade–off of longer oven time for lower temperature; perhaps doubling for each 20°C reduction.

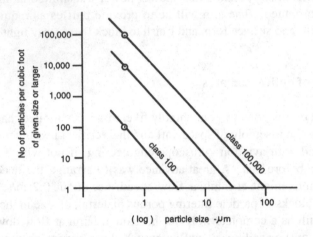

Fig. 6.6.   Dust concentration allowed in clean rooms.

The following lists the panoply of contamination control techniques deployed for a sensitive space optics payload (ROSAT; Sims *et al.,* [1993]).

- sealed compartments with dust close–out doors, pyrotechnically released in orbit; purged with dry nitrogen before launch; vented through 2–way breather valves with mesh filters
- activated charcoal dessicators (removed before launch)
- ultrasonic cleaning in degreasing solution, followed by cascade rinsing in water
- bagging of parts for transport between clean rooms, in clean anti–static polythene sheet; sometimes double bagging; and for storage after bake–out
- sealed (but vented with filter) transport container, with purging connections.

Clearly contamination control requires a lot of expensive equipment. But it is all for nothing without an earnest and intelligent attitude towards each operation by dedicated laboratory staff.

7

Project management and control

7.1   Preamble

The selection and operation of an appropriate and efficient project management scheme is as necessary to the success of a space project as the selection of the correct electronic components or the execution of a competent thermal design. Unlike pure engineering tasks, project management concerns the engineering of complex systems, the components of which are individual human beings. These sometimes exceed their specifications and sometimes fail to meet them, but they are always different. The presentation here of a structured approach to the creation of a project management scheme does not imply that one structure will suit all projects or all individuals. Considerable effort and care is needed to ensure that the management plan is efficient, appropriate and agreeable to all parties. The task of designing a project management plan can be seen as a method of ensuring that the project meets its time and cost budgets, in the same way as the electronic and mechanical engineering described previously in the book seek to meet power and mass budgets.

7.2   Introduction

Management is a very common word in everyday life but an accurate definition of it is difficult to get agreement on amongst practitioners. One definition that can be considered is:

> ' Management is the task of deciding what should
> be done and then getting other people to do it '

Although there are some unfortunate overtones in the last part of the definition, this description is a good one for endeavors like space projects which require a team of people with a mix of skills. The two important components of the job are identified–the need to decide what should be done and then the need to organise people to do it. If the management task does not include the stage of decision taking then it becomes more akin to administration, whereas if the organisational requirement is not included then the role described is one of unconstrained and perhaps rather haphazard leadership.

Clarity is at the heart of good management. Clarity in the overall objectives, clarity in the individual task assignments and clarity in the schedules and resource allocations. Clarity, though, can only be radiated downwards from the top of a management team. One of the most effective diagnoses of a space project with considerable problems was once carried out by asking the most junior members of the project team (as they stood at their machines, test rigs or drawing boards) ;

'when does the project require the item you are working on ?'

After receiving the same answer ('I don't know') from all those interviewed, the simple and correct conclusion was that the team was not managed with anything approaching the clarity required. An image frequently used in emphasising the need for clear objectives in order for a team to function effectively is that of each team member contributing to the project's progress by pulling on a rope attached to the project 'load'. (see Fig. 7.1)

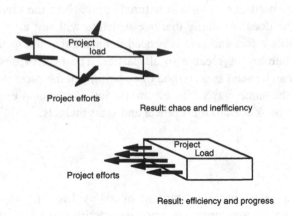

Fig. 7.1. Clarity of the project objectives.

If everyone is pulling in the same direction then a team of twenty can achieve twenty times the work of a single person. Where the clarity of their objective is absent and the team all pull in different directions then the effectiveness of the team can be reduced to zero, indeed they may lurch off in the direction of negative advancement by accident. The requirement in setting up a project management scheme is therefore to provide clarity of leadership and information to the team members, and to other participants in the project.

Objectives have been mentioned several times already in this chapter. It is good practice for a manager to enunciate the project objectives in a visible and hopefully inspiring way for the whole project team and this means down to the most junior member. The process of doing so will both confirm in the manager's mind what his / her objectives are (is it managing within cost or managing within schedule for example?) as well as providing a most powerful means of informing the team of the overall objective of their work. Clear and effective management, especially when combined with good decision making and a high profile style of behaviour is better described as leadership, something which is frequently aimed at or even emulated but not so often achieved.

The issue of objectives is not simple. Conflicts can arise between managing a project to a fixed schedule and managing to a fixed total cost. Tension between reliability and quality control on one hand and schedule on the other is very common. Which of these objectives is paramount may change with time throughout the project and the project leader, or principal investigator (PI), must make sure that the issues are well discussed among his co-investigators (Co-Is), agreed and promulgated throughout the team. These may be seen as tactical objectives and none the less important for that. However even the most junior team member must have in his/her mind a sentence or two to describe what the overall objective of the project is in everyday scientific terms. Not only is this inspirational to the person concerned but there are many routes by which a simple statement enthusiastically expressed can influence others to greater or more focussed efforts to the benefit of the project.

The main part of this chapter describes a fairly formal system for providing the management needed in a project but the reader must not confuse or misunderstand the objective here. The scheme to be described in some detail will, if properly used, provide the basis for an effective management system for a medium sized space instrument project. It can also be easily extended to cope with larger undertakings and therefore has the advantage that the concepts and nomenclature will be familiar to large aerospace companies, or even to the larger parts of the same project of which the instrument is a subsystem. However, if the scale of the task demands only two engineers and a draughtsman, all in the same room, then a blackboard with a list of tasks and delivery dates may well be very effective and clear. The unthinking application of the management scheme to be described might in that case obstruct the desired clarity of purpose and organisation. The management process must be designed to fit the overall project objectives and scope every bit as carefully as the thermal blanket or power supply is designed to play its part in the instrument operation. The next section describes the main organisational elements in a space project, and their relationship to the project team carrying out the instrument design and preparation.

7.3   The project team and external agencies

The size and scope of space missions demands that several organisations are involved, often in more than one country. The international aspect of space is increasing as budgets are squeezed and the sophistication of the individual missions increases. A successful project leader, or PI as he/she is frequently named at the instrument level, therefore needs to keep two or three (or more) organisations, each of which have different objectives, policies and funding cycles, in step with his requirements. This is an important task for the PI and in most project structures no-one else is in a position to carry out this duty.

The main elements of organisation in such a project are :

    the project team

    the funding agency

    the mission agency

    the launch agency.

Before analysing their roles individually it is necessary to understand that, usually as a result of the international dimension referred to above, there may be more than one funding agency. In many cases there are several, one for each country represented in the Co–I team or hardware team. The fact that so many science missions have been joint projects between NASA and ESA has meant that there are often two mission agencies. Developments in Eastern Europe may increase this complexity even more in the future. The end result is that a PI may need to keep several agencies informed of the project's progress. He may need to coordinate bids for funding in several countries, each using different conventions on costing, different timescales for approvals and each facing different financial crises as the project proceeds. Playing chess with 10 people simultaneously is a good analogy.

7.3.1   The project team

The project team is the element which designs the instrument, oversees its preparation, calibration and use and wishes in some sense to benefit from the scientific outcome. It is useful to define the role and structure of this important element before proceeding to analyse the interfaces to the other agencies involved in the project. In this section the relationship of the PI to the project team is outlined.

The PI and the project manager have to lead the project team as their most important and immediate task. If the project team loses enthusiasm or commitment then no-one else but they can supply it. If the project team loses effectiveness through lack of resources or lack of clear instructions then the PI must detect this before others if he is to remove the problem before it causes a serious threat to the project.

There are as many different styles in carrying out the leadership of the project team as there are PIs, but the ultimate test of leadership is the evidence that the leader is being followed. In every aspect of the project task the PI has two alternatives, either to carry out a task himself or to delegate it to someone well capable of doing it for him. A regular and vital part of the PI function is the conscious decision to take one of these alternatives or the other. Such decisions should not be made by default. In many ways, then, the style of the PI is determined by those tasks which he chooses to execute himself and those that he delegates. There are therefore 'science PIs' who rely strongly on their managers and engineers, 'management PIs' who delegate the engineering and the science tasks and 'hardware–instrument PIs' who concentrate on the design and qualification tasks. A relatively few individuals have enough breadth and depth of experience to lead in all these areas though there are many more who try this approach and fail in some regard. Even when dealing with the reported result of a delegated task the PI should be in a position to lead a project team discussion centered on the question :

'What does this test result or calculation mean for our project?'

After the discussion, the PI must be able to answer this question on even the most esoteric subject within the project boundary or he has failed in his task and begins to rely irrationally on luck for a successful outcome.
The PI's leadership tasks therefore include :

    a) Defining the scientific objectives of the project and the technical performance required to achieve those objectives,

    b) Chairing regular Co–I meetings,

    c) Ensuring a clear assignment of tasks to individuals and institutes,

    d) Defining the overall project strategy, in scientific, technical, political and programmatic matters,

    e) Adjudicating trade–off studies,

    f) Maintaining adequate visibility of the project in each participating country,

    g) Ensuring that the technical progress made by the team will result in the achievement of the scientific objectives.

The detailed structure of project teams will be analysed further in section 7.4.

7.3.2   The funding agency

In each country the funding agency occupies a crucial position in the project. Interaction between the funding agency and the project team can reach a peak at several phases of the project. The most obvious periods of activity are at the proposal stage and immediately after selection when the mission agency is very keen to know that the instruments it has selected will be funded by the participating countries. These phases are simplified greatly when the funding agency and the mission agency are the same, i.e. the mission agency pays directly for the procurement of the instruments. For reasons mentioned before, the international nature of space science has led to there being several funding agencies involved in most instruments that are selected.

The lead Co–I in each participating country will be in a position to advise the PI on how best to approach the funding agency over any appropriate issue. Following the early phases of selection, the most frequent circumstance in which this will be necessary is when either the project or the funding agency run into financial problems. The causes that lead to the project finding itself short of funds to complete on schedule and to the agreed specification generally derive from either some technical setback which requires extra work, or a schedule problem which necessitates the use of additional manpower or some kind of parallel working arrangement. Of course, the project may have been undercosted to begin with, in which case the problem will become apparent at some stage and will require either a descope or more funds. On many occasions, financial problems in a remote part of the agency budget can cause financial pressure on the project itself. When this happens the PI will need to be active in defence of the project. Political support from the other participating countries, either from the relevant Co–Is, from their funding agencies and even from their embassies, needs to be marshalled to support the case for the project continuing at its previously approved level. Good financial and technical progress as well as a high scientific priority can be useful factors in convincing the appropriate agency officers of the benefits of continuation.

Each international funding agency has its own characteristics and these need to be borne in mind when approaching it for support. Two of the most important aspects of the policy used by the agency to define its operations are :

a) The governmental purpose of the funding for the project (is it for technology, science, education, to encourage foreign relations or to provide underpinning for commercial or military technology ?). Knowing the basis for funding the mission, the PI and Co–Is can better target their argument for approval, increased funding or project continuation.

b) The costing methods used for the programme, in particular whether projects have their funds allocated in one block for the whole duration of the project, or whether the project is funded on an annual basis. Such differences become crucial when major project delays are announced by the mission agencies.

7.3.3   The mission agency

Most of the world's space mission agencies are household names, at least in their own country: NASA, the European Space Agency ESA, the two Japanese agencies NASDA and ISAS, the Indian agency ISRO and the Russian Space Agency RKA. All these organisations use their funds to :

a)   define space missions or projects

b)   procure instruments, satellites, space stations

c)   organise, fund or carry out the launch of space hardware.

Some of them fund the process from beginning to the end result while others, ESA for example, have a policy of requiring the ESA member states to fund the preparation of instruments and the data analysis phase.

7.3.4   The launch agency

The process of defining and engineering space science satellites and instruments has become refined to a sufficient extent that the interface between the instrument scientist and the launch agency is nowadays remote. The handling and access requirements of the instrument as part of the payload are generally passed on to the launch agency by the mission agency which is itself responsible for the contractual interface to the launch agency. The increased availability of cheap commercial launchers and the promise of small institute–scale satellites may result in the evolution of direct purchasing of launches by the science project teams but this is not currently the case.

Having outlined the major organisational elements in the project, most of whose management structures are beyond the influence of the PI, it is appropriate to consider the management structure within the project team. This is something which has to be designed by the PI and his/her colleagues to meet many requirements.

7.4   Management structure in the project team

The success of a management scheme for space instruments depends on two factors: the people chosen and the management structure created. Obviously the two factors can interact to a greater or lesser extent. At the outset of the project some of the key staff will already be chosen and there is usually, anyway, only a limited set of staff who are available and qualified. The choice of personnel for the key tasks is either obvious or extremely difficult and this textbook (or any other) is unlikely to help much. The only guidance which a formal text can offer is some definition of the tasks to be carried out, leaving those involved in real decisions about real people to imagine how the individual candidates might perform.

One factor in assessing the performance of an individual is graphically illustrated in section 7.6 which outlines the various project phases. There are a large number of different phases, spread over a period of five to ten years. While some individuals may relish the task of setting up projects and engaging in the early and very political battles for funding, the issue of their long term interest in the mission and indeed their endurance must also be considered if they are to be valuable long–term members of the team.

The key organisational elements of a space instrument management system will be outlined in some detail, followed by a discussion of the issues which arise when these elements are incorporated into an actual management structure. What then is a 'management structure' and what is its purpose? The kind of organisation charts of which Figs. 7.2–7.4 in this chapter are examples, attempt to set out the paths by which authority and decisions flow in the project team. It is sometimes difficult to represent on a single chart both financial authority pathways and the chain of technical responsibility. Line management issues (who is senior to who, and who controls the work of others) within an individual institute should not be shown unless they are also active routes of responsibility in the project. While the organisational chart shows the flow of authority downwards, responsibility flows upwards and this is an important feature which extends the usefulness of the chart, especially for the more junior members of the team and for outsiders seeking a point of contact at an appropriate level. The management structure or project organogram attempts to answer the following important questions:

a)   Who is in overall charge?

b)   Who does what?

c)   Who has control over which parts of the team?

Let us begin with the two most important posts; the PI and the project manager.

7.4.1   *The principal investigator*

In almost all space instrument projects with a high degree of scientific innovation, the PI is the most important person in determining the success of the project. Very frequently the PI chooses him or herself by proposing the instrument concept and the scientific objectives in the first place. On other occasions a PI has to be selected from a group of scientists at different institutes which wish to collaborate in the project. In some ways it is difficult to define the PI task and it might be fairer to those considering taking it on, to simply state that the PI must:

> ' Do everything necessary to ensure that the project team achieves the objectives of the project.'

Almost always the PI is a scientist of high standing, usually with at least some experience as a junior member of a previous project team. He or she will be recognised by the funding agency in the relevant country as someone in a position suitable to receive and properly deploy resources on a space mission. The main functions of the PI are :

a)   Scientific leadership of the project team.

b)   Strategic leadership of the various technical groups.

c)   Political leadership of the project team including the important tasks of interfacing to the several agencies involved. (see section 7.3).

d)   Maintenance of a plan, probably several, for completing the project on–time and within budget.

e)   Decision making and arbitration on major issues as the final authority within the approval status of the project.

7.4.2   *The project manager*

An industrially based project would receive the scientific performance specification from the PI and appoint a project manager on the other side of the contract interface to ensure the completion of the contractual requirements on–time and within budget. Most space instruments are not built this way and the manager is frequently part of the project team, often a Co–I and located at the PI institute. Co–location with the PI is by no means a necessity and some very good management teams have an off–site PI at the end of a good communication link to the manager's office. In this case the management centre is enshrined in a project office serving both the PI and the day to day management needs. In projects with a low level of scientific innovation or development risk it may be the project manager who is the senior interface with the funding authority, above the PI. This is the usual practice in some countries.

The tone or style of the project team is often set by this PI–manager interface.

Some PIs are skilled managers and play an important role in management decisions and the setting of technical standards while others act purely in a scientific and political role delegating all technical, financial and day to day activities to the project manager. The composition of the project team and the personal skills and tastes of the staff involved will determine this. Obviously the PI will have a huge influence on these matters since the mission agency may view him as the person with formal responsibility for the project. Other agencies see the project manager in this light and the PI again as a science adviser. The issue of the mission agency interface to the PI and project manager is important to explore and agree at an early stage. There can only be one person in overall charge.

The project manager functions include :

a) Creating and maintaining all the technical specifications from the requirements documents.

b) Defining and maintaining a realistic schedule consistent with the mission agency's launch dates and model philosophy.

c) Organising and monitoring the assignment of tasks to institutes .

d) Monitoring the progress of the project and directing resources to overcome problems before they become critical.

e) Setting up and monitoring the appropriate technical standards for the design, manufacture and test of the instrument.

f) Acting as final arbiter on technical issues which do not affect the scientific performance.

g) Acting as final arbiter in the case of conflict between the quality assurance/product assurance (QA/PA) function and the schedule, cost or reliability requirements.

h) Financial reporting and planning of resource requirements at the overall project level.

7.4.3   The co-investigators

The Co-I team often provides the highest level forum for advising the PI on a regular and frequent basis. Membership of the Co-I team confers some status on the personnel concerned and members are often appointed on the basis of certain technical or scientific skills or as the senior scientist or engineer representing each collaborating institute. The Co–I team can, however, grow too large! It then becomes unwieldy, troublesome and devalued. The PI can use the Co–I team as an advisory body meeting every three to six months, or as a link in the management chain with individual members having subsystem responsibility and reporting progress, problems etc. to the meetings chaired by the PI.

7.4.4   The local manager

Each participating institute may wish to appoint a local manager instead of using the lead Co–I from that institute in the management function. The local manager to Co–I interface then follows many of the issues raised for the project manager to PI interface but on a correspondingly smaller scale. Normally the local manager reports to the overall project manager directly on issues which do not impact the scientific performance of the instrument.

7.4.5   The steering committee

Many PIs find it helpful to set up a committee of the heads of the various institutes in the collaboration to discuss overall progress and strategy, especially on political and national funding issues. This can interest and involve these influential people in the project which may otherwise be just one of many in which their institute is participating. The PI can use this committee as another avenue to support a Co–I experiencing difficulties in extracting resources from his or her institute director. National funding issues are often very effectively discussed at such committees which by their membership have leading space administrators present to advise and help in the case of difficulties. Membership of a steering committee can generate a sense of identity with the project for senior, and influential, scientists and administrators. Some funding agencies would wish to place a representative on the steering committee to gain detailed insight into the progress and state of the project. The steering committee should meet regularly but not frequently, say twice a year, so as not to interfere with the function of the normal management committees.

7.4.6   The project management committee

Some national funding agencies require a project to be reviewed or managed in a strategic financial sense by a committee of authorised officers of that agency. In some cases there can be real advantages in such a committee being chaired by an independent scientist who then represents the project at various important funding stages as a non-participating advocate.

7.4.7   Management structures

Figs. 7.2 to 7.4 give some simple examples of organisational structures which have in the past proved successful in managing space projects.

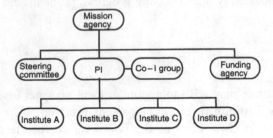

Fig. 7.2.   The PI model.

Fig. 7.2 is the PI model. Essentially the whole project is led in scientific and technical decision making by the PI, with the assistance of the Co–I group. For this to be successful the PI must have a wide range of real (i.e. not imagined) skills, considerable personal authority and a substantial amount of time free of other commitments.

Fig. 7.3 is the model used where the manager has a high profile with the mission agency or funding agency and the PI confines him or herself to scientific issues, perhaps by chairing the 'science team' of Co–Is.

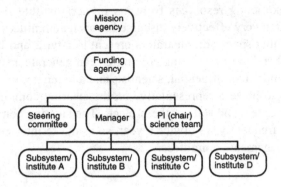

Fig. 7.3.   The manager model.

Fig. 7.4 is the strong PI plus well delegated manager model. If the PI can resist bypassing his manager and perturbing the project below him and if the PI and manager have confidence in each other this can be very successful.

The most important aspects of these organisational charts are that they must be unambiguous, they must be agreed by all parties (signed for in blood if necessary) and they must be adhered to.

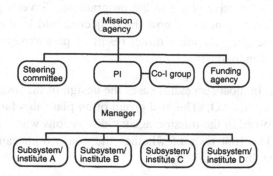

Fig. 7.4.   The delegated authority model.

7.5   Project phases

No two projects proceed along the same lines, or even to the original plans. A list of the 'normal' project phases can be misleading therefore. Despite the differences from one project to another however, there is an overall shape to the development of an instrument to fulfill a specified function on a space mission.

7.5.1   Normal phases

The main phases are listed below. Some of them need little in the way of explanation.

a)   Conception. This is perhaps the most idiosyncratic of events. Sometimes it is triggered by unfolding inadequacies in the previous datasets, while in other missions the announcement of opportunity for the mission itself triggers the appropriate thinking which leads to the instrument or scientific concept.

b)   Mission concept. This can be a separate phase in which the scientific objective is offered up to a funding or mission agency as the proposal for a whole new scientific mission. The mission agency may conduct an Assessment study of the project at this stage, or, if the project has reached an advanced stage of definition, a phases A study will be carried out.

c)   Announcement of opportunity (AO). When the mission agency has selected the mission as a new start in its space programme then an announcement of opportunity for instrument proposals is made public.

d) Formation of a collaboration. The point has been made several times that many, if not most, space instruments are made by consortia of institutes. The formation of these consortia is a sensitive phase in the programme. Several groups may be involved each with its own opinions of how the project should be done and who should lead it. Often these opinions are a direct result of past experience of working together as institutes and individuals.

e) Instrument design. In most successful cases the design of the instrument is started one or two years before the AO. The lead group often plans this far ahead because they are personally involved in the mission agency discussions which lead to the mission being selected. The first tests of whether the consortium can work successfully together are made at this stage.

f) AO response. As with the design phase, the process of agreeing, writing and delivering the proposal on time to the mission agency is a useful practice for later events.

g) PI selection. At some point in the previous three stages it will either become apparent who is the correct person to lead the consortium, or there will be a specific moment at which that decision must be taken.

h) Instrument selection. This is done by the mission agency, sometimes after interviews with the PI and his Co–I team. There may be some iteration over issues of mass, power, size etc over a period of weeks or months. If the PI has included options in his proposals then the mission agency will wish to decide between these as part of its selection process. The selection at this stage will be conditional on the funding agencies agreeing to fund the various teams involved.

i) Funding approval. This may be simple if the funding agency requires few financial details from its scientists or if the work is to be supported from some institutional budget. Some agencies require large volumes of financial and other management data in a different form from the mission agency proposal and this can cause a great deal of extra work, beyond the response to the AO.

j) Detailed design. This is termed the phase B by industry and has the objective of completing the interface definition with the spacecraft and agreeing the subsystem specifications. Since these interfaces are extremely difficult or expensive to change, a great deal of detailed design must have been completed by this point in the project if the interface agreements are to be formally entered into. phase C includes the fabrication and assembly of the space components and their functional and environmental testing. There may be several models to be delivered in phase C.

k) Size and mass model. Some mission agencies require a simple model to represent the size, i.e. shape, and the mass of the instrument for use by the spacecraft manufacturers as a check on the drawn and written interfaces.

l) Prototype model. The project team, formed from the consortium who proposed the instrument, may wish to build a laboratory–standard model of the whole or parts of the instrument at an early stage to check certain critical functions or obtain technical performance data.

m) Structural and thermal model (STM). Many agencies demand this as the first deliverable item. Sometimes it also fulfills the functions of the size and mass model. The STM is often non–operational in an electrical sense except insofar as the power dissipation is accurately simulated with dummy loads. It is used to verify the structural and thermal designs and confirm the accuracy of the mathematical models developed to describe the response of the instrument to thermal and mechanical stress.

n) Engineering model (EM). This is usually the first fully functional model of the instrument and is the first time that all the subsystems are interconnected. Problems are often uncovered at this point as a result of incompatibilities of the interfaces. The EM is frequently used to environmentally qualify the design. In instruments where there are several identical units (for instance many channels of identical electronics), cost savings can be made by including just one representative unit. There are dangers in this because the interaction of these units cannot then be modelled but the risks have to be traded–off against savings.

o) Flight model (FM). In some programmes the flight model is the first model to undergo a full environmental test and a full calibration. Leaving these important events to this stage can be dangerous in a schedule sense. Every attempt should be made to qualify all the design aspects at the STM or EM stages and run through a trial calibration. Calibration facilities are notoriously time consuming to commission and an accurate estimate of the time required for a full flight calibration can only be obtained via a partial trial in the real facility.

p) Flight spare. The flight spare, if it is required at all, follows on unexpectedly quickly after the flight model and often benefits in terms of the lessons learnt on the flight model. It can be cleaner and better calibrated than the FM for this reason. In order to save costs it is becoming increasingly common to provide only subsystem replacements as spares instead of a complete flight unit.

q) Launch and operations. These phases mark the need for a changeover from purely technical and hardware skills in the team. A handover of responsibility in managing the project can occur at this point and this transfer needs to be planned carefully and clearly. The launch itself is sometimes defined as phase D, with the operations as phase E.

r)   Data analysis and interpretation. This phase of the project usually does arrive though many doubt it during the earlier phases. In a perfect science programme it should spawn the ideas which lead to the next project.

7.5.2   Project reviews

From time to time either the funding agency or the mission agency may require a review of the project. The thoughtful PI or project manager will have already planned a series of project reviews. These are much more helpful and effective if they are chaired and carried out by independent scientists and engineers. Reviews are a non–negligible resource drain on the project, requiring much preparation by the project team. In a well run project much of the necessary documentation will already exist and its preparation will not represent a new task. Even so there are performance reports to prepare, diagrams to redraw to illustrate a particular aspect and narrative text to write in order to educate the review panel from a starting point of little detailed knowledge. It is all very worthwhile if the review panel is truly independent and thinks through the project tasks with a fresh mind using its, hopefully considerable, experience. Review panels which are not independent and truly critical do not repay the effort of the review to nearly the same extent and can provide a misleading sense of reassurance.

The principal points in a programme at which reviews are helpful are as follows:

a)   Conceptual review. Held at an early stage in the design, soon after instrument selection, this review exposes the solidity of the instrument concept, the design and development status, the calibration plans and the preparation for data analysis.

b)   Development review. These may be called for at various points to assess the status of critical subsystems.

c)   Critical design review. A lengthy and searching review, subsystem by subsystem including design details, materials choice, manufacturing methods, QA/PA practices, assembly techniques, test philosophy and calibration plans. The realism of the schedule may be an important concern for the review panel.

d)   Engineering model review. This accepts the basis of the design from the previous reviews but assesses the EM programme either before delivery to the spacecraft or after EM testing, in which case a thorough analysis of the test results can be expected.

e) Flight model review. This in many ways parallels the EM review except that the full QA/PA procedures will have been operative and the non–conformance reports and test failures will be available for review in a formal and rigorous manner. In theory no delivery should take place until all the non–conformance reports (NCRs) and review recommendations have been signed off by the review chairman.

f) Pre–shipment review. This provides an opportunity to close off any remaining NCRs and make the formal decision that the unit is in a condition which justifies shipping to the launch site. This is the last time to get things right. Things stay wrong for a long time in orbit if they malfunction or underperform. Despite the obvious pressure to agree that all is well, this opportunity to assess, review, retest or even redesign should not be missed.

g) Flight readiness review. (FRR) Much the same objectives as in the pre–ship review but the results of post transportation testing will be available and must be checked. This event is often carried out as part of the spacecraft and launcher FRR.

h) Failure review (or Materials review board, MRB). This should be held whenever a serious subsystem or component failure requires comprehensive analysis to ascertain the cause of the failure and the necessary remedial action. An independent chairman and team is essential for this to be effective.

The overhead on training the review panel from scratch at each review can be much reduced by obtaining the agreement of the review panel and chairman at the start of the project to continue their personal involvement throughout the project lifetime and take part in all the relevant reviews of the project.

7.6   Schedule control

The main purpose of a schedule is to provide some kind of 'ruler' against which to judge progress. Schedules are seldom very accurate to begin with but if it is found that progress is being made at only half the rate required then action, of one kind or other, has to be taken. A schedule on its own is therefore no good unless supported by some means of progress measurement and reporting. There exist various different kinds of schedule display, some simple and some so complex that computers are required to analyse and print them. A decision should be made early on as to which form of schedule analysis will be used. The effort expended on the more complex scheduling tools is not always rewarded in modest, instrument sized projects.

7.6.1   *Progress reporting*

It is vital that the project management set up an agreed and efficient system of progress reporting, right from the start of the project. Intervals between reports must depend upon the project requirements and the state of the project but a regular, short, written report on a monthly basis, together with a progress meeting every three months is a reasonable norm. In periods of crisis management a weekly or even daily meeting may be required. Such meetings have other benefits in terms of team identity and coherence beyond the basic aim of information dissemination. Setting deadlines for the completion of activities within the spacing of a month or so allows the most effective oversight of the project progress. If deadlines are spaced at longer intervals there is a greater time delay before the schedule problem is identified.

7.6.2   *Milestone charts*

At the very beginning of a project there are a number of fixed time frames specified, mainly in terms of the delivery of individual models and, of course, the launch date itself. These dates, perhaps eight to twelve in number, form the project milestones and the necessary project activities can be arranged between them as in Fig. 7.5.

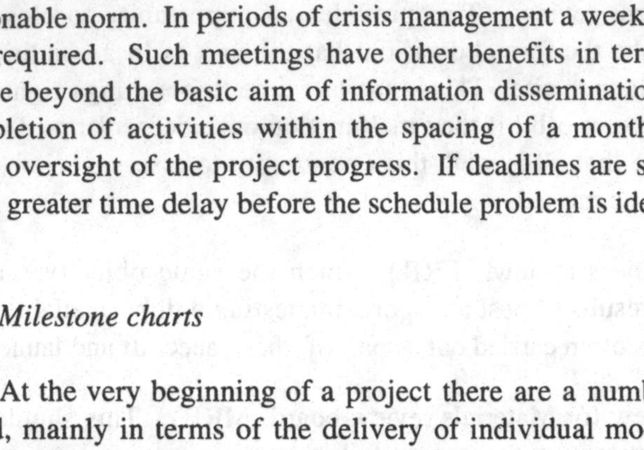

Activity	Year 1		Year 2		Year 3		Year 4	
Phase B review	▼							
Phase C/D		▼						
Design review	▼							
STM delivery			▼					
Thermal tests				▼				
Structural test				▼				
EM start				▼				
EM delivery					▼			
EM test review						▼		
FM start				▼				
FM delivery						▼		
FM test review							▼	
Delivery to range								▼

Fig. 7.5. Milestone chart.

It is quite usual to include the major review dates in the set of milestones. Modest projects with small, closely organised teams can be run quite effectively by means of milestone charts alone. This is especially true if the resources needed are easily available and fairly constant in profile. Even if a more complex method of schedule control is being used, displaying the project plans and achievements as a milestone chart is the simplest and most effective way of communicating with the project team and the funding agency.

7.6.3   *Bar charts or waterfall charts*

A slight increase in complexity from a simple milestone chart is the bar chart or waterfall chart. As the example in Fig. 7.6 shows, this method of schedule representation allows each separate major activity to be planned both in terms of its relationship to the major milestones and in respect of its logical connection to preceding and following activities. As with the milestone chart, the bar chart is simple enough to be readily understood by all interested parties. Many of the complex computer packages for schedule control include the option to print out a bar chart representation of the schedule as well as other displays. Bar charts, which clearly can be quite complex if a large number of separate activities are represented, frequently satisfy the needs of modest projects. In addition to being easy to understand and assimilate, bar charts require only a low level of activity to generate them and keep them up to date.

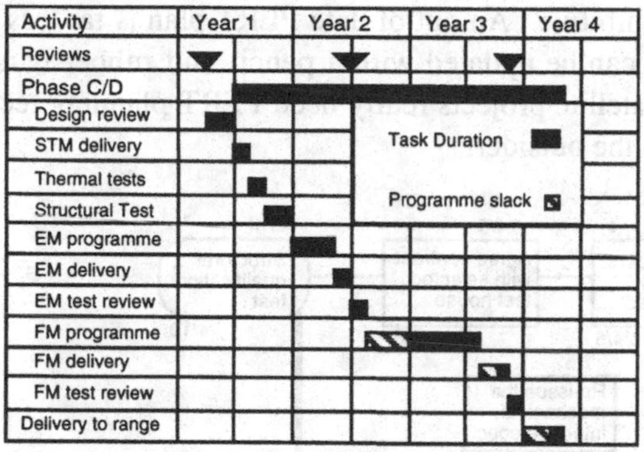

Activity	Year 1	Year 2	Year 3	Year 4
Reviews	▼	▼	▼	
Phase C/D				
Design review	■			
STM delivery	■	Task Duration ■		
Thermal tests	■			
Structural Test	■	Programme slack ▨		
EM programme	▬			
EM delivery		■		
EM test review		■		
FM programme		▨▬		
FM delivery			▨■	
FM test review			■	
Delivery to range				▨■

Fig. 7.6.   Bar chart.

7.6.4   *PERT charts*

The generic name for the most complex method of schedule control is the 'Project Evaluation and Report Technique', known commonly by its acronym –PERT. A wide range of mini–computer software products is available under various trade names offering the capability of carrying out a PERT analysis and displaying the results in several ways. The power of the PERT technique is two–fold. Firstly the computer, when provided with the logical relationship of each activity to the others together with the earliest start date, latest finish date and minimum duration for each task, can calculate which sequence of activities is on the 'critical path'. These are the activities which define the path to the earliest completion date for the project in the sense that shortening or lengthening the duration of any task on the critical path shortens or lengthens the overall project duration. Tasks off the critical path have a measure of slack in them or in the tasks which are logically related to them. Identification of the critical path is very important because the project completion date can only be affected by changes to tasks on the critical path. It is these tasks to which extra resources must be directed if the completion date is to be brought forward. The second important attribute of PERT scheduling software is that most implementations of it (but check this before buying it) will allow the optimisation of the schedule to be done taking account of any imposed limitations on the available human resources. Since almost every project team is of finite size the preparation of a project schedule which recognises this in seeking the optimum path through the various activities brings a useful level of realism to the plan.

The disadvantages of the PERT technique are also two–fold. While at the beginning of the project the manager may have the money to buy exciting new software packages which promise the ability to plan his or her way out of any problems, and the time to use it, as the project proceeds the effort required to maintain the PERT plan can become prohibitive. An out of date PERT plan is far less use than an out of date Bar chart which can be updated with a pencil and rubber if needs be. Few instrument–size or small satellite projects really need PERT planning techniques, impressive though they seem to the outsider.

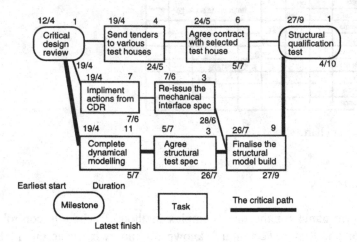

Fig. 7.7. A small section of a PERT chart.

The availability of PERT programmes in PC form has made them more accessible but a rational decision has to be taken early on as to who will maintain the plan and what this effort will cost the project. The second disadvantage of these more complex methods is that the graphical display of the output is by no means easy to absorb, or even print on reasonable size paper! If a project is complex enough to need the PERT technique then the number of tasks will be large enough to make presentation of the output a problem in itself. When all the work has gone into specifying the project plan to the software but the user is only printing out a bar chart of the results, the question must be asked whether a bar chart analysis would not have sufficed in the first place, at much less overall cost.

7.7   Documentation

The choice of the level of documentation appropriate for the project is an important one. Factors which need consideration are the size and geographical spread of the project team, the level of formal QA required or available, the complexity of the technical tasks and the formality of the hardware procurement process.

Of all the documents in a project none is more important than the proposal since without this the project never gets going. In addition to its importance in gaining funds to do the work, the proposal lasts for the duration of the project as a comprehensive source of information on what the objectives of the project were at the time it was conceived.

7.7.1   The proposal

The audience at which the proposal is aimed is a wide one It includes highly skilled and experienced scientists working in relevant fields, as well as science administrators, aerospace engineers and space policy makers. Many of the subsequent members of the project team will obtain their first impression of the project from the project proposal. In order to attract the favourable attention of this highly inhomogeneous group of people the proposal must be well written and address a number of key issues, such as:

What is the broad scientific need for this investigation?
What specific science will be accomplished?
What will the overall benefit be?
What is actually proposed?
What will its lifetime be?
What data will it produce?
Who will analyse the data, using what methods?
What is the minimum acceptable scientific performance?
How easily does the proposed instrumentation meet this specification?

What are the likely sources of error, both statistical and systematic?

What other, similar ,experiments are planned in other countries?

What models will be built, what will they be used for?

How much will it cost to build?

How much will it cost to operate?

Who will own the hardware and the data?

Who will own the intellectual property rights?

Who are the key staff, what experience do they have and how much time will they commit to the project?

7.7.2   Project documentation

A good analogy can be made between documentation and glue. In both cases where there is an excess above the necessary level there will be a 'gumming up' of the process and a reduction in progress or efficiency. If there is not enough of the required documentation, as is the case with glue, the whole system begins to fall apart. In particular, a lack of appropriate documentation can result in a great deal of wasted effort working to the wrong specifications or to the wrong schedule or technical standards.

Documentation is not a favourite topic with scientists. The natural inclination is to reduce the requirements and to give the subject a low priority in their time allocation. This is an understandable tendency and can be effective if not carried to extremes. The list of documents given below has been derived in response to this minimal approach. It attempts to set a lower, safe limit on the number of separate documents required to define a space science instrument sufficiently that Co–Is, subsystem engineers, contractors and independent review teams can all work on a defined basis towards the same goal. The 'minimum' list is :

a)   Science requirements specification

b)   Instrument performance specification

c)   Interface specification (to the spacecraft)

d)   Mechanical interface drawing(s)

e)   Workpackage assignment list, including a formal statement of responsibilities

f)   Schedule in a network form

g)   QA/PA plan

h)   Inter–institute agreement on data rights

i)   In–orbit operations plan

j)   Environmental test specification

k)   Transportation/handling specification

l)   Test plan, including details of tests and normal response to stimuli.

These documents define the project and must be kept up to date at each institute with proper issue numbers and formal sign–off sheets to establish beyond doubt their status in the project. As the project proceeds it will become necessary to change designs and even details of the specifications. Normally such decisions will be taken as part of the regular Co–I meetings chaired by the PI. For this reason these meetings should be minuted at least to the extent that decisions are recorded formally. This enables the results of the decisions to be put into action before the next issue of the relevant document.

7.8   Quality assurance

Quality assurance (QA) and Product assurance (PA) are so closely related in the space instrumentation field that a distinction between them is not useful. It is always necessary to produce a QA/PA plan to demonstrate the standards to be employed in organising the work and carrying it out. Nothing strikes terror into the heart of the novice as much as the request to write a QA/PA plan but in fact the process can be relatively simple once access to professional advice is available. Often the agency requesting the plan will help to a very large extent by defining the standards it wishes to be used. One then has to tell them in a formal manner what it is they wish to hear. However, writing an acceptable QA/PA plan is only one issue. A structure of QA/PA procedures has to be set up, including the necessary monitoring and reporting procedures, in order to comply with the proposed requirements. While overall PA systems such as ISO 9000 or BS 5750 can be a help, every project is different and has specialised requirements.

In summary, the objective of the QA/PA plan is to state what you are going to do, how you are going to do it and how you will prove to the mission agency that you have done it.

7.8.1   Specification of the project

The QA/PA plan is the place where the documents describing and defining the project are listed, together with those official documents which specify standards or processes. The statements needed are simple and straightforward references to the applicable documents which either exist in the project or are available in the public domain. Generally little justification of the choices is needed except where an unusual or controversial choice is made.

7.8.2   *Manufacturing methods and practices*

Again, simple statements of the processes of component selection, testing procedures, design factors, derating policies, test philosophy and model programme are all that are required. More explanation than in the previous sections may be helpful in this part because the choices are more closely allied to the actual design and function of the instrument.

7.8.3   *Monitoring and reporting of results*

For the plan to be put into action, various staff will need to measure, record and report the results of the programme of work. What steps are carried out in this process need to be specified. A definition may be helpful of the relationship between the QA staff and the project team and the steps that will be taken to resolve disputes. Of special importance is the reporting of malfunctions or failures. All the agencies involved will wish to see a clear proposal of what happens in the event of a failure, how it is recorded and who will judge what action should be taken. Some formal method of collecting and reviewing NCRs will be needed. The relationship of these reports to the normal project reviews can also be proposed.

7.8.4   *Samples of documents and reports*

It can be very useful to include as an annexe to the QA/PA plan, blank copies of the reporting forms that are proposed for various recording and reporting purposes in the project. These may be forms which already have some status in the institutes QA procedures or they may be designed for the project in question. Since they will be needed anyway, it is quicker to include an example than to try and describe the form in words.

7.8.5   *Detailed contents*

The actual contents of a QA/PA plan are obviously project specific but a starting point for a new plan is provided below in list form which includes the main sub–headings that might be needed :

a)   General PA requirements and management

> Project organisation.
> Contractor and supplier monitoring
> Status reviews
> Agency participation in inspection and test
> PA progress reporting
> Performance verification

b)   Quality Assurance

Procurement controls
Incoming inspection
Surveillance of manufacture and integration
Test witnessing
Records and log books
Cleanliness and contamination control
Non–conformance control
Metrology andcalibration
Handling, storage, marking, packaging and transportation
Component alerts
PCB manufacturing standards
Wiring standards
Electrostatic discharge control

c)   Reliability assurance

Failure modes, effects and criticality analysis (FMECA)
Single point failure and critical items list
Numerical reliability assessment
Component stress analysis and derating policy
Worst case analysis

d)   Safety assurance

Applicable requirements
Safety assurance tasks

e)   Component selection and procurement

Prohibited materials and components
Radiation sensitive components
Component derating
Component approval and preferred parts List
Non–qualified components
Component screening and burn–in
Lot Acceptance procedures
Declared components list
Manufacturer surveillance
Destructive physical analysis
Storage

f)   Materials and process selection

Materials
Stockists
Conformance documentation
Contamination and corrosion
Storage

g)   Configuration management and control

Configuration management
Configuration control system
Configuration identification and baseline
Documentation numbering scheme

h)   Acceptance reviews and data packages

NCR procedure flow chart
List of controlled documents
Approval procedure flow chart
Logbook/factoring record hierarchy
Standard report forms

7.9   Financial estimation and control

In the real world projects are run with a fixed financial envelope. The estimation of the overall project cost is therefore an important activity at the early stage of the project approval. Errors either way can come back to haunt the project team. Once the financial approval is given the problem is to complete the project within the allowed funds, but also to time and in a manner which meets the performance requirements. Different mission agencies have different attitudes to cost overruns but even those which were indulgent in the 1970s and 80s are feeling the pressure from their funding sources to manage programmes more effectively and that means managing them to budget.

7.9.1   Work breakdown schemes

The most straightforward means of estimating the cost of a project, as with estimating the cost of anything, is to break the project down into its constituent 'work packages', cost each one separately and find the total. This is usually done by means of a 'Work breakdown scheme', or WBS, which provides a nomenclature which can be used everywhere in the project to identify and document the various activities or tasks.

There are some fairly standard WBS formats which can be used on most projects or a new one can be generated from the specific project requirements. In essence the WBS provides a logical structure for the subdivision of the whole project activity into its constituent parts.

An important issue when setting up a WBS is that one group or person must have responsibility for each work package. This is the person who has the responsibility for seeing that the work is properly specified, carried out and reported on to the overall manager, it is not necessarily the person or institute which actually carries out the work, although it would be odd if they were not very closely involved in its execution. There can be only one person or institute responsible for each work package. One of the benefits of agreeing the work breakdown in this formal way is that the process of reaching an agreement will automatically result in a clear understanding of who is responsible, provided that the content of each work package is well defined. It is customary to generate a single sheet work package description to assist in the definition of the work content of each work pack. Such a work package description will contain the name of the responsible person or institute, together with the start date and end date of the activity and a specification of the inputs required to start the work and the output required from it.

7.9.2   Cost estimates

The basic cost ingredients of each work package are the materials required, the staff effort necessary to complete it on time, the funds needed for any associated travel and the financial provision for any consumables involved.

Estimating the materials costs can only be done after defining the items required and obtaining quotations or catalogue prices. If a similar task has been performed before then a comparison with the previous spend may provide a useful check. The issue of what quality of product is needed in various phases can be important. Flight standard electronic components can be a factor of thirty times more expensive than functionally similar units which are appropriate for prototypes or even for Engineering models.

It is notoriously difficult to estimate the staff effort required for various tasks. Comparison with previous work is especially valuable at this point. It is better to estimate, and record for further projects estimation, the staff effort in staff–months rather than in cash, as accounting systems differ widely between institutes and different countries. An issue which needs to be agreed at the outset is whether the overall management effort should be included, *pro rata*, in each work package or whether a single, separate, management work package is easier to deal with.

A similar issue arises with the estimate of travel costs. Some work breakdown schemes have a single work package for travel, others allocate a travel budget to each work package. The estimate of travel costs can only be done by assuming a pattern of meeting and work visits and costing each separately. Decisions (guesses ?) have to be made as to how many people will travel, how many nights they will stay away and how many hire cars will be needed. Once these parameters are estimated then costs can be generated from the relevant air fares, subsistence rates and car hire prices.

Project teams need pencils, computer paper, stationery as well as lighting, heating, telephone and fax services etc. There are a myriad of items which will need funding via the consumables line in the work package costs unless these are provided by means of institutional overheads. Even if the institute does provide these at the start of the project, beware of efficiency exercises over the ten year project time span which change the internal institute rules for infrastructure support. Such changes can leave a project with new charges to face and no appropriate budget from which to pay them.

It is extremely valuable to compile the costings for each work package on a year by year basis throughout the project duration. When fed into a suitable spreadsheet this allows the project finances to be totalled for each year, for each work package and as a complete total for the whole project. Funding agencies are frequently interested in the spend profile year by year whereas work package managers need to know what their total budget is to achieve the work package objectives.

7.9.3   *Financial reporting*

As with the reporting of technical progress, judgements have to be made of the frequency and content of financial reporting which is necessary. Whilst these requirements may sometimes be laid down by outside agencies, the project is usually left to define the procedures for itself and a balance between excessive detail on one hand and a lack of data for proper control on the other is needed. Financial reporting is not undertaken simply for accountancy purposes. If a project is not spending at the rate which was planned then one reason may be that adequate technical progress is not being achieved. Monitoring the spend can therefore provide an additional and very quantitative input to the management of the schedule.

Reports of the commitment and spend broken down by work package are usually requested on a three monthly basis. This is frequent enough to allow some insight into the annual financial processes but not so heavy a burden that it dominates other management tasks. Recording commitment is very important in judging the carry over from one financial year to the next. Commitment of funds by employing staff, placing contracts etc. is vital if the project spend is to be maintained. Studying the graph of commitment and spend therefore provides an insight into whether the level of commitment is sufficiently far above the total spend to be able to generate the required send rate. A gap of three to five months is common between the commitment of funds and the actual spend in the sense of paying out money from the project coffers.

7.9.4   Financial policy issues

Although the design and manufacture of space instruments is supposed to be a technical issue, the majority of the team are only free to concentrate on the science and technology if someone is looking after the money properly. A number of policy issues arise which are best sorted out at the start of the project to ensure that the flow of funds is both adequate and continuous.

Since space projects generally take a long time, say five to ten years, the question should be raised with the various funding bodies as to whether adjustments in the overall cost are allowed to account for inflation. Inflation is a fact of life and either some allowance has to be built into the original costings to provide a buffer against future inflation, or an inflation index must be chosen which is published regularly and is agreed by all parties to represent a reasonable assessment of inflation in the relevant business sectors. Using the index, the unspent balance of funds can be retrospectively adjusted. A new total cost to completion must then be periodically recognised by the funding agency in an appropriately formal way.

It would be foolish to imagine that projects can be costed in advance with complete accuracy. Prudence demands that some level of contingency is included, say 10 per cent of the unspent funds, to provide for completely unexpected circumstances. It is usually more effective if this contingency is rigorously excised from individual work packages and held centrally. This practice creates a large pressure on each work package manager to keep the costs of his or her work package within budget by seeking savings internally before coming forward to the project with bids against the central contingency. It can be effective for the steering committee or other senior body to hold the contingency. Releases of funds are then very carefully scrutinised by all the interested parties as part of a very public process.

There are certain financial issues which are always left for others to deal with and hence run the risk of being overlooked. These include VAT or local taxes, import duties, software licenses and overhead contributions. The list of these and other items which may benefit from consideration when costing a project is given below :

a)   Overall project costs

    Staff costs
    Materials
    Travel costs
    Consumables
    Consultancies (for expert opinion, safety approval etc.)
    Secretarial / administration charges
    Licenses for design rights
    Software licenses

b)   Overheads

            Lighting
            Heating
            Computer access
            Library access
            Safety services
            Pension schemes
            Site overheads

c)   Special expenditure

            Import/export licensing
            Transit containers
            Special jigs and fixtures
            Electrical support equipment
            Alignment equipment
            Lifting and handling equipment
            Calibration facilities
            Transportation charges
            Civil engineering charges
            Insurance premiums

d)   Post Launch Costs

            Operations staff
            Travel
            Computing equipment
            Software licenses
            Networking charges
            Data media
            Postal charges

7.10   Conclusions

No scientist or engineer starts a project because he loves management. Years of observation show that the people who spend a small amount of time developing a clear and efficient management process actually spend very little time managing the projects they care about. Scientists who despise management seem to spend more time doing it, but doing it badly, to everybody's discomfort.

8

Epilogue: space instruments and small satellites

8.1   The background

At the end of the nineteen sixties satellites characteristically had masses in the range 150 to 300 kg. By the end of the nineteen nineties many, if not most , satellites are being designed to mass budgets between one and ten metric tonnes. Satellites have expanded to fill the launchers available would be one conclusion. In fact, many changes have taken place during the past thirty years and most of these have led to a growth in satellite masses. The scientific problems being solved from space have grown more and more exacting in terms of the equipment required. As more is discovered, seemingly, more is left to be discovered and ever more sensitive instruments are demanded. Greater sensitivity usually requires large collecting areas, cooled telescopes or massive detectors, all strong factors in determining the mass of the payload. In the fields of Astronomy and Earth Observation the scientific problems seem best tackled by 'Observatory Class' missions in which a payload of five or ten separate instruments is compiled to provide the varied individual measurements necessary to address the mission objectives. In some cases these instruments could be launched on separate platforms if their observations could be properly coordinated, in other cases the full set of measurements must be simultaneous in space and time. Launch vehicles which are principally designed to meet the growing needs of geostationary communications satellites are available with the lift capability of many tonnes and so there has been little pressure to identify missions which can achieve the highest quality science from small, and hence inexpensive, satellites.

The financial pressures which most governments now experience are leading to a radical reassessment of the value of space programmes and the costs of access to space. Following the decline in the defence equipment industry as a result of the cessation of the cold war, the space sector has a lower throughput and the management overheads in large aerospace companies are being distributed across a smaller range of projects, leading to further cost increases in some cases. All these factors bring with them increased project durations and a longer gap between a conceptual idea and a resultant mission.

All these factors might point to a terminal decline in the use of space were it not for the fact that space engineering can now be rated as a mature discipline and hence accessible to small and medium sized companies, universities and research institutes, in addition to the major aerospace companies who were the only credible providers of space hardware in the past. The improvements in the size, power consumption and reliability of electronic components has played a large role in making the design and manufacture of space instruments a realistic possibility for organisations with modest technical resources. Indeed, it is the aim of this book to bring such opportunities to the widest possible community. It is then obvious that only a small extension of the ideas presented in the preceding chapters is necessary to design, not a space instrument but a whole satellite.

8.2   What is small ?

Since the size or mass of an instrument or a spacecraft has been very closely related in the past to the cost of designing and manufacturing it, the word 'small' has come into use as a euphemism for low–cost, inexpensive or cheap. In fact neither the size nor the mass of a 'small' spacecraft would have any special significance in a world of unlimited funds. Since that world has not yet been discovered, the term 'small' must be seen as synonymous with a low–cost approach to design, manufacture, test and operation. But low–cost does not have to mean unreliable or modest in outcome. Various attempts to define a size range that may be considered 'small' are therefore doomed from the outset and miss the whole point of the exercise. Nevertheless, the possibility of adopting a low–cost approach to the creation of a very large and sophisticated 'Observatory–Class' spacecraft must be considered unlikely and so it is realistic to expect that 'small' spacecraft would be below one tonne in mass. There will always be a scientific and technical need for 'large', as opposed to 'small' spacecraft to carry out such missions.

The real difference between a 'normal' spacecraft and a 'small' one lies in the management and technical approach taken by the project team. Basically one has to replace 'money' with 'ingenuity' and replace 'planners' with 'do–ers'. Neither of these objectives can be taken to extremes: money and good plans are both needed. However, the project cost can only be kept small and within the original estimate if the channels of influence between the budget, the objectives, the design and the procurement are all kept very clear and efficient. A cost increase in one part of the project has to be balanced by a reduction somewhere else before the money is committed. With large and complex teams this process of the feedback of the implications of high level decisions is so slow and inefficient that the project motors on with an inertia which inevitably defaults to the more expensive solution. The more layers of management, the worse this tendency is. It seems on the basis of current experience that the preparation of 'small' spacecraft can only be done by rather small, experienced teams, freed from onerous reporting and monitoring requirements and relying very heavily on the integrity and commitment of the team members rather than on excessive written guidance. There are many risks in such an approach, depending so critically on the individual, and it can only be justified if the cost reductions are sufficient in scale that the overall cost benefit balance is positive. With factors of two or three already achieved by such management techniques, the small reduction in reliability seems at present a worthwhile sacrifice, particularly if coherent programmes of small satellite launches are organised to reap the full benefit of the cost reductions. Since failures occur from time to time in programmes run in the traditional way with full QA/PA and management procedures, and since the most reliable launch vehicles only achieve success rates of 95–97% the philosophy of adopting the small satellite approach is sound if it is carried out thoughtfully and carefully.

8.3   The extra tasks

Numerous examples now exist where a small satellite has been constructed and launched successfully to fulfill a well specified mission objective by deploying the principles of space instrument design as outlined in this book. There need be no difference in the project philosophy between instruments and spacecraft, the extension of these methods to satellite design simply requires the addition of a few new subsystems and the inclusion of effort to cover a small number of extra activities. In most cases there exist specialist texts covering the extra subsystems where they are not readily available 'off–the–shelf'.

Power generation has already been discussed in Chapter 4 which also includes a section on attitude sensing, and one simple form of attitude control. Wertz has written an excellent text: *Attitude Measurement and Control* and this should be consulted if required. Chapter 4 includes a section on data handling which describes the main types of telemetry systems but from the users point of view rather than the designers perspective. In fact telemetry transmitters and receivers are rather easy to procure to space standards as a result of their widespread use both for missiles and remote monitoring in hostile environments. Without specialist skills in RF design and manufacture it would not be efficient to attempt a new design. Of these extra subsystems it is the attitude control system which is the most challenging for the non–expert. However, some industrial concerns will make available a completed cold gas control system only requiring electrical command input from an on–board computer responding to the attitude sensors. Another, simpler, route which is relevant where the absolute pointing accuracy requirement is not so severe is to use a magnetorquer which uses an electrical coil to create torques on the spacecraft from the Earth's magnetic field. Such systems are rather easy to design and operate and are outlined in Chapter 4.

There are a number of tasks which are necessary in the design and testing of a spacecraft but which are not required at the instrument level. Obviously the interfaces between the spacecraft and the launch vehicle become the responsibility of the project team as do those to the ground segment. Both these tasks require serious amounts of effort. The interfaces to the launcher cover mechanical issues, the dynamic inputs to the spacecraft, any aerodynamic loading after fairing release, the keep alive electrical power requirements and a range of thermal conditions which may be encountered during the various launch phases and orbit injection. Interfaces to the ground segment will require not only planning, but also real tests of the spacecraft telemetry and command functions over RF links to suitable ground stations.

Close interaction will be necessary with the launcher team to plan the activities during the launch campaign, especially if there is a need for access to the spacecraft or instruments at a late stage in the launch sequence. Project teams managing the preparation of spacecraft will need to ensure that the transmission and reception frequencies are properly authorised with appropriate bodies, and that the spacecraft is registered with the relevant United Nations conventions on the peaceful use of outer space.

8.4   Conclusion

Despite the added responsibility of providing the whole space platform, the intellectual and scientific rewards that result from a successful project of this nature are considerable. A central advantage of small missions is that their timescale for completion is much better adapted to the duration of training courses, both for those who wish to use the data and those who see the project as a realistic and valuable training exercise, than the traditional space projects which last ten years or more. If a small satellite programme is not complete in four years it does not deserve the title 'small'.

Histories of the world will be written in hundreds or thousands of years time commenting on the attitudes, objectives and achievements of the present age, an age in which mankind discovered the ability to move away from the Earth and shape its destiny with the benefit of that new vantage point. How will we be judged? The authors of this book hope that its pages may inform and possibly inspire young scientists and engineers to make good use of the space opportunity to assist them in the exploration of the universe and the management of our presence on this fragile planet. The future of mankind might just depend on those skills.

Appendixes

Appendix 1   List of symbols

The symbols are aggregated in blocks associated with each chapter. The number given in the right column shows the chapter, section/subsection in which the symbol first appears with that meaning.

A	cross-section area	2.2.1
A	displacement amplitude	2.6.1
a	receptance $\alpha\,(i\omega)$	2.6.2
α_n	phase angle in Fourier series expansion	2.6.5
B	bandwidth of resonant response	2.6.6
b	breadth of cross-section	2.2.2
c	damping constant	2.6.2
C_n , C_n'	coefficients of a Fourier series $(n = 0, 1...$etc.$)$	2.6.5
D	abbreviation for $c/2\,m$; identical to $(1/2\,h\,p)$	2.6.2
d	average depth between skins of honeycomb plate	2.2.2
d	depth or diameter of bar	2.2.2
d	diameter	2.3.4
D	mean diameter of a thin tube	2.2.2
ε	strain, (increase of length per unit length)	2.2.1
E	Young's elastic modulus	2.2.1
$\varepsilon_x; \varepsilon_y$	direct strain	2.2.1
ϕ	phase angle of a response	2.6.2
F	shear force	2.2.2
f	vibration frequency (Hz)	2.3.2
f_n	natural frequency of oscillator or normal mode	2.6.6
g	acceleration of free fall $(9.81\ \mathrm{ms^{-2}})$	2.3.1
G	elastic modulus for shear	2.2.1
γ	shear strain angle	2.2.1
η	damping loss coefficient, $c/\surd\,(km)$	2.6.2
I	second moment of area of cross-section of a beam	2.2.2
J	polar second moment of area of a bar	2.2.2
k	stiffness	2.6.2
L	length	2.2.1
l	length of twisted bar	2.2.2
M	bending moment	2.2.2
m	mass	2.5.1
n	number of cycles	2.6.1
ν	Poisson's ratio	2.2.1
p	pressure	2.3.4
P	proof factored load	2.5.1
p	rms sound pressure	2.6.1
p	*see* ω_n	2.6.2

$p\,(x)$	probability of (x); probability density	2.6.5
P_0	amplitude of (sinusoidal) exciting force	2.6.2
θ	angle of twist (of a bar)	2.2.2
q	dynamic pressure in a fluid flow ($= 0.5\rho V^2$)	2.3.1
Q	resonant magnification ; (gain; quality factor')	2.6.2
ρ	atmospheric density at the altitude of launch vehicle	2.6.4
ρ	density	2.2.1
R	local radius of curvature of bending of a beam	2.2.2
r	mean radius of a tube	2.5.1
r	radius from axis of torsion (of a bar)	2.2.2
ρ	radius of gyration of pin−ended strut	2.5.1
σ	rms value of random vibratory acceleration	2.6.6
s	sound power spectral density	2.6.10
S	stress	2.6.1
σ	stress, (load per unit area of cross−section)	2.2.1
σ	yield stress	2.5.1
σ_c	critical stress	2.5.1
σ_t	buckling stress of a thin−walled tube of thickness t	2.5.1
$\sigma_x; \sigma_y$	direct stress	2.5.1
T	half of loop period	2.6.5
τ	shear stress	2.2.1
t	thickness	2.3.4
T	twisting torque	2.2.2
V	velocity	2.3.1
ω	(circular) frequency of (sinusoidal) exciting force	2.6.2
W	spectral density	2.6.6
ω_n	circular frequency of a natural vibration	2.6.2
ξ	time parameter	2.6.5
y	distance of particle or fibre from neutral plane of zero bending mass	2.2.2
y	transverse deflection of neutral plane	2.2.2
Z	section modulus $= I/y_{max}$	2.2.2
A	absorption	3.3.1
A	cross-sectional area	3.3.2
A	radiating area	3.1.2
a	radiation constant	3.3.3
α	absorptivity	3.1.3
α	angle	3.3.3
B	Planck function integrated over frequency	3.3.3
B_v	Planck function	3.3.3
C	capacitance	3.2
C	centigrade degrees	3.1.2
c	specific heat capacity	3.2
c	velocity of light	3.3.3
d	distance	3.3.3
d	length	3.3.2

E	energy	3.2
e	exponential	3.3.3
ε	emissivity	3.1.3
ε_e	effective emissivity	3.5.2
ε_{eff}	effective emissivity	3.3.3
ε_b	equivalent emissivity	3.5.2
F_{ab}	view factor between surfaces a and b	3.3.3
H	dimensionless height	3.3.3
H_{ab}	heat exchange between surfaces a and b	3.3.3
H_C	conducted heat	3.4
H_R	radiated heat	3.4
h	Planck constant	3.2
h	height, length	3.3.3
I_ν	specific intensity	3.3.3
I_{ab}	heat exchange between surfaces a and b	3.3.3
∞	infinity	3.8
K	Kelvin	3.1.1
k	Boltzmann constant	3.2
k	conductance	3.3.2
k_C	conductance	3.4
k_{12}	conductance between nodes 1 and 2	3.3.2
k_R	equivalent radiative conductance	3.4
l	length	3.3.3
L	luminosity	3.3.3
l_{12}	conductance between nodes 1 and 2	3.4
λ	wavelength	3.2
λ	thermal conductivity	3.3.2
m	mass	3.2
$n(E)$	number of particles with energy, E	3.2
ν	frequency	3.2
P	radiated power	3.1.2
P_ν	specific power	3.3.3
π	pi (= 3.142)	3.3.2
Q	charge	3.2
Q	energy flux	3.3.2
Q	heat load	3.4
r	radius	3.1.2
R	reflection	3.3.1
R	impedance	3.3.2
R	dimensionless radius	3.3.3
σ	Stefan's constant	3.1.2
Ω	solid angle	3.1.2
Ω	ohms	3.3.1
S_ν	specific source function	3.3.3
T	mean temperature	3.4
T	temperature	3.1.2
T	transmission	3.3.1
t	time	3.4

Appendixes

Appendix 2   List of acronyms and units

Acronym	Meaning
ADC	Analogue to Digital Converter
AO	Announcement of Opportunity
AU	Astronomical Unit
AWG	American Wire Gauge
BCH	Bose Chauderi and Hochegem
BOE	Back Of Envelope
BOL	Beginning Of Life
CAP	Count Acceptance Period
CCD	Charge Coupled Device
CCL	Control Line
CMOS	Complementary MOS
CMR	Common Mode Rejection
CSA	Charge Sensitive Amplifier
CVCM	Collected Volatile Condensable Materials
DAC	Digital to Analogue Converter
DAP	Di – Allyl Phthalate
DMS	Data Management System
DOD	Depth of Discharge
ECL	Error Control Line
EMC	Electromagnetic Compatibility
EMR	Electromagnetic Radiation
EOL	End Of Life
EOR	Exclusive OR
ESA	European Space Agency
ESTEC	European Space Technology Centre
FEE	Front End Electronics
FET	Field Effect Transistor
FIFO	First In First Out
GSFC	Goddard Space Flight Centre
Hi-Rel	Highly Reliable
IC	Integrated Circuit
IR	Infrared
ITO	Indium Tin Oxide
JAN	Joint Army Navy
LEO	Low Earth Orbit
LLD	Lower Level Detector
LLLD	Lower Lower Level detector
MCP	Micro – Channel Plate
ME	Main Electronics
MGSE	Mechanical Ground Support Equipment
ML	Memory Load
MLI	Multilayer Insulation

MOS	Metal Oxide Semiconductor
MSB	Most Significant Bit
MTBF	Mean Time Between Failure
NASA	National Aeronautics and Space Administration
NTP	Normal Temperature and Pressure
OBDH	On – Board Data Handling
PCB	Printed Circuit Board
PSD	Position Sensitive Detector
PSK	Phase Shift Key
PSS	Procedures, Standards & Specifications
PWM	Pulse Width Modulation
RAM	Random Access Memory
RCL	Rotational Control Line
RF	Radio Frequency
ROM	Read Only Memory
ROSAT	Rontgensatellit
RPM	Revolutions per minute
RTG	Radio – isotope Thermal Generator
RTU	Remote Terminal Unit
S/N	Signal to Noise ratio
SCC	Space Components Co-ordination group
SEU	Single Event Upset
SMPS	Switched Mode Power Supply
SR	Shift Register
SSM	Second Surface Mirror
TFE	Tetra FluoroEthylene
TML	Total Mass Loss
TMM	Thermal Mathematical Model
ULD	Upper Level Detector
US	United States of America
VCHP	Variable Conductance Heat Pipes

Symbol for unit	Name of unit	Multiple	Prefix
A	amp	10^{-12}	pico (p)
dB	decibel	10^{-9}	nano (n)
F	farad	10^{-6}	micro (μ)
g	gram	10^{-3}	milli (m)
H	henry	10^{3}	kilo (k)
h	hour	10^{6}	mega (M)
Hz	hertz	10^{9}	giga (G)
J	joule		
K	kelvin		
l	litre		
m	metre		
N	newton		
Pa	pascal		
s	second		
V	volt		
W	watt		
Ω	ohm		

Notes to the text

Chapter 2 Mechanical design

Page 22: Calculation of sections in mechanical design; section 2.2.2:
The given equations show the relevance of bending moment M to both beam stress and local radius R of curvature of deflection. Calculus, or graphical integration, may be used to establish M, or shear force F. Many examples are given in Benham, Crawford, and Armstrong [1996].

Page 43, Multi–freedom systems:
As an illustration of the technique of modelling, for dynamic analysis, the structure of a launch vehicle carrying a spacecraft within its nosecone, see the schematic drawing (Fig. 9.3, page 270) in G.P. Sutton's book 'Rocket Propulsion Elements' (5th edition, Wiley, 1986). The rocket is reduced to 20 lumped masses, of which the spacecraft, as payload, is but one. Such a model would be of limited usefulness, which is why the recent practice has been to assemble a model of hundreds of nodes, of which the spacecraft and its component units would be a small percentage.

Page 50, Fig. 2.29 and section 2.6.7:
Damping and Q data can be tabulated as the dimensionless damping parameter $\zeta = D/p = D/\omega_n = 1/2Q$. Figures collected by Simonian and Demchak (Sarafin, [1995]) from spacecraft structures support the wide variations of damping shown in Fig. 2.29, from 0.001 for welded metal to 0.02 (Q=25) for honeycomb structures.

Page 58, Fig. 2.32:
The epoxy adhesive Scotchweld 1838, which outgasses less than Araldite Av100/HV100, has begun to replace Araldite for space instrument use. But the parameters TML, CVCM at the point (0.06, 0.03) for Scotchweld 1838 (key Sw) have regrettably not been plotted on this graph.

Page 66:
In proof tests of structures, elastic response to a load–unload cycle is demonstrated by sensitive measurement of deflections, mainly in the load direction, of extremities, relative to a base or interface, as the load is stepped up. A linear increase of deflection with load is looked for, with the line retraced during unloading steps. There should be only negligible permanent set at the end. Strain gauges may be applied to confirm stresses.

Chapter 5 Mechanism design and activation

Page 296:
The figure (5.1b) of Laboratory stands, exhibiting the Kelvin 'clamp' version of the principle of kinematic location, is not, of course, an illustration of an instrument assembly which could be flown in weightlessness. Without preloaded springs to maintain the defining contacts, it would

come apart. But this principle has been applied for the setting and re–setting of demountable alignment reference mirrors in pre–launch testing.

Page 302, Fig. 5.6:
In design of a parallel–motion flexure, beam theory allows the interplay of the dimensions length, width, thickness and maximum motion with the available deflecting force and the material properties of elastic modulus and yield strength.

Page 302, Fig. 5.7:
The isometric sketch of the cross–flexure, hinge suggests an idealised model. Such a hinge, made at MSSL for an infrared space spectrometer, had blades at 45° to its base, and proved amply rugged when vibration tested. As for the parallel–motion flexure, beam theory should be applied during design to ensure that strength meets requirements, and flexure stiffness is not too much for the means of actuation.

Note on the cover drawing

The isometric drawing on the cover of this book illustrates the Optical Monitor Telescope built for ESA's X–ray Multimirror Mission, a scientific spacecraft planned for orbital operation from the year 2000.

The tube comprises a stiff central structure surrounding the Ritchey–Chretien optics, preceding a detector interchange mirror, selectable filters or grisms, and CCD detectors. The tube opening reveals baffle vanes. Heat pipes convey electrical dissipation from electronic units at the far end. Mechanisms unlatch and open the cover after injection into orbit, respond to filter commands, and interchange detectors if need be. Some thermal blankets will surround the tube when installed alongside X–ray mirror modules in the spacecraft.

References and bibliography

Acton, L.W., Culhane, Bentley, Patrick, Rapley, Sheather & Turner. (1980). The Soft X – ray Polychromator for the Solar Maximum Mission. *Solar Phys.*, 65, 53 – 71.

Agarwal, D.A. & Broutman. (1990). *Analysis and Performance of Fiber Composites.* 2nd edition. Wiley Interscience, N.Y.

Agrawal, B.N. (1986). *Design of Geosynchronous Spacecraft.* Prentice–Hall Inc. Englewood Cliffs, N.J.

Ariane 4 User's Manual . (1985). Issue 1, rev. 1, Jan. [Supplied by Arianespace, Evry, France, for launch customer's use. Revisions of data obtainable from Ariane Customer Service].

Barnes, C. *et al.* (1993). Microelectronics Space Radiation Effects Program at JPL: Recent Activities. *Proceedings 2nd ESA Electronic Components Conference.* ESA WPP–063 May, 277 – 283.

Barron, R.F. (1985). *Cryogenic Systems.* 2nd edition. Oxford Univ. Press.

Bendat, J.S. & Piersol, A.G. (1971). *Random Data : Analysis & Measurement Procedures.* Wiley –Interscience, N.Y.

Benham, P.P., Crawford, R.J., & Armstrong, C.G. (1996). *Mechanics of Engineering Materials.* Longman Scientific, Harlow.

Beynon, J.D.E & Lamb, D.R. (1980). *Charge Coupled Devices & their Applications.* McGraw Hill , London.

Berlin, P. (1988). *The Geostationary Applications Satellite.* Cambridge University Press.

Betz, F.E. (1982). Real & Potential Nickel Hydrogen Superiority. *GSFC Battery Workshop.* NASA Conference Publication 2263, N.T.I.S., US Department of Commerce, Springfield, VA. 22161. 416 – 21.

Bhargava, V.H., Haccoun D., Matyas, R. & Nnuspl, N. (1981). *Digital Communications by Satellite.* John Wiley & Sons, N.Y.

Bonati, B. *et al.* (1993). Acoustic Microscopy Analysis of Hi–Rel Electronic Components for Space Applications. *Proceedings 2nd ESA Electronic Components Conference.* ESA WPP–063 May, 207 – 212.

Bose, R.C. and Chauderi. (1960). On a Class of Error Correcting Binary Group Codes. *Information Control*, 3.

Bruhn, E.F. (1965). *Analysis & Design of Flight Vehicle Structures.* Tri–State Offset. Cincinatti, Ohio.

BS 9000. *General Requirements for Electronic Components of Assessed Quality*, British Standards Institute, London.

BS 9400. *Specification for Integrated Electronic Circuits and Micro – Assemblies of Assessed Quality*, Generic Data & Methods of Test, British Standards Institute, London.

Cable N. (1989). Computer simulation of space mechanisms. *ESA Bulletin*, 60, p. 46, Nov.

Cable N. (1988). Pyrotechnic devices. *ESA Bulletin*, 59, No. 54, p. 66, May.

Callister, W.D. (1997). *Materials Science and Engineering.* 4th. edition. Wiley, N.Y.

Campbell, W. A. Jr. *et al.* (1993). *Outgassing Data for selecting Spacecraft Materials.* NASA Reference Publication, 1124 Revision 3, Sep.

Chobotov. (1991). *Spacecraft Attitude Dynamics & Control.* Orbit. Krieger, Florida.

Coates, A.J., Bowles, Gowen, Hancock, Johnstone & Kellock. (1985). The Ampte UKS Three–Dimensional Ion Experiment. *IEEE Transactions on Geoscience and Remote Sensing*, Vol GE – 23, No 3, May, 287 – 92.

Couch, L. W. (1983). *Digital & Analog Communication Systems.* Macmillan, N.Y.

Crabb, R.L. *et al.* (1991). Power system performance prediction for the Hipparcos spacecraft. *European Space power conference*, ESA SP-320, ESA, Noordwijk.

Cruise, A.M., Barnsdale, Bowles & Sheather. (1991). A Wide Angle Earth Albedo Sensor for Spacecraft Attitude Determinations. *Jrnl of Brit Interplanetary Soc.*, Vol 44, 142 – 4.

Cuevas, A. *et al.* (1990). Point & planar junction solar cells for concentration applications: fabrication, performance, & stability. *Proceedings of 21st IEEE Photovoltaic Specialists' Conference*, Kissimee, FLA, I.E.E.E. Press, N.Y. 327 – 32.

Davidge, R.W. (1979). *Mechanical Behaviour of Ceramics.* Cambridge Univ. Press.

Davis, G.R., Furniss, Patrick, Sidey & Towlson. (1991). Cryogenic Mechanisms. *NASA 25th. Aerospace Mechanisms Symposium*, 1991, CR 3113.

Drago, R.J. (1988). *Fundamentals of Gear Design.* Butterworths, London.

Dunn, B.D. (1997). Metallurgical Assessment of Spacecraft Parts, Materials and Processes. Wiley – Praxis, Chichester.

Elishakoff, I. & Lyon. (1986). Random Vibration : Status and Recent Developments. *Studies in Applied Mechanics*, 14 . Elsevier.

Erdman, A.G. & Sandor. (1991). *Mechanism Design: Analysis and Synthesis.* Prentice Hall, Englewood Cliffs, N.J.

ESA (1989). *Fourth European Space Mechanisms & Tribology Symposium*, Cannes, ESA SP – 299.

ESA PSS – 01 – 201. (1987). [Eaton D. (ed.)] : *Structural Acoustics Design Manual.*, issue 1.

ESA PSS – 01 – 301. (1992). *Derating Requirements and Application Rules for Electronic Components.* Product assurance Division, ESA Publications Division, ESTEC, Noordwijk, Netherlands. Apr.

ESA PSS – 01 – 603. (1995). *ESA Preferred Parts List.* Product Assurance Division, ESA Publications Division, ESTEC, Noordwijk, Netherlands. Sep.

ESA PSS – 01 – 609. (1993). *Radiation Design Handbook.* Product Assurance & Safety Dept, ESA Publications Division, ESTEC, Noordwijk, Netherlands, May.

ESA PSS – 01 – 701. (1990). *Data for Selection of Space Materials.* Product Assurance & Safety Dept, ESA Publications Division, ESTEC, Noordwijk, Netherlands. Jun.

ESA PSS – 01 – 708. (1985). *ESA Manual of Soldering of High-Reliability Electrical Connections.* Product Assurance Division, ESA Publications Division, ESTEC, Noordwijk, Netherlands, May.

ESA PSS – 01 – 710. (1985). *Qualification & Procurement of Double sided PCBs.* Product Assurance Division, ESA Publications Division, ESTEC, Noordwijk, Netherlands. Oct.

ESA PSS – 01 – 726. (1990). *The Crimping of High-Reliability Electrical Connections.* Materials & Process Division, ESA Publications Division, ESTEC, Noordwijk, Netherlands, Dec.

ESA PSS – 01 – 728. (1991). *Repair & Modification of PCB Assemblies for Space Use.* Materials & Process Division, ESA Publications Division, ESTEC, Noordwijk, Netherlands. Mar.

ESA PSS – 03 – 108. (1989). *Spacecraft Thermal Control Design Data.* Vols 1 – 5.

ESA PSS – 03 – 1101. (1986). [Eaton D.C.G. (ed.)] : *Composites Design Handbook for Space Structure Applications* (2 vols.), issue 1, Dec.

ESA PSS – 04 – 103. (1989). *Reed Solomon Coding Standard.* Telemetry Channel Coding Standard. Standard Approvals Board for Space Data Communications. Sep, 13 – 8.

ESA PSS – 04 – 106. (1988). *Packet Telemetry Standard.* Standards Approval Board for Space Data Communications. ESA Publications Division, ESTEC, Noordwijk, Netherlands. Jan.

ESA PSS – 04 – 107. (1992). *Packet Telecommand Standard.* Standards Approval Board for Space Data Communications, ESA Publications Division, ESTEC, Noordwijk, Netherlands. Apr.

ESA RD – 01 Rev 3. (1992). *Outgassing and Thermo-Optical Data for Spacecraft Materials.* ESA Publications Division, ESTEC, Noordwijk, Netherlands. Apr.

ESA TTC – B – 01. (1979). *Spacecraft Data Handling Interface Standards*, issue 1, Data Handling & Signal Processing Division, ESTEC, Noordwijk, Netherlands, Sep.

ESA WPP 115. (1996). Reinhard R., Ed., STEP Testing the Equivalence Principle in space, *Pisa symposium,* 1993. ESA. July.

Fano, R.M. (1961). *Transmission of Information.* John Wiley & Sons, N.Y.

Forney, G.D. Jr. (1973). The Viterbi Algorithm. *Proc I.E.E.E.*, Vol 61, No 3, Mar, 268 – 78.

Fortescue, P. & Stark. (1996). *Spacecraft Systems Engineering.* Wiley, Chichester.

Gonzales del Amo, J. A. (1990). GaAs/Ge Solar Cells for Space Applications. *ESA Electronics Components Conference*, ESA SP – 313. Nov, 21 – 4.

Gordon, J.E. (1988). *The Science of Structures and Materials.* Scientific American Library, HPHLP, N.Y.

Hamming, R.W. (1950). Error Detecting & Error Correcting Codes. *Bell Systems Tech. J.*, 29, 147 – 60.

Hanna, M.S. & Beshara. (1993). Orthotropic behaviour of shells & stability of uniaxial compressed cylinders. *Aeronautical Journal*, 97, pp 111 – 7, March.

Hocquenghem, A. (1959). Codes Correcteurs d'Erreurs. *Chiffres*, 2, 147 – 56.

Holmes-Siedle, A. & Adams, L. (1993). *Handbook of Radiation Effects.* Oxford University Press.

Huffman. (1952). A method for the Construction of Minimum Redundancy Codes. *Proc. I. Radio Eng.* Vol 40, 1098 – 101, Sep.

Kallmann – Bijl, H. [1961]. COSPAR International Reference Atmosphere, edited by H. Kallmann – Bijl, North Holland Pub Co.

Kawakami, H., Thomas, P., Bowles, J.A. (1992). 3.3 – MHz charge sensitive preamplifier. *Rev. Sci. Inst.*, Vol 63, No 1. Jan , 820 – 3.

Kenjo, T. (1984). *Stepping Motors and their Microprocessor Controls.* Oxford.

Kenjo, T. (1991). *Electric Motors and their Controls*. Oxford.

Kent, B.J., Swinyard & Hicks. (1992). Contamination effects on EUV optics in the space environment. *SPIE Conference 'The Expanding Spectrum'*.

Khemthong, S. *et al.* (1990). Large Area Space Solar Cells Si or GaAs? *Proc. 21st I.E.E.E. Photovoltaic Specialists' Conference*, 1374 – 7.

Lapington, J.S *et al.* (1986). The Design & Manufacture of Wedge & Strip Anodes. *I.E.E.E. Trans. Nuc Sci.*, Vol 33, No1, Feb, 288 – 92.

Leuenberger, H. and Person, R.A. (1956). Compilation of Radiation Shape Factors for Cylindrical Assemblies. Paper No 56 – A – 144, *Am. Soci.Mech. Eng.*

Lin, S. *et al.* (1983). *Error Control Coding: Fundamentals & Applications*. Prentice Hall, Englewood Cliffs, N.J.

Malacara, D. (ed.) 1978 & 1993 (2nd edition) . *Optical Shop Testing*. Wiley, Chichester.

Martin, C. *et al.* (1985). *Rev .Sci. Inst.*, Vol 52, 1067.

McConnell K. (1995). *Vibration Testing Theory and Practice*. Wiley, Chichester.

Morgan, W.L. and Gordon, G.D. (1989). *Communications Satellites Handbook*. Wiley & Sons, N.Y.

NASA Materials Selection Guide, Revision A. (1990). Edited by Romer E Predmore & Ernest W Mielke. Materials Assurance Office, Materials Branch, Office of Flight Assurance, GSFC, Greenbelt, Maryland.

Newland, D.E. (1990). *Mechanical Vibration Analysis & Computation*. Longman, Harlow.

Newland, D.E. (1993). *Introduction to Random Vibrations, Spectral & Wavelet Analysis*. Longman, Harlow .

Niu, M.C.Y. (1988). *Airframe Structural Design*. Conmilit Press Ltd, HongKong.

O'Shea, D.C. (1986). Annotated bibliography of optomechanical design. SPIE paper No. 2255 770, 3 – 13.

O'Sullivan, D. (1994). Space Power Electronics–Design drivers. *ESA J.* Vol 18, No 1, 1 – 23.

Peterson, W.W. *et al.* (1972). *Error-Correcting Codes*. 2nd edition, MIT Press, Cambridge, MA.

Pickel, J.C. & Blanford, J.T. (1980). Cosmic Ray induced errors in MOS devices. *I.E.E.E. Transactions on Nuclear Science*, Vol 27, No 2, Apr, 1006 – 15.

PPL – 20. (1993). *GSFC Preferred Parts List*. NASA GSFC, Greenbelt, MD 20771. Mar.

QPL – 19500 – 129. (1994). *Qualified Products List of Products Qualified under Military Specification MIL-S-19500* – General Specifications for Semiconductor Devices. Jul..

QPL – 38510 – 94. (1993). *Qualified Products List of Products Qualified under Military Specification Mil-M-38510* – General Specifications for Microcircuits. Sep.

Ristenbatt, M.P. (1973). Alternatives in Digital Communication. *Proc. I.E.E.E.*, Vol 61, No 6, June, 703 – 21.

Rowntree, R.A. (editor). (1994). Tribology for Spacecraft. *AEA Technology, European Space Tribology Laboratory*, 8 – 9 June.

Sarafin, T.P. (Editor). (1995). *Spacecraft Structures & Mechanisms*. Microcosm/Kluwer Academic, Torrance, CA.

Schmidt, G.E. Jr. (1978). Magnetic Attitude Control for Geosynchronous Spacecraft. *Proc. AIAA Communications Satellite Systems Conference*, San Diego, CA, Apr, 110 – 2.

Shannon, (1948). A Mathematical Theory of Communications. *Bell Systems Tech. J.*, Vol 27, 379 – 423 & 623 – 56.

Shigley, J.E. & Mischke. (1989). *Mechanical Engineering Design*. (5th edition), McGraw-Hill International Edition. N.Y.

Sidi, M.J. (1997). *Spacecraft Dynamics & Control*. Cambridge Univ. Press.

Siegel, R and Howell, J.R. (1992). *Thermal Radiation Heat Transfer*. 3rd Edition. Washington DC. Hemisphere Pub. Corp.

Sims, M.R. *et al.* (1993). Contamination and cleanliness control of the ROSAT Wide Field Camera, *J.B.I.S.*, 46, p. 483.

Smith, A.D. (1986). A Study of Microchannel Plate Intensifiers. *I.E.E.E. Transactions on Nuc. Sc.*, Vol 33, No1, Feb, 295 – 8.

Smith, S.T. & Chetwynd, D.G. (1992). *Foundations of Ultraprecision Mechanism Design*. Gordon & Breach Science Publishers, Montreaux, Switzerland.

Spilker, J.J. Jr. (1977). *Digital Communication by Satellite*. Prentice Hall, Englewood Cliffs, N.J.

Stein, B.A. & Young. (1992). *LDEF Materials Workshop 1991*. NASA Conf. Publication 3162, part 1.

Stimpson, L.D. and Jaworski, W. (1972). Effects of Overlaps, Stitches and Patches on Multilayer Insulation. *AIAA Thermophysics Conference*, San Antonio, Texas.

Tascione, T.F. (1994). *Introduction to the Space Environment, 2nd Ed.* Kreiger Publishing Co., Malabar, FLA.

Thierfelder, H.E. (1982). Battery Charge Control with Temperature Compensated Voltage Limit. *The 1982 GSFC Battery workshop.* NASA CP – 2263, N.T.I.S., Springfield, VA. 259 – 69.

Tribble. (1995). *Space Environment.* Princeton University Press.

Trubert, M. (1986). Loads Analysis for Galileo Spacecraft. *Proc. Conf. Spacecraft Structures,* 1985, ESA SP 238, (pp 293 – 8).

Trudeau, N.R., Sarles, F.W. Jr., Howland, B. (1970). Visible Light Sensors for Circular Near Equatorial Orbits. *AIAA Third Communications Satellite Systems Conference,* Paper 70 – 477. L.A., CA.

Turner, R.F. & Firth. (1979). *13th Aerospace Mechanisms Symposium,* 1979. NASA Conference Publication 2081.

US 311 – HNBK – 001. (1990). *Handbook for the selection of Electrical, Electronic & Electromechanical Parts for GSFC Spaceflight Applications.* Parts Branch of Office of Flight Assurance,U.S. Department of Defense, Washington DC, 20301, Jun.

US Mil Spec Handbook Mil – S – 2000. *Standard Requirements for soldered electrical and electronic assemblies.* U.S. Department of Defense, Washington DC, 20301.

US MIL – HDBK – 217F. (1991). *Reliability Prediction of Electronic Equipment.* U.S. Department of Defense, Washington DC, 20301. Dec.

US MIL – HDBK – 338. (1984). *US Electronic Reliability Design Handbook.* Department of Defense, Washington DC, 20301. Oct.

US MIL – I – 38535 (1995). *Military Specification for Microcircuits, General Specification for.* U.S. Department of Defense, Washington DC, 20301.

US MIL – M – 38510J. (1993). *Military Specification for Microcircuits, General Specification for.* U.S. Department of Defense, Washington DC, 20301, Jul.

US MIL – STD – 19500. (1993). *Military Specification for Semiconductors, General Specifications for.* U.S. Department of Defense, Washington DC, 20301, Jul.

US MIL – STD – 202F. (1993). *Test Methods for Electronic and Electrical Component Parts.* U.S. Department of Defense, Washington DC, 20301, Jul.

US MIL – STD – 461D. (1993). *Requirements for the Control of Electromagnetic Interference Emissions and Susceptibility.* US Dept of Defense, Washington DC, 20301, Jan.

US MIL – STD – 462D. (1991). *Measurement of Electromagnetic Interference Characteristics.* US Dept of Defense, Washington DC, 20301. Jan.

US MIL – STD – 750. (1991). *Test Methods for Semiconductors.* U.S. Department of Defense, Washington DC, 20301.

US MIL – STD – 883D. (1991). *Test Methods & Procedures for Microelectronics.* US Dept of Defence, Washington DC, 20301. Nov.

US MIL – STD – 975M. (1994). *NASA Standard Electrical, Electronic & Electromechanical Parts List.* US Dept of Defense, Washington DC, 20301. Aug.

US MIL – STD – 978. (1988). *NASA Parts Applications Handbook.* Dept of Defense, Washington DC, 20301. Mar.

Viterbi, A.J. (1971). Convolutional Codes & their Performance in Communication Systems. *I.E.E.E. Trans. Comm. Tech.,* Vol COM – 19, Oct, 751 – 72.

Ward, A.K., Bryant, Edwards, Parker, Patrick, Sheather, & Cruise. (1985). *The Ampte UKS Spacecraft.* I.E.E.E., Trans. Geoscience GE – 23, 3, 203 – 11.

Weidmann, G., Lewis & Reid. (1990). *Structural Materials.* Open University, Butterworths, London.

Weinberg, I. (1990). Radiation damage mechanisms in silicon and gallium arsenide solar cells. *Current topics in photovoltaics,* Vol 3, Academic Press, London.

Wertz, J.R. (1978). *Spacecraft Attitude Determination & Control.* Kluwer Academic Press, Dordrecht.

Wertz, J.R. & Larson. (1991). *Space Mission Analysis & Design.* Kluwer Academic Press, Dordrecht.

White, D.R. (1980). *Electromagnetic Shielding Materials & Performance.* Library of Congress Catalogue Card No 75 – 16592.

Yoder, P.R. (1992). *Opto-Mechanical Systems Design.* (2nd ed.), Marcel Dekker, N.Y.

Yoshikawa, T. (1990). *Foundations of Robotics.* MIT Press, Cambridge, MA.

Young, W.C. (1989). *Roark's Formulas for Stress & Strain.* (6th edition),. McGraw – Hill, N.Y.

Index